AF344525

Climate Adaptation and Flood Risk
in Coastal Cities

Climate Adaptation and Flood Risk in Coastal Cities

Edited by

Jeroen Aerts, Wouter Botzen, Malcolm J. Bowman, Philip J. Ward and Piet Dircke

London • New York

First published 2012
by Earthscan
2 Park Square, Milton Park, Abingdon, Oxon OX14 4RN

Simultaneously published in the USA and Canada
by Earthscan
711 Third Avenue, New York, NY 10017

Earthscan is an imprint of the Taylor & Francis Group, an informa business

British Library Cataloguing in Publication Data
A catalogue record for this book is available from the British Library

Library of Congress Cataloging in Publication Data
A catalog record for this book has been applied for

ISBN: 978-1-84971-346-7 (hbk)

Typeset in Times
by JS Typesetting Ltd, Porthcawl, Mid Glamorgan

Printed and bound in Great Britain by the MPG Books Group

Contents

List of Figures, Tables, Boxes and Plates

Figures

Tables

Boxes

Plates

List of Contributors

Jeroen Aerts, VU University Amsterdam, The Netherlands.

Edvin Aldrian, Badan Meteorologi Klimatologi dan Geofisika (BMKG), Jakarta, Indonesia.

Muh Aris Marfai, Faculty of Geography, Gadjah Mada University, Yogyakarta, Indonesia.

Daniel Bader, Columbia University New York City, New York, US.

Wouter Botzen, VU University Amsterdam, The Netherlands.

Hamish Bowman, University of Otago, Dunedin, New Zealand.

Malcolm J. Bowman, Stony Brook University, New York, US.

Philip Bubeck, Helmholtz Centre Potsdam, GFZ German Research Centre for Geosciences, Germany.

Brian A. Colle, Stony Brook University, New York, US.

Piet Dircke, Rotterdam University of Applied Sciences, The Netherlands.

Maria Francesch-Huidobro, Department of Public and Social Administration, City University of Hong Kong, Hong Kong.

Lidia Gaslikova, University of Bern, Switzerland.

Vivien Gornitz, NASA Goddard Institute for Space Studies (GISS), New York, US.

Susan Hanson, University of Southampton, Hampshire, UK.

Kristina Hill, University of Virginia, Charlottesville, Virginia, US.

Radley Horton, NASA Goddard Institute for Space Studies (GISS), New York, US.

Peter Jansen, Rijkswaterstaat, Department of Infrastructure, The Netherlands.

Tom Jongeling, Deltares, The Netherlands.

S. N. (Bas) Jonkman, Royal Haskoning, The Netherlands; Delft University of Technology, The Netherlands; University of California Berkeley, US.

Eric Koomen, VU University Amsterdam, The Netherlands.

A. J. (Joost) Lansen, Royal Haskoning, Coastal and Rivers Division, Rotterdam, The Netherlands.

David C. Major, Columbia University New York, US.

Arnoud Molenaar, City of Rotterdam, The Netherlands.

Robert J. Nicholls, University of Southampton, Hampshire, UK.

Pieter Pauw, VU University Amsterdam, The Netherlands.

Poerbandono, Faculty of Earth Sciences and Technology, Bandung Institute of Technology, Indonesia.

Jennifer K. Poussin, Institute of Environmental Studies, VU University Amsterdam, The Netherlands.

Christoph C. Raible, University of Bern, Switzerland.

Cynthia Rosenzweig, NASA Goddard Institute for Space Studies (GISS), New York, US.

Aurel Schwerzmann, Swiss RE, New York, US.

William Solecki, Department of Geography, City University of New York–Hunter College, New York, US.

Bianca Stalenberg, Technical University Delft, The Netherlands.

Michiel van Drunen, VU University Amsterdam, The Netherlands.

Susan van 't Klooster, VU University Amsterdam, The Netherlands.

Philip J. Ward, Institute for Environmental Studies (IVM), Faculty of Earth and Life Sciences (FALW), VU University Amsterdam, The Netherlands.

Acknowledgements

We would like to thank all of the authors who contributed to this book and others who helped us to make it a success. In particular, we would like to thank the Connecting Delta Cities network for their support. We gratefully acknowledge The New York City Department of City Planning (DCP), the New York City Department of Buildings (DOB) and the New York City Mayor's Office for supporting and reviewing research related to Chapter 9. We are also grateful to Arnoud Molenaar from the City of Rotterdam and Piet Dircke from ARCADIS for financial support; the Knowledge for Climate programme (HSINT02, HSGR06 and Themes 1 and 6) and the Climate Changes Spatial Planning programme (A20) for assisting with the flood risk modelling in Chapter 6; and both the Royal Netherlands Academy of Arts and Sciences Mobility Programme (09-MP-10) and the Knowledge for Climate programme (HSINT02 and Theme 1) for supporting the work on Jakarta in Chapter 14.

Preface

Climate change will have a severe and inevitable impact upon the flood risk of global coastal cities – even if we do succeed in substantially reducing its causes and mitigating its effects. As a consequence, the vulnerability of infrastructure, people, nature and economic sectors is expected to increase in the decades to come. Both scientists and policy-makers have addressed the issue of adapting to the challenge of climate change, and both call for embedding long-term scenarios in city planning and investments in all sectors. Since the choices we make today will influence vulnerability to climate risks in the future, it is important to link adaptation measures to, and benefit from, on-going investments in infrastructure and spatial planning, and to draw up detailed estimates of the benefits of adaptation. In this way, adaptation becomes a challenge rather than a threat, and climate adaptation may initiate opportunities and innovations for investors and spatial planners.

This book explores the different aspects of climate adaptation in coastal cities. It provides an overview of how large coastal cities can deal (and are dealing) with the problem of flooding, and how they can improve their adaptation strategies for the future. Methods and instruments are described for assessing the social, physical and economic causes and impacts of flooding, and adaptation strategies are presented that are being used to cope with and manage flood risk. Flood risk is defined here as the probability of flooding multiplied by its consequences. Through in-depth case studies of the cities of New York, Jakarta, Hong Kong and Rotterdam, the book will link knowledge and research in these cities.

The initiative for this book has been supported by the C40 Global Cities Network (see www.c40cities.org) and the underlying Connecting Delta Cities initiative (see www.deltacities.com). In October 2008, a C40 meeting in Tokyo on the topic of climate adaptation officially adopted the Connecting Delta Cities initiative (CDC) put forward by the city of Rotterdam. It was addressed as Joint Action 8: Climate Adaptation Connecting Delta Cities. The C40 agreed that the network should (initially) consist of a small number of cities that are front-runners in climate adaptation, with the objective of exchanging knowledge on climate adaptation and sharing best practices.

Together with C40, we continue to work on connecting global cities in order to support developing climate adaptation strategies. For this, the Connecting Delta Cities initiative serves as a platform for sharing (novel) experiences and making these available to professionals in the public and private sectors.

List of Acronyms and Abbreviations

ADCIRC	Advanced Coastal Circulation and Storm Surge model
AFD	adaptable flood defences
amsl	above mean sea level
AR4	IPCC's *Fourth Assessment Report*
ARK	National Programme on Spatial Planning and Adaptation to Climate Change (The Netherlands)
BAKORNAS	National Coordinating Board for the Management of Disaster (Indonesia)
BFE	baseline flood elevation
BMKG	Badan Meteorologi, Klimatologi, dan Geofisika (Meteorological, Climatological and Geophysical Agency)
C	current city
CAC	current city, all changes
CCBF	Climate Change Business Forum
CCC	current city, climate change
CcSP	Climate Changes Spatial Planning programme
CDC	Connecting Delta Cities initiative
CDF	cumulative distribution function
CDM	Clean Development Mechanism
CER	Certified Emissions Reduction
CLPH	China Light and Power Holdings
CO_2	carbon dioxide
CPB	The Netherlands Bureau for Economic Policy Analysis
CRS	Community Rating System
CSA	Combined Statistical Area
3D	three dimensional
DEM	Digital Elevation Model
DIVA	Dynamic Interactive Vulnerability Assessment model
DSS	decision support system
ECC	Environmental Campaign Committee
EFD	European Flood Directive
EPA	Environmental Protection Agency
ERPG	*Emergency Response Planning Guidelines*

EST	Eastern Standard Time
EV	electric vehicle
FAC	future city, all changes
FCC	future city, climate change
FEMA	Federal Emergency Management Agency
FIRM	Flood Insurance Rate Map
FNC	future city, no environmental change
GCM	Global Climate Model
GDP	gross domestic product
GE	global economy scenario
GEO-3	*Global Environment Outlook 3*
GFS	Global Forecast System
GHG	greenhouse gas
GIA	glacio-isostatic adjustment
GIS	geographic information systems
GPS	global positioning system
HEMP	Hazard and Effects Management Process
HKO	Hong Kong Observatory
hPa	hectopascal
HSBC	Hong Kong Shanghai Banking Corporation
ICT	information and communications technology
IMF	International Monetary Fund
IPCC	Intergovernmental Panel on Climate Change
IWGCC	Inter-departmental Working Group on Climate Change
KNMI	Royal Netherlands Meteorological Institute
kt	knot
kWh	kilowatt hour
KvR	Climate Changes Spatial Planning programme
LIDAR	light detection and ranging
LOOP	Low-Carbon Office Operations Programme
LPG	liquid petroleum gas
masl	metres above current mean sea level
MNP	The Netherlands Environmental Assessment Agency
mph	miles per hour
MSA	Metropolitan Statistical Area
NCEP	US National Centers for Environmental Prediction
NDBC	National Data Buoy Center
NFIP	National Flood Insurance Program
NGO	non-governmental organization
NGVD	National Geodetic Vertical Datum
NOAA	National Oceanic and Atmospheric Administration
NOS	National Ocean Service

NPCC	New York City Panel on Climate Change
NWS	National Weather Service (US)
NYC	New York City
NYCDEP	New York City Department of Environmental Protection
NYCOEM	New York City Office of Emergency Management
NYHOPS	New York Harbor Observing and Prediction System
NYS	New York State
OECD	Organisation for Economic Co-operation and Development
PCB	polychlorinated biphenyl
PDSI	Palmer Drought Severity Index
PMI	pluses, minuses and interesting issues
PPP	purchasing power parity
PRD	Pearl River Delta
RAS	Rotterdam Adaptation Strategy
RC	regional communities scenario
RCI	Rotterdam Climate Initiative
RCP	Rotterdam Climate Proof plan
RIVM	Dutch National Institute for Public Health and the Environment
RMS	root mean square
RoRo	roll on/roll off
RPB	The Netherlands Institute for Spatial Research
SAR	Special Administrative Region
SARS	severe acute respiratory syndrome
SBSS	Stony Brook Storm Surge Model
SFHA	Special Flood Hazard Area
Sida	Swedish International Development Cooperation Agency
SIT	Stevens Institute of Technology
SLOSH	Sea, Lake and Overland Surges from Hurricanes model
SLR	sea-level rise
SOP	Standard of Protection
SRES	*Special Report on Emission Scenarios*
TCG	The Climate Group
UK	United Kingdom
UKCIP	UK Climate Impacts Programme
UNFCCC	United Nations Framework Convention on Climate Change
US	United States
USACE	US Army Corps of Engineers
UTC	coordinated universal time
WHO	World Health Organization
WLO	*Welvaart en Leefomgeving*
WWF	World Wide Fund for Nature (*formerly* World Wildlife Fund)
WWTP	wastewater treatment plant

1

Introduction: Coastal Cities and Adaptation to Climate Change

Jeroen Aerts, Wouter Botzen, Malcolm J. Bowman, Philip J. Ward and Piet Dircke

1.1 Setting the Scene

Currently, more than half of the world's population live in cities, especially in vulnerable coastal cities. It is estimated that many of the world's large cities are vulnerable to rising sea levels and climate change, with millions of people being exposed to extreme floods and storms (Aerts et al, 2009). By the middle of this century, the majority of the world's population will live in cities in or near deltas, estuaries or coastal zones, resulting in even more people living in highly exposed areas. Such socio-economic trends amplify the possible consequences of future floods, as more people move towards urban delta areas and capital is continuously invested in ports, industrial centres and financial districts in flood-prone areas. It is also expected that in many regions in the world, the frequency, intensity and duration of extreme precipitation events will increase as a result of climate change, as well as the frequency and duration of droughts. At the same time, many coastal cities suffer from severe land subsidence. As a consequence of these urban developments and projected land subsidence and climate change, the vulnerability of our coastal cities is expected to increase in the decades to come (Nicholls et al, 2008; Rosenzweig et al, 2010).

It is increasingly recognized that cities need to adapt in order to moderate the harm or exploit the beneficial opportunities resulting from changes in climate and other physical or socio-economic factors. However, the issue of climate adaptation is very complex, and there is no single readily available adaptation solution applicable to all delta cities. Adaptation is partly a matter of learning by doing, or allowing experiments and innovation. There is also the need to keep all options open because of the uncertainty of how the future will unfold: one can never predict exactly how the future will develop and what measures will be needed. This uncertainty is dealt with by using scenarios of the future, which represent different possible storylines of the future. Hence, climate-robust and flexible no-regret or low-regret measures should be considered. In addition, complicated issues such as policy-making, stakeholder involvement and financing new measures may hinder the speedy implementation of adaptation measures, and may cut ambitious plans to more modest levels. It is, therefore, important to consider a variety of possible measures in the planning process of climate adaptation, and to learn from the experiences of other areas and coastal cities.

Cities play an important role in the climate adaptation process since they have already developed the ability to adapt continuously to change and attract economic activity and investments. One could say that cities have already been adapting to changing conditions for many years or even centuries, and climate change is an additional challenge that needs to be addressed in cities' planning, investments and regulations. Many cities are gradually taking on the issue of climate adaptation, and there is a growing interest in sharing and exchanging experience and knowledge between cities. Since the choices made today will influence vulnerability to climate change in the future, it is important to link adaptation measures to on-going investments in infrastructure and spatial planning, and to draw up detailed estimates of the benefits of adaptation. In this way, adaptation becomes a challenge rather than a threat, and climate adaptation may initiate opportunities and innovations for investors and spatial planners.

This book explores the different aspects of climate adaptation in coastal cities. It provides an overview of how large coastal cities can deal, and are dealing, with the problem of flooding, and how to improve adaptation strategies for the future. Methods and instruments are described for assessing the social, physical and economic causes and impacts of flooding, and adaptation strategies are presented that are being used to cope with and manage flood risk. Flood risk is here defined as the probability of flooding multiplied by its consequences. Through in-depth case studies of the cities of New York, Jakarta, Hong Kong and Rotterdam, the book will link knowledge and research in these cities. This comparison is of special interest since these cities all have similar climate change risks and invest considerably in adaptation, while optimal policies are likely to be different in cities in developing countries than capital-intensive developed countries. Active knowledge exchange will be beneficial to these cities and provide examples of how to, and how not to, adapt to flood risk around the world.

1.2 Future Trends and Coastal Cities

The Intergovernmental Panel on Climate Change's (IPCC's) *Fourth Assessment Report* states that it is inevitable that flood risks and other climate change impacts will continue to increase, and that adaptation measures and policies need to be developed parallel to mitigation efforts (IPCC, 2007). The question is not if, but how quickly, societies and cities will need to adapt. Adaptation to changing climatic and socio-economic conditions is not new; cities have been adapting to societal and environmental changes for centuries. However, the world of today is much more complex than it was in the past, and interventions taken to adapt to climate change in one sector have significant impacts upon other economic sectors and upon the environment. Adaptation to climate change, therefore, requires a holistic approach, where all sectors and stakeholders participate in order to include long-term adaptation planning in their daily operations. Existing climate policy documents state that long-term planning is the key to successful adaptation. Effective land-use planning is crucial for enhancing cities' adaptive capacities to climate change. Effective adaptation requires the local implementation of measures, and requires collaborating with non-governmental organizations (NGOs) to improve interrelationships with local institutions. A participative approach ensures that stakeholders can express their objectives, concerns and visions, and stimulates the development and implementation of innovative ideas in the adaptation process. An adaptation process also increases the commitment of stakeholders to ensure that new measures are accepted and implemented.

Urban population growth and, as a consequence, urban development have an enormous impact upon land use. Studies carried out to assess the effects of population growth and land-use change in the lower Netherlands show that flood risks have increased by a factor of seven over the last 50 years due to urbanization and land-use change (Aerts et al, 2008). Thus, even without climate change, flood risk in urbanized deltas will increase simply because residents and businesses continue to settle in vulnerable locations. Furthermore, Bouwer et al (2007) show that by 2025, loss potentials among the world's ten largest cities are projected to increase by at least 22 per cent (Tokyo) to more than 50 per cent in Shanghai and Jakarta. Since economic growth and urban development in these areas are inevitable, and the economic impacts of climate change may not be limited to the city boundaries alone, rising sea levels could have devastating effects on the worldwide population and economic activity in the future.

1.2.1 Climate change, subsidence and sea-level rise

Sea-level rise is partly a natural phenomenon, and historical measurements in several delta cities such as New York and Rotterdam show an increase in mean sea-level rise of 17cm to 22cm over the last 100 years (Aerts et al, 2008). Prior to the Industrial Revolution, sea-level rise in New York and Rotterdam could be attributed mainly to regional subsidence of the Earth's crust, which is still slowly readjusting to the melting of ice sheets since the end of the last ice age (Aerts at al, 2009; NPCC, 2009). For New York and Rotterdam, land subsidence accounts for 3mm to 4mm per year, mainly due to these post-glacial geological processes. But much higher subsidence rates occur as well. For example, in Jakarta, parts of the city are sinking at a rate of 4cm per year, mainly due to groundwater extraction and construction loading.

Climate change, however, will accelerate natural sea-level rise through the thermal expansion of the oceans, melting of glaciers and ice sheets such as in Antarctica and Greenland, and changes in the accumulation of snow. It may also change the paths and speeds of major ocean current systems. The *Fourth Assessment Report* of the IPCC projected an increase in global temperature of between 1.1°C and 6.4°C over the next century (IPCC, 2007). As a result, average sea levels could rise by up to 59cm by 2100. There are regional differences in projected sea-level rise, and it is expected that sea levels in the north-east of the Atlantic Ocean will rise by 15cm more than the world average by 2100. This can be explained through the weakening of the warm Gulf Stream, gravitational effects and the extra warming of seawater at greater depths. The projected sea-level rise for Rotterdam and New York, for instance, is estimated at around 50cm to 85cm by 2100. The most extreme low-probability scenarios indicate a sea-level rise of 108cm to 140cm. Substantial uncertainty exists about the future behaviour of the large ice sheets in Greenland and Antarctica. Although it is not well understood how quickly the ice sheets will melt, a theoretical collapse of the Greenland and West and East Antarctica ice sheets through accelerated glacier flow would lead to a rise in sea level of several metres over the coming centuries.

1.3 Flood Risk Vulnerability

Flood risk is defined as the probability of flooding multiplied by the potential consequences, such as economic damage or loss of lives (Smith, 1994). The level of flood risk therefore depends upon:

- the hazard characteristics, such as flood depths and extent, flood duration or flow velocity (Milly et al, 2002; Kundzewicz and Schellnhuber, 2004);
- the exposure characteristics in flood-prone areas, such as number of people, land use and value of assets (Kundzewicz and Schellnhuber, 2004); and
- the vulnerability of the exposed assets and population to the hazard, which can vary largely depending upon the location (e.g. in developed or developing countries) (Kron, 2005).

Estimates of flood risk can be further disaggregated into coastal floods, river floods and extreme rainfall. In all three cases the impact can be very high, with numerous casualties and much damage to property. Extreme flood events are relatively rare, with typical return periods of 100 years and higher. Extreme precipitation events in non-tropical cities rarely cause casualties, but frequently cause damage to property and infrastructure. Tropical cities such as Hong Kong, however, have interesting historic events recorded where extreme rainfall has caused flash flooding and mud flows, leading to casualties and flood damage in parts of the city. It should be noted that risk is not a static concept. If flood protection is improved or evacuation plans are developed, then risk can be decreased. With expected advances in scientific modelling and prediction of storms and storm surges, improved warnings can be brought to bear in alerting communities at risk and in managing evacuations.

The amount of damage from a flood is dependent upon, among other factors, the size of the flooded area and water depth. Other factors include the duration of the flood and flow velocities. Furthermore, the rate at which the water rises and the time allowed for evacuation largely determine the number of casualties (Jonkman et al, 2008). Looking at the most important consequences of a flood for different economic sectors, it appears that most coastal cities are subject to similar threats from flooding, both from oceanic storm surges and from inland sources. For most ports, both land-based transportation and the use of inland waterways are of importance to connect the port areas with surrounding regions. These connections may be threatened as the clearance levels of bridges decrease during a flood. Train and subway stations may be flooded, coastal highways inundated, emergency and hospital services curtailed, and communications disrupted. Furthermore, floods cause direct economic damage to infrastructure and property, with the magnitude of the damage depending upon the depth and duration of the flood. Most estimates of flood damage rely on studies that quantify only the direct economic damages. However, other non-flooded areas may also be affected, as the supply of goods and services from (and to) the flooded area may be hindered. Production loss due to floods, however, is difficult to quantify at present. Indirect flood damage may be twice as high as the direct economic damage.

1.4 Organization of the Book

As each global city is unique in its climate change and adaptation challenge, there is no unique solution or approach that applies to all cities. What we can offer in this book are the most important building blocks that are needed to develop a flood adaptation plan for a coastal city. The book follows a so-called scenario analysis whereby solution trajectories are analysed and compared under the assumption of various long-term scenarios. The scenario method is derived from research in the area of scenario and policy analysis (Findeisen and Quade, 1985; Aerts, 2002), and is explained in detail by Aerts and Droogers (2004).

At the core of the scenario approach are the stakeholders, such as water managers, urban planners, engineers and NGOs at different levels. Their agenda, institutional setting and the way in which they communicate and interact largely determine the potential for climate adaption and whether adaptation measures can be successfully implemented. The governance aspects of climate adaptation are described in Chapter 12. Furthermore, the adaptation planning process also requires a multidisciplinary systems approach with the full participation of stakeholders. If stakeholders and the scientists and engineers involved are aware of the risks of climate change, it will be much more feasible to jointly develop a workable adaptation plan for a city. An example of how stakeholders can be involved in developing adaptation strategies is provided in Chapter 11.

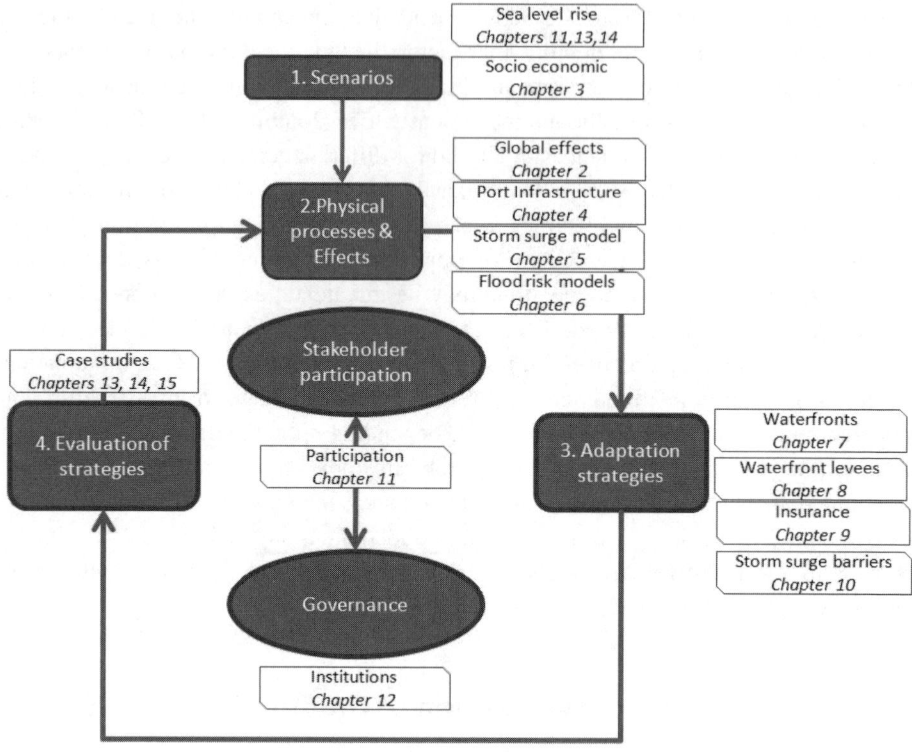

Figure 1.1 *Main building blocks required to develop a flood risk adaptation plan for a coastal city*

Stakeholders involved with developing a flood adaptation plan are confronted by four main building blocks. These building blocks (numbers 1, 2, 3 and 4) are addressed in Figure 1.1. We now briefly describe each building block:

1   *Scenarios.* In order to know what adaptation measures are needed, one should first assess the potential effects of future developments. As the future is inherently uncertain, this can be done by using a range of scenarios. A scenario is a combination of internally consistent assumptions with regards to future socio-economic (Chapter 3) and climatological developments (Chapters 11, 13 and 14). Here, scenarios are seen as external variables and cannot be influenced by the stakeholders in a city. One can think of sea-level rise, population growth or changes in national water policies. These are often developments that have an international dimension. Feedback mechanisms can be important, but are not addressed in this book. Most scenarios have a time horizon (e.g. the year 2050 or 2100).

2   *Processes and effects.* Based on the scenarios, several effects will occur in urban water systems around the globe (Chapter 2) and, more specifically, on port infra-

structure (Chapter 4). Effects can be calculated qualitatively using expert knowledge or quantitatively using numerical models. In this book we describe two types of models: storm surge models to simulate surge heights (Chapter 5) and their probability flood-risk models to estimate flood damage and their probability (Chapter 6). Flood risk is here defined as flood probability multiplied by flood damage.

3   *Adaptation strategies.* An adaptation strategy is here defined as a set of measures that share a logical cohesion. In this book we describe several strategies: in Chapters 7 and 8, innovations in waterfront architecture are addressed and how waterfront designs can be used to lower flood risks. Chapter 9 takes New York City as a case study to examine the role of insurance policies in adaptation and how insurance both covers residual flood risk and (indirectly) lowers flood risk through damage mitigation requirements. In Chapter 10, engineering options are discussed by evaluating different (existing) storm-surge barrier types and their pros and cons.

4   *Evaluation.* In this book we qualitatively describe adaptation strategies and their effectiveness for three cities: New York City (Chapter 13), Jakarta (Chapter 14) and Rotterdam (Chapter 15).

References

Aerts, J. (2002) *Spatial Decision Support for Resource Allocation: Integration of Optimization, Uncertainty Analysis and Visualization*, PhD thesis, University of Amsterdam, The Netherlands

Aerts, J. and Droogers, P. (2004) *Climate Change in Contrasting River Basins: Adaptation Strategies for Water, Food and the Environment*, CABI, Wallingford, UK, p261

Aerts, J., Sprong, T. and Bannink, B. (2008) *Attention for Safety*, Report for the Dutch Delta Committee, www.adaptation.nl, p258

Aerts, J., Major, D., Bowman, M. and Dircke, P. (2009) *Connecting Delta Cities: Coastal Cities, Flood Risk Management and Adaptation to Climate Change*, VU University Press, Amsterdam, The Netherlands

Bouwer, L. M., Crompton, R. P., Faust, E., Höppe, P. and Pielke, Jr., R. A. (2007) 'Confronting disaster losses', *Science*, vol 318, p753, http://dx.doi.org/10.1126/science.1149628

Findeisen, W. and Quade, E. S. (1985) 'The methodology of systems analysis: An introduction and overview', in H. J. Miser and E. S. Quade (eds) *Handbook of Systems Analysis: Overview of Uses, Procedures, Applications, and Practice*, John Wiley and Sons, New York, NY, pp117–150

IPCC (Intergovernmental Panel on Climate Change) (2007) *Climate Change 2007: The Physical Science Basis. Contribution of Working Group I to the Fourth Assessment Report of the Intergovernmental Panel on Climate Change*, Cambridge University Press, Cambridge, UK

Jonkman, S. N., Bo karjova, M., Kok, M. and Bernardini, P. (2008) 'Integrated hydrodynamic and economic modeling of flood damage in the Netherlands', *Ecological Economics*, vol 66, no 1, pp77–90

Kron, W. (2005) 'Flood risk = hazard x values x vulnerability', *Water International*, vol 30, no 1, pp58–68

Kundzewicz, Z. W. and Schellnhuber, H. J. (2004) 'Floods in the IPCC TAR perspective', *Natural Hazards,* vol 31, pp111–128

Milly, P. C. D., Wetherald, R. T., Dunne, K. A. and Delworth, T. L. (2002) 'Increasing risk of great floods in a changing climate', *Nature,* vol 415, pp514–517

Nicholls, R. J., Hanson, S., Herweijer, C., Patmore, N., Hallegatte, S., Corfee-Morlot, J., Château, J. and Muir-Wood, R. (2008) *Ranking Port Cities with High Exposure and Vulnerability to Climate Extremes Exposure Estimates,* OECD Environment Working Paper no 1, 19 November 2008

NPCC (New York City Panel on Climate Change) (2009) *Climate Risk Information,* NPCC, www.nyc.gov/html/om/pdf/2009/NPCC_CRI.pdf

Rosenzweig, C., Solecki, W., Hammer, S. A. and Mehrotra, S. (2010) 'Cities lead the way in climate-change action', *Nature,* 21 October, vol 467, no 7318, pp909–911

Smith, D. I. (1994) 'Flood damage estimation: A review of urban stage-damage curves and loss functions', *Water SA,* vol 20, pp231–238

2

Planning for Changes in Extreme Events in Port Cities throughout the 21st Century

Susan Hanson and Robert J. Nicholls

2.1 Introduction

There is a strong correlation between the world's largest coastal cities and the location of important international ports, with 13 of the world's 20 largest cities (based on UN, 2005) being port cities. As key nodes in transport and trade networks, ports represent critical infrastructure within the local, regional and global economy. A total of 80 per cent of world freight moves by ship, and (although recent developments in trade routes and maritime technologies have allowed ports to distance themselves from associated urban areas) port cities remain concentrations of people and locations of high economic value along the coast, which makes them particularly susceptible to climate change (see Box 2.1).

While all cities face the challenge of planning for long-term climate change and climate variability, few have systematically considered the potential effect of global climate change, either economically or environmentally, within the decision-making process. For cities located on the coast, this is particularly important given their high exposure to flooding from extreme water-level events today (Nicholls et al, 2008; Hanson et al, 2011). Moreover, it is important to remember that future impacts of such events will not only be affected by climate change, but also by land movements, exacerbated in some cases by human-induced subsidence, and global-to-local socio-economic changes, such as continued urbanization and growth in trade. The range and magnitude of impacts will therefore vary according to local conditions, economic networks, local-to-global policies, as well as the ability to adapt and minimize losses. Assessing current and potential flood and storm damage is therefore an important issue for port cities.

The scale of human and economic effects of extreme water-level events can be extensive, as shown by Hurricane Katrina in the US (see Box 2.2); while local consequences for both people and infrastructure were substantial, the economic implications extended beyond the local scale to national and global scales (e.g. RMS, 2005). Other studies, based on the analysis of land elevation data using geographic information systems (GIS), also show that such effects are likely to be substantial into the future. For example, many elements of the transportation system in the Greater New York region lie 2m to 6m above current sea level; this is within current worst-case storm modelling for extreme water levels, and the number of elements within these extreme levels will increase as global sea levels rise (Jacob, 2001). More recently, for the same region, Jacob et al (2007) estimated that a 1m global sea-level rise would increase the frequency of coastal storm surges and flooding incidences of between a factor 2 to 10, with an average of 3.

Box 2.1 The world's port cities

There are over 2500 coastal or offshore ports recorded in *Ports of the World 2009* (Lloyds List, 2009), with the majority of coastal countries around the world having at least one port (see Plate 1). Despite differing in size, these play a significant role in the country's economic welfare, and historically have a strong relationship with a neighbouring town or city – their presence promoting the concentration of population (largely associated with inward migration) and the development of supporting economic activities and infrastructure.

Maritime trade and ports infrastructure have always responded to the economic and technological development of sea trade and this will continue; but an additional future concern for both the port and its associated city is the potential consequences of climate change. For port cities, this is of particular significance as they have characteristics which make them additionally susceptible to the effects of climate change. These include:

- low elevations around estuaries or river deltas, including reclaimed ground near sea level, making them prone to flooding due to storm surge and sea-level rise;
- high populations and population densities;
- high concentrations of economic activity, often centred around the port;
- difficulty in relocating due to the requirement of deep water with a protective harbour and the historical development of supporting infrastructure.

If cities remain in these areas, the exposure to such events needs to be carefully evaluated. The costs (human and economic) of extreme events, and how they may be altered under climate and other change, are key to maintaining acceptable standards of risk management and will be an important aspect of port city governance during the 21st century. As the majority of coastal extreme-event damages and fatalities are caused by flooding, the main manifestations of climate change for port cities are relative sea-level change and the potential increase in frequency and/or intensity of storm surges. Understanding how these (and their impacts) may change over time can help in determining acceptable risk levels and the development of appropriate policies and plans to improve preparedness and response systems, whether at the global (by reducing greenhouse gas emissions) or city scale (via flood risk management).

This chapter reviews current and future exposure to flood events under a range of climate and non-climate scenarios. It demonstrates that this exposure will grow substantially during this century, and the associated growth in flood risk can only be managed by mitigation and adaptation strategies.

Box 2.2 The Gulf Coast, New Orleans and Hurricane Katrina

New Orleans, Louisiana, has always faced major flooding risks as it is located in the low-lying Mississippi Delta, is surrounded by water and is subject to high river flows and frequent hurricanes. The city, situated largely below sea level, is surrounded by levees and depends upon pumped drainage to remain dry.

Hurricane Katrina was the 11th tropical cyclone of the 2005 season, recording storm surge heights of up to 9m, which allowed floodwaters to reach at least 10km inland along sections of coastal Mississippi and up to 20km inland along bays and rivers. As a result, Hurricane Katrina is acknowledged as the most destructive US storm in terms of economic losses, with thousands of homes and businesses affected (see Table 2.1). While the surge at New Orleans was smaller, at about 4m, most of the storm's damage occurred here due to the widespread failure of the levees, which led to 80 per cent of the city being subject to flooding. The number of fatalities was also the largest of recent US hurricane events, but was substantially fewer than the 1900 hurricane, which killed in excess of 5000 people and led to almost complete destruction of Galveston, Texas.

Table 2.1 *Top five US hurricanes in terms of economic damage with associated fatalities*

Date	Event	Port cities affected	Severely affected areas	Overall economic losses (US$ millions, 2010)	Attributed fatalities
August 2005	Hurricane Katrina	New Orleans	Louisiana, Mississippi, Alabama	142,000	more than 1300
September 2008	Hurricane Ike	Houston	Texas, Louisiana, Cuba	39,300	around 200
August 1992	Hurricane Andrew	Miami	South Florida, Louisiana, Bahamas	41,800	around 60
September 2004	Hurricane Ivan	n/a	Caribbean Islands, Alabama, Florida, Louisiana, Texas	26,800	around 120
October 2005	Hurricane Wilma	Miami	Cuba, Florida, Bahamas	25,000	around 25

Note: n/a = not available.
Source: Linham et al (2010)

Despite this record of major hurricane-related flooding, post-event analysis exposed failures in defence structures and emergency management procedures for New Orleans. The pre-Katrina defences were clearly inadequate, as shown by the widespread overtopping and breaching. The concept, design criteria, construction and maintenance of the levee system were particularly criticized as was the siting of residential areas, over 130 petroleum and chemical plants, and other infrastructure within the Mississippi River floodplain. New Orleans' flood defences are now largely rebuilt and upgraded to a much higher standard (between 1/100-year and 1/500-year standard) than before Katrina at a cost of US$15 billion. It should be noted that if New Orleans was located in The Netherlands, defence standards would be even higher: currently 1/10,000, with a recommended increase to 1/100,000 in the coming decades (cf. Deltacommissie, 2008). Recent reports have also recommended that potential hurricane impacts are accorded a higher priority than in the past and be central to future development plans and decisions (e.g. Committee on New Orleans Regional Hurricane Protection Projects, 2009).

2.2 Coastal Extreme Water Events

Extreme coastal water levels are produced by the combination of astronomical tides, storm surge heights and mean sea level, each of which may change over time (see Box 2.3). Changes in mean sea level due to relative sea-level change (the combination of global and regional sea-level change and vertical uplift/subsidence) are the best understood of these components, and projections to the end of this century are not uncommon. Current understanding of potential changes in storm surges is limited; intensification is thought possible in some regions, especially for tropical storms (Meehl et al, 2007). For many areas, observed increases in extreme water levels are consistent with increases in global mean sea level (e.g. Menéndez and Woodworth, 2010). However, while it is not possible to predict with absolute certainty how any of these components will change into the future, it is possible to create scenarios (or plausible futures) that can provide a first approximation for impact and adaptation assessments, supplying a robust basis for planning decisions. It is also important to remember that flood levels will increase and become more frequent where sea levels rise even if storm intensity and behaviour remains unchanged; this indicates that erosion and flooding incidences will become increasingly relevant for coastal regions around the world and need to be considered in terms of their potential damage, both human and economic.

Box 2.3 Components of extreme water levels

The magnitude of extreme water-level events at any particular time or place is influenced by tidal conditions, storm severity and regional mean sea level. These will change over time and can be predicted to varying degrees of certainty:

- *Storm surge.* Storm surge is the temporary rise in water level due primarily to variations in wind and air pressure. Changes in storm surge height and frequency due to climate change are difficult to predict given the large uncertainty surrounding future wind patterns, although significant differences in intensity might be expected depending upon the region.
- *Astronomical tides.* Astronomical tides are the response of the oceans to the gravitational attractions of the sun and moon. They are relatively well understood and can be easily predicted. In future projections they are often an implicit component of total water levels.
- *Climate-induced mean sea-level change (average elevation).* Changes in sea level are a result of the change in the global volume of the ocean. In the 20th/21st century, this is expected to be primarily due to:
  - thermal expansion of the ocean as it warms;
  - the melting of small glaciers and ice caps due to human-induced global warming; and
  - changes in the mass balance of the Greenland and Antarctic ice sheets, which is less certain.

  The Intergovernmental Panel on Climate Change's (IPCC's) *Fourth Assessment Report* (AR4) (IPCC, 2007) contains estimates of the global mean sea-level change, with likely rises of between 0.18m and 0.59m projected by 2099. Regional rises in sea level can vary significantly, with substantial departures from the global average value for the thermal expansion component of sea-level change (see Lowe et al, 2009). This is largely related to changes in meteorological conditions and the gravity field of the Earth.
- *Land subsidence/uplift.* Changes in sea level due to vertical land movement also occur in most places and are particularly relevant in deltaic environments. Natural causes include neotectonics, glacio-isostatic adjustment (GIA) and sediment compaction/consolidation. These changes can be regional, slow and steady, as in the case of GIA, but also localized, large and abrupt (e.g. as associated with earthquakes). Human activity can exacerbate this movement – for example, the extraction of groundwater, causing accelerated subsidence in large cities such as New Orleans and Jakarta.

2.3 Exposure to Coastal Extreme Water-Level Events

Recent work carried out by the Organisation for Economic Co-operation and Development (OECD) (Nicholls et al, 2008) produced a first estimate of the exposure of the world's 136 large port cities (population exceeding 1 million inhabitants in 2005) to coastal flooding due to sea-level rise and storm surge in 2005 and during the 2070s. The report also recognized that the socio-economic reasons why cities were located at the coast are important and included scenarios of socio-economic changes. The report took a global overview of coastal flood exposure in the 136 world port cities and produced rankings based on physical exposure and socio-economic vulnerability to climate extremes – in particular, storm surges, the effects of relative sea-level rise due to global climate change, and local subsidence, both individually and in combination. The study assessed two indicators of exposure to flooding – population and assets – using population distributions as a function of elevation to estimate the population and assets below a 1/100-year extreme water level for both the current situation and in the 2070s, excluding the influence of any defences (in effect, a worst-case scenario, which can translate into major losses during extreme events when any defences may fail due to breaching or overtopping). By excluding the effect of defences, exposure discloses to a large extent the reliance that people and infrastructure located in the floodplain place on any formal or informal flood defences, and allows for comparison of the potential impact of flooding between cities. Over longer timescales, it also includes the uncertainty around whether or not defences will be constructed or whether they will be sufficiently maintained to be fully effective.

The analysis suggests that about 40 million people (0.6 per cent of the global population, or roughly one in ten of the total port city population in the cities considered) were living at elevations below the 1/100-year coastal flood event in 2005 (see Figure 2.1a). By the 2070s, total population exposed could grow more than threefold (see Figure 2.1b) due to the combined scenarios of sea-level rise, human-induced and natural subsidence, population growth and urbanization. While the assumptions within these scenarios mean that these estimates can be regarded as being at the high end of projections, the pattern in exposure and concentration of exposure in a relatively low percentage of the cities considered (see Figure 2.2) illustrates that many cities retain high levels of exposure irrespective of the magnitude of change. These cities include Mumbai, Guangzhou, Shanghai, Miami, Ho Chi Minh City, Kolkata, New York, Osaka-Kobe, Alexandria, New Orleans, Tokyo, Tianjin, Bangkok, Dhaka and Hai Phong.

A similar city-level concentration occurs for asset exposure, although the focus for the cities concerned moves from developed towards developing countries by the 2070s. The total value exposed in 2005 across all cities considered is estimated to be US$3000 billion, corresponding to around 5 per cent of global gross domestic product (GDP) in 2005 (both measured in international US dollars). In 2005, the top ten cities in terms of assets exposed were Miami, Greater New York, New Orleans, Osaka-Kobe, Tokyo, Amsterdam, Rotterdam, Nagoya, Tampa-St Petersburg and Virginia Beach.

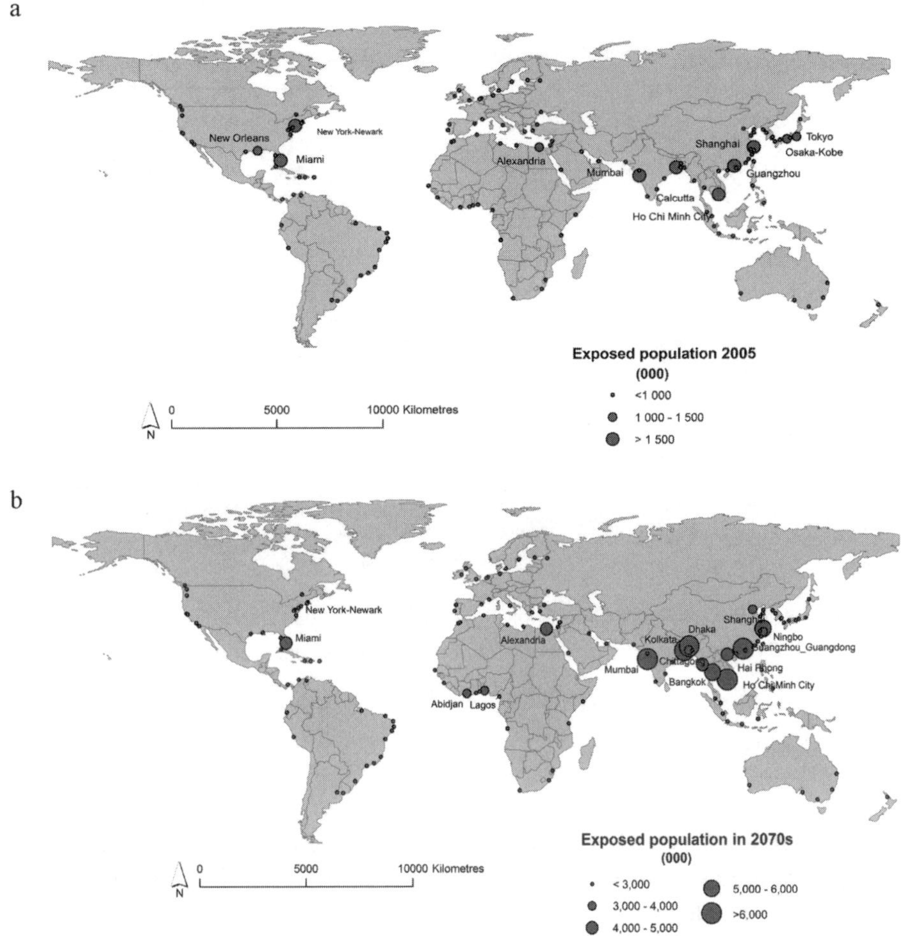

Figure 2.1 *(a) Spatial distribution of population exposed to the 1/100-year storm event in 2005 for the world's large port cities; (b) population exposure to the 1/100-year flood event in the 2070s under a scenario which combines 0.5m sea-level rise, a 10 per cent increase in surge height for those regions considered plausible in the IPCC's* Fourth Assessment Report *(IPCC, 2007), natural land movement with additional human-induced subsidence for those cites known to be subject to its effect, and population growth, including urbanization (note difference in scale)*

Source: Nicholls et al (2008)

These cities contain 60 per cent of the total exposure, but are from only three (wealthy) countries: the US, Japan and The Netherlands. By the 2070s, asset exposure increases to more than ten times current levels, approximately 9 per cent of projected global GDP during this period, with Miami, Guangzhou, New York–Newark, Kolkata, Shanghai,

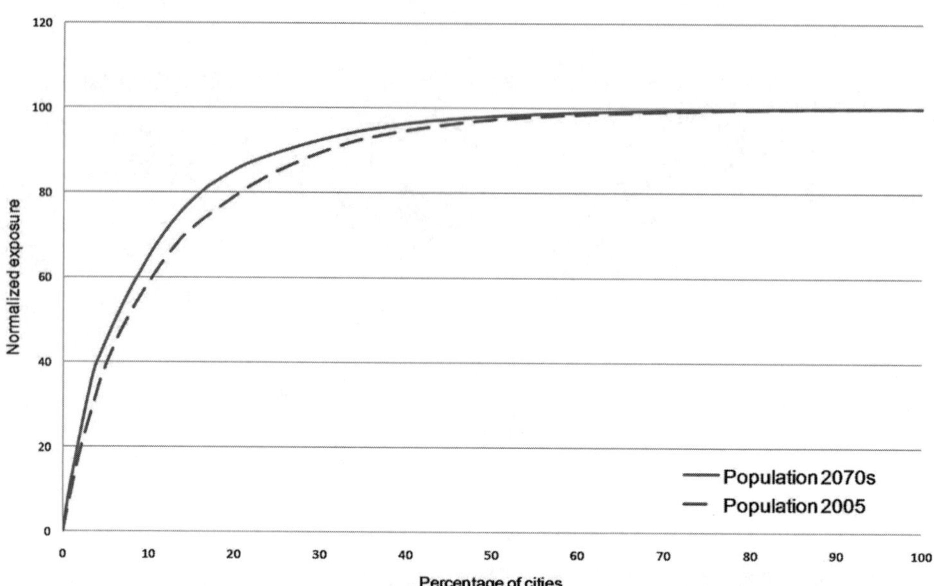

Figure 2.2 *Cumulative distribution of total exposure for population in 2005 and during the 2070s illustrating the concentration of exposure in a small percentage of the 136 cities considered*

Source: Nicholls et al (2008)

Mumbai, Tianjin, Tokyo, Hong Kong and Bangkok forming the top ten cities. This higher proportion of exposure found in Asia corresponds with projected increases in both population and levels of per capita GDP.

In addition to absolute exposure levels, some cities show proportionately high increases in levels of exposure to sea-level rise and storm surge. These are mainly found in rapidly growing cities in developing countries in Asia, Africa and, to a lesser extent, Latin America for both population (see Figure 2.3) and assets (all but one of the top 20 cities for percentage increase in asset exposure during the 2070s is an Asian city; the exception is Miami). This indicates that, on the global scale, population growth, socio-economic growth and urbanization are important drivers of the overall increase in exposure, particularly in developing countries, as low-lying areas are urbanized. For developed countries (where population and economic growth are expected to be smaller), climate-related factors are proportionately more important.

While these exposure levels are unlikely to happen as the result of a single event and even if all cities are well protected against extreme events, large-scale city flooding may remain frequent at the global scale as so many cities are located in low-lying areas. For instance, assuming that flooding events are independent, there is a 74 per cent chance of having one or more of the 136 large port cities affected by a 100-year event every

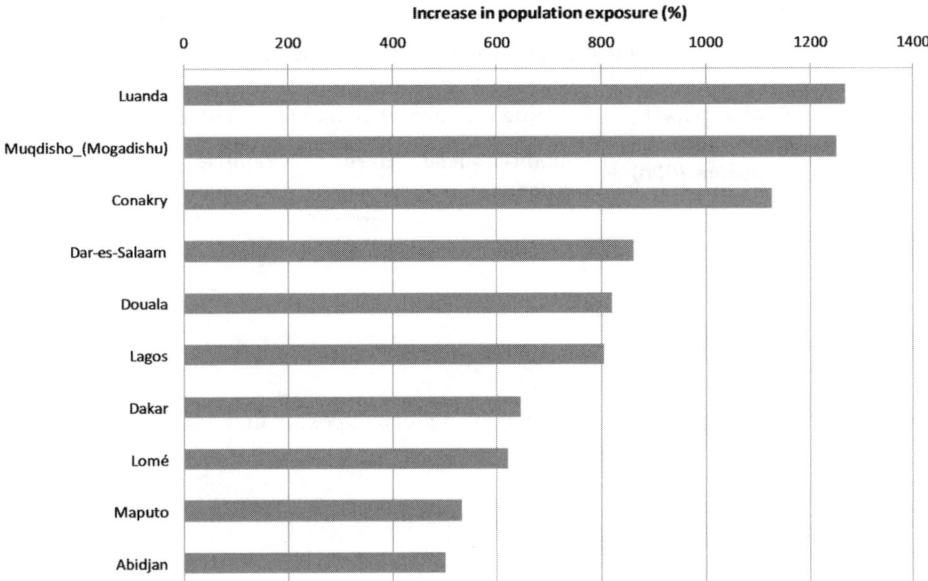

Figure 2.3 *Top ten cities in Africa with the highest proportional increase in exposed population by the 2070s under the same scenario as Figure 2.1(b) relative to 2005*

Source: Nicholls et al (2008)

year, and a 99.9 per cent chance of having at least one city affected by such an event over a five-year period. At the global scale, 100-year and even 1000-year events will frequently affect individual large port cities.

2.4 Reducing Exposure Levels

Examination of the proportion of exposure growth associated with the components of extreme coastal water levels can highlight the most effective methods of reducing exposure levels – those that can be most immediately beneficial – and whether decisions need to be made at international (global climate mitigation) or local (climate mitigation, adaptation and flood risk management) levels.

For the 136 port cities in the OECD report (Nicholls et al, 2008), research carried out under the UK AVOID programme (Hanson et al, 2010) indicated that, if no reduction in global greenhouse gas emissions is anticipated, the majority of any increase in exposure levels is associated with increases in city populations (see Table 2. 2). Nearly one quarter is directly related to the global climate (sea level and storm enhancement), with a similar amount linked to human-induced subsidence, despite this being a factor in only a limited number of cities. This indicates that a combination of policies addressing

Table 2.2 *Growth in global exposure from 2005 to the 2070s and its relative contribution by extreme water-level component*

Climate scenario	Global growth in population exposure (000)	Extreme water-level component contribution (%)			
		Human-induced subsidence*	More intense storms**	Relative sea level	Urbanization
Unmitigated	85,525	20	11	13	56

Note: * Applies to 37 and ** applies to 55 cities of the 136 cities considered.
Source: Hanson et al (2010)

the different components is the ideal way to reduce exposure and, consequently, flood risk levels during this century.

2.4.1 Global climate mitigation

Hanson et al (2010) investigated the potential benefits of global climate mitigation on exposure to coastal flooding by 2070 using a range of greenhouse gas (GHG) scenarios based on the IPCC's A1B storyline (Nakicenovic and Swart, 2000). Sea-level and storm surge projections were calculated based on temperature changes associated with the following climate scenarios (Gohar and Lowe, 2010):

- peak GHG emissions in 2016 with post-peak annual decreases in emissions of 5 per cent;
- peak GHG emissions in 2030 with post-peak annual decreases in emissions of 2 per cent;
- unmitigated A1B climate scenario.

This replaced the global sea-level scenario used in the OECD study with spatially variable sea levels for the 136 cities of between 0.12m and 0.49m across the scenarios. In addition, changes in storm surge heights (for those parts located in tropical cyclone-prone areas) were scaled with temperature change during the century.

Unsurprisingly, the study showed that exposure levels are reduced if GHG emissions are reduced; the magnitude of the decrease is greater for a more rapid initiation of GHG reductions and for larger total reductions. However, it highlighted that, due to the time-delayed response of sea levels to climate mitigation (see Nicholls and Lowe, 2004), the most substantial benefits of climate mitigation will only become apparent towards the end of the century and nearly 90 per cent of the growth in exposure found under the unmitigated scenario still occurs by the 2070s. Despite this, reductions in exposure of up to 9 million people within the 136 cities under the most highly mitigated scenario were still found by the 2070s.

The pattern of population exposure reduction is consistent across mitigation scenarios due to the global nature of emission reductions. While all port cities show a reduction in exposure, the main regional benefits are found in the rapidly developing areas of Asia (see Figure 2.4). This reduction in exposure would be amplified if the population growth and urbanization rates of the cities are also reduced (see Figure 2.5). This effect would be particularly noticeable in rapidly developing Asian countries such as China. For asset exposure by the 2070s, the benefits of climate mitigation are more evenly distributed, with the more developed countries such as the US, Japan and countries in Europe benefiting along with China and other Asian countries. This reflects the convergence of per capita GDP projections, as well as the population increases expected in Asia.

In common with other cities, port cities have an important role to play in achieving reductions in emissions, although the relationship between population growth, urban growth and greenhouse gas emissions is complex. However, even taking an optimistic view and assuming that significant global mitigation is achieved at a near point in the future, substantial adaptation will still be needed. The large exposure in terms of population and assets is still likely to translate into recurring city-scale disasters at the global scale. This makes it essential to consider both adaptation as well as disaster planning and management strategies in order to determine acceptable levels of risk and, perhaps more importantly, to address what happens when adaptation and, especially, defences fail.

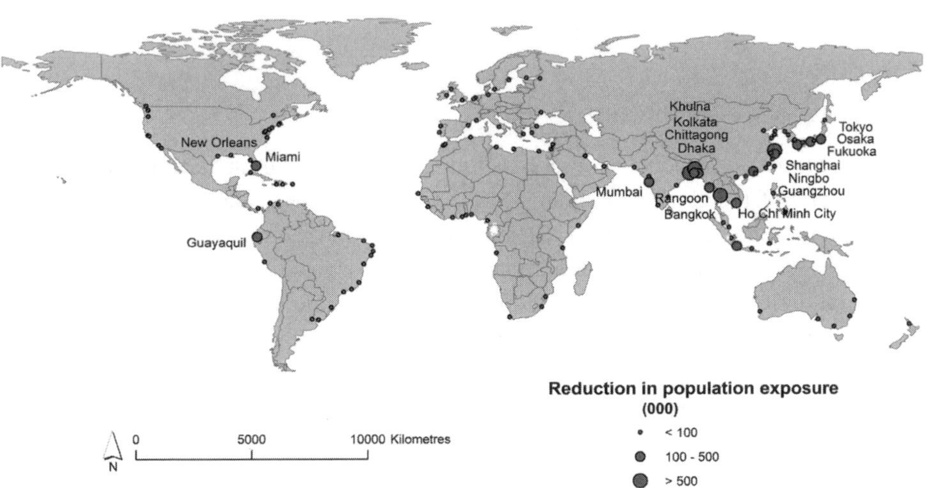

Figure 2.4 *Reduction in population exposure by 2070 with an annual reduction of greenhouse gases of 5 per cent from 2016 and a rapid urbanization scenario*

Source: Hanson et al (2010)

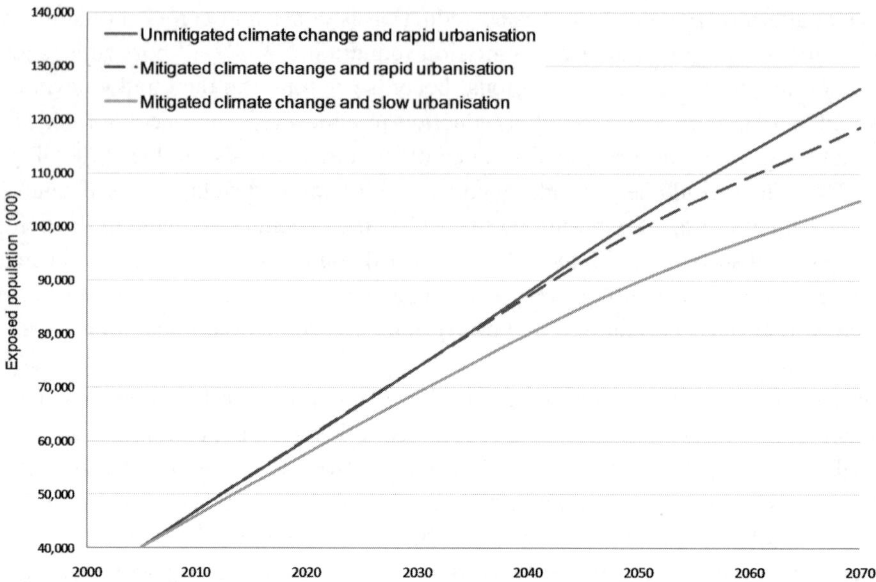

Figure 2.5 *Reduction of population exposure under a mitigated climate scenario (maximum emissions in 2016) when compared to an unmitigated climate, and the further reduction in exposure achieved if a reduced urbanization rate is included*

Note: Urbanization rates are based on United Nations 2005 to 2030 urbanization rates, which were extended to 2070 for the rapid scenario and reduced by 25 per cent for the slow scenario.
Source: Hanson et al (2010)

2.4.2 City-scale adaptation

Given that most of the benefits of global climate mitigation will not become appreciable until the end of the century, reducing flood risk at the city scale needs to encompass more than measures to address climate change mitigation; coastal adaptation also needs to be considered. Although there are many available coastal adaptation techniques, no individual option should be relied on to manage and reduce risks to acceptable levels; the most effective adaptation policy option, in addition to long-term planning to ensure viable development and settlement in low-lying areas, is a portfolio of the approaches described in Box 2.4. This means that successful adaptation (and mitigation) is largely dependent upon the integration of planning scales, temporally and spatially.

Planning for climate change usually requires consideration of a timescale of between 50 to 100 years, and spatial scales from the local to global. Development decisions usually take place on a much shorter, often reactive, economic and political timescale with a more local emphasis. Additionally, effective disaster management strategies, safer land-use choices, more resistant infrastructure and protection investments commonly take time to develop. For example, major coastal defence projects, such as the Thames

Barrier in the UK, have shown that implementing coastal protection infrastructure typically has a lead time of 30 years or more, although it is notable that the upgrade of defences in New Orleans post-Katrina has occurred over only six years!

At each level, the actions of decision-makers may have significant implications decades into the future; including the consideration of long-term issues such as climate change would therefore encourage robust and adaptable solutions, broadening the management options available to future generations rather than focusing on optimal strategies for the current situation. The consequences of not considering long-term planning in this way are events such as that experienced in New Orleans where local impacts have had, and continue to have, important ramifications.

Policy development to address climate change depends upon many parameters, including population risk aversion and broader policy goals (e.g. priority attributed to poverty reduction and disaster risk reduction). However, port cities are, to a large extent, dependent upon their own actions to reduce the impacts of future climate change. Urban growth, given the population numbers and densities involved, and the importance of ports in future global economic and social systems, offer these cities the opportunity to play a key role in developing climate mitigation and adaptation techniques. In particular, the cities in Asia and Africa that are undergoing rapid urban growth (see Figure 2.3) have an opportunity to plan well-designed adaptable cities, while also contributing to global environmental climate mitigation efforts. Taking advantage of this opportunity will require the adoption of proactive effective and participatory strategies to urban planning aimed at improving energy efficiency, and reducing emissions while providing adequate economic and living conditions. The scale of the potential exposure indicated by the OECD and AVOID research argues that the benefits of a focused effort across all scales of management would be well worth the effort and create a knowledge base that could help to advance action in many locations in the coming decades.

Of the adaptation shown in Box 2.4, the improvement in hard-engineered defences, whether by upgrades or new construction schemes, is an immediate adaptation option that is widely adopted but this requires significant economic input. A global assessment of the costs of climate change (Nicholls et al, 2010) indicated that the cost of maintaining port areas at their current height relative to current mean sea levels by 2050 with a global sea-level rise of 0.4m would, for 108 of the ports represented in Plate 1, be in the region of US$19.6 billion (at 2005 prices with no discounting). While still nominal in comparison to the construction and maintenance of sea walls in the same countries, this cost can only be expected to increase during this century if climate mitigation is not achieved. In addition, while the extent of flooding from storm surges can be reduced by the repair and enhancement of current flood defences, continual maintenance and improvements will be needed to maintain, or enhance, protection levels in the future (Ward et al, 2007). They may also promote a false sense of security that areas behind structures are safe for habitation and development, with consequent high losses in the event of future failure. This means that the issue of residual risk and its management must always be considered during the planning process.

Box 2.4 Adaptation options for port cities

- A long-term approach to planning for climate change and urban growth.
- Relocation, redesign, construction and maintenance of coastal protection schemes (e.g. levees, sea walls, dikes, raising dock and wharf levels, and elevating other infrastructure) in order to maximize and maintain protection levels.
- Consider rapidly deployable temporary barriers and protection structures.
- If appropriate, reduce any human-induced subsidence (e.g. groundwater extraction).
- Relocation of existing infrastructure, commercial properties and associated populations.
- Building regulations (e.g. flood-proof buildings) to reduce the consequences of flooding.
- Integrated emergency evacuation and disaster management procedures.
- Contingency plans to cover delays or cancellations in port operations and alternative transport routes.
- New and improved technologies for extreme event detection (early warning systems).
- Risk-sharing through taxes, insurance and reinsurance.

Relocation seems unlikely for valuable city infrastructure, considering the cost of rebuilding infrastructure and buildings (e.g. more than US\$400 billion in Miami today), and the political difficulties to do so (illustrated by the public objections after the French government suggested the destruction of 1500 houses in the highest-risk areas affected by winter storm Xynthia in 2010). Some ports also face difficulties as the adaptation or expansion of infrastructure or facilities can be limited by surrounding residential and other city development. Transport infrastructure networks that have built up around ports can also add to the restriction of relocation as an adaptation option.

For human-induced subsidence, the increased risk could be mitigated to some degree by avoiding the processes that lead to shallow subsidence, such as groundwater withdrawal, alongside urban water-demand management. Several Asian cities appear to have successfully implemented such water management policies during the 20th century, including Tokyo, Osaka-Kobe and Shanghai, with others, such as Jakarta, beginning to address the issue. Exchange of experience between these and other subsidence-prone cities could be beneficial.

Flood warnings, evacuation plans and disaster management are strategies aimed at minimizing risks to human life. The presence of a port within the city presents opportunities for the development of these strategies, which are not available elsewhere, and these should be considered within the city-wide planning process. However, these strategies will do little to change risks to assets. Building regulations, including flood-proofing and flood-resistant utility systems, have the potential to reduce these losses, as well as enabling a swifter recovery period.

Adapting to climate change will not only require responding to the physical effects of global warming, but will also require adapting the way in which risk is conceptualized, measured and managed (Kousky and Cooke, 2009). Growing levels of exposure may have important consequences on the insurance and reinsurance markets. For example, a huge increase in the amount of capital needed to insure these risks at the global scale may be needed, and changes in risk modelling to include the consideration of secondary consequences, such as the unanticipated failure of the pumping system in New Orleans, could change risk appreciation (RMS, 2005). However, along with taxes, insurance policies can also encourage or discourage development and behaviour to improve the economic, social and environmental components of the city (Bagstad et al, 2007).

2.5 Port Cities in the Future

Port cities in the future will need robust mitigation and adaptation plans that can reduce the possible impacts of climate change and other factors which increase exposure to flooding. The benefits of global climate change mitigation policies, while significant in the long term, are not manifest at the city scale until towards the end of the 21st century. Hence, city-scale adaptation must become a core element of long-term urban planning. Interactions between national- and city-level decision-makers, public and private, as well as national and often international policy-makers (i.e. where relevant, official development assistance) inevitably shape the way in which cities and city infrastructure develop. This is particularly true in port cities where urban growth influences the space available for port development, while port development affects the urban structure in terms of economic health, environmental impacts, and cultural and social implications. By working in partnership, local, regional and national decision-makers will bring greater resources and expertise to bear on the adaptation problem, including policies which establish incentives for public and private investors to develop adaptation plans. National governments are well placed to assist port city adaptation efforts by bringing available research to bear on specific locations in order to better understand the nature of the risks in local contexts and the costs and benefits of adaptation, and to facilitate the development of risk-sharing approaches and insurance markets. Local governments, on the other hand, will need to work closely with local stakeholders and decision-makers to assess and choose amongst available adaptation options to reflect acceptable risk levels and balance the interests of those most directly affected. This broad engagement across scales of governance and different types of actors will be necessary to protect against, and to manage, coastal flood risk, especially if cities do expand into high flood-risk areas.

However, it is important to remember that climate change offers opportunities as well as challenges. Sea-level and storm pattern changes may make some existing areas less viable for commercial port activities, but may open new routes (such as the opening of a North-West Passage) or development locations. Larger 'super-ports' or 'hubs', which require space and economies of scale, may move away from the traditional port areas or even offshore, requiring new supporting infrastructure and population centres.

This, in addition to almost all future population growth expected to occur in urban areas, mostly in developing countries, as well as growth in the economic significance of ports and new global markets, means that changes in the port and shipping network are almost inevitable.

In effect, a key question for determining the impact of climate change for current port cities is what will the maritime transport network look like in 50 to 100 years' time?

Acknowledgements

Much of the work discussed here was carried out in conjunction with J. Corfee-Morlot of the OECD; Dr S. Hallegatte from the Centre International de Recherche sur l'Environnement et le Développement (CIRED); and colleagues based at Risk Management Solutions (RMS), London. It was largely funded by the OECD, supported by the UK's AVOID programme (DECC and Defra) under contract GA0215.

References

Bagstad, K. J., Stapleton, K. and D'Agostino, J. R. (2007) 'Taxes, subsidies, and insurance as drivers of United States coastal development', *Ecological Economics*, vol 63, pp285–298

Committee on New Orleans Regional Hurricane Protection Projects (2009) *The New Orleans Hurricane Protection System: Assessing Pre-Katrina Vulnerability and Improving Mitigation and Preparedness*, National Academies Press, Washington, DC

Deltacommissie (2008) *Working Together with Water: A Living Land Builds for Its Future*, Deltacommissie, The Netherlands

Gohar, L. K. and Lowe, J. A. (2010) 'Summary of the emissions mitigation scenarios: Part 2', in *AVOID: Avoiding Dangerous Climate Change through Stabilising Greenhouse Gas Concentrations Report AV/WS1/D1/03*, Department of Energy and Climate Change (DECC) and Department for Environment, Food and Rural Affairs (Defra), London

Hanson, S., Nicholls, R. J., Hallegatte, S. and Corfee-Morlot, J. (2010) 'The effects of climate mitigation on the exposure of worlds large port cities to extreme coastal water levels', in *AVOID: Avoiding Dangerous Climate Change Report AV/WS2/D1/07 (DECC: GA0215 / GASRF123)*, Department of Energy and Climate Change (DECC) and Department for Environment, Food and Rural Affairs (Defra), London

Hanson, S., Nicholls, R., Patmore, N., Hallegatte, S., Corfee-Morlot, J., Herweijer, C. and Chateau, J. (2011) 'A global ranking of port cities with high exposure to climate extremes', *Climatic Change,* vol 104, p89

IPCC (Intergovernmental Panel on Climate Change) (2007) *Climate Change 2007: The Physical Science Basis. Contribution of Working Group I to the Fourth Assessment Report of the Intergovernmental Panel on Climate Change*, Cambridge University Press, Cambridge, UK

Jacob, K. (2001) 'Infrastructure', in C. Rosenzweig and W. D. Solecki (eds) *Climate Change and a Global City: The Potential Consequences of Climate Variability and Change – Metro East*

Coast. Report for the U.S. Global Change Program, National Assessment of the Potential Consequences of Climate Variability and Change for the United States, Columbia Earth Institute, New York, NY

Jacob, K., Gornitz, V. and Rosenzweig, C. (2007) 'Vulnerability of the New York City metropolitan area to coastal hazards, including sea-level rise: Inferences for urban coastal risk management and adaptation policies', in L. McFadden, R. J. Nicholls and E. Penning-Rowsell (eds) *Managing Coastal Vulnerability*, Elsevier, Oxford, UK

Kousky, C. and Cooke, R. M. (2009) *Climate Change and Risk Management: Challenges for Insurance, Adaptation, and Loss Estimation*, Resources for the Future, Washington, DC

Linham, M., Green, C. and Nicholls, R. J. (2010) 'Costs of adaptation to the effects of climate change in the world's large port cities', in *AVOID: Avoiding Dangerous Climate Change Report AV/WS1/D1/02*, Department of Energy and Climate Change (DECC) and Department for Environment, Food and Rural Affairs (Defra), London

Lloyds List (2009) *Ports of the World 2009*, Informa and Maritime, London

Lowe, J. A., Howard, T., Pardaens, A., Tinker, J., Holt, J., Wakelin, S., Milne, G., Leake, J., Wolf, J., Horsburgh, K., Reeder, T., Jenkins, G., Ridley, J., Dye, S. and Bradley, S. (2009) *UK Climate Projections Science Report: Marine and Coastal Projections*, Meteorological Office, Hadley Centre, Exeter, UK

Meehl, G. A., Stocker, T. F., Collins, W. D., Friedlingstein, P., Gaye, A. T., Gregory, J. M., Kitoh, R., Knutti, R., Murphy, J. M., Noda, A., Raper, S. C. B., Watterson, I. G., Weaver, A. J. and Zhao, Z.-C. (2007) 'Global climate projections', in S. Solomon, D. Qin, M. Manning, Z. Chen, M. Marquis, K. B. Averyt, M. Tignor and H. L. Miller (eds) *Climate Change 2007: The Physical Science Basis. Contribution of Working Group 1 to the Fourth Assessment Report of the Intergovernmental Panel on Climate Change*, Cambridge University Press, Cambridge, UK, and New York, NY

Menéndez, M. and Woodworth, P. L. (2010) 'Changes in extreme high water levels based on a quasi-global tide-gauge data set', *Journal of Geophysical Research. Oceans*, vol 115, p15

Nakicenovic, N. and Swart, R. (eds) (2000) *Emissions Scenarios: Special Report of the Intergovernmental Panel on Climate Change*, Cambridge University Press, Cambridge, UK

Nicholls, R. J. and Lowe, J. A. (2004) 'Benefits of mitigation of climate change for coastal areas', *Global Environmental Change-Human and Policy Dimensions*, vol 14, pp229–244

Nicholls, R. J., Hanson, S., Herweijer, C., Patmore, N., Hallegatte, S., Corfee-Morlot, J., Chateau, J. and Muir-Wood, R. (2008) *Ranking Port Cities with High Exposure and Vulnerability to Climate Extremes: Exposure Estimates,* Environmental Working Paper no 1, Organisation for Economic Co-operation and Development (OECD), Paris

Nicholls, R. J., Brown, S., Hanson, S. and Hinkel, J. (2010) *Economics of Coastal Zone Adaptation to Climate Change, Development and Climate Change*, Discussion Paper no 10, International Bank for Reconstruction and Development/World Bank, Washington, DC

RMS (2005) *Hurricane Katrina: Profile of a Super Cat – Lessons and Implications for Catastrophe Risk Management*, Risk Management Solutions, Newark, CA

UN (United Nations) (2005) *World Urbanization Prospects: The 2005 Revision*, United Nations, New York, NY

Ward, R. E., Muir Wood, R. and Grossi, P. (2007) 'Flood risk in megadelta coastal cities: Lessons from New Orleans', Paper presented to the AGU 7th Joint Assembly, American Geophysical Union, 22–25 May, Acapulco, Mexico

Socio-Economic Scenarios in Climate Adaptation Studies

*Susan van 't Klooster, Michiel van Drunen
and Eric Koomen*

3.1 Introduction

Climate change and sea-level rise projections are important factors in the estimation of future flood risk. Recently, a large number of studies have evaluated the conditions and time by which impacts of climate change would become apparent and their potential effects (e.g. Meehl et al, 2007; Schneider et al, 2007). Such effects include, for example, increased damages and other losses as a result of flooding.

In addition to physical boundary conditions, socio-economic changes also affect the risk of flooding. For example, population growth, increase in the value of assets and land-use changes (industrial expansion, urbanization, use of different crop types, nature development) may change the expected losses in the case of flooding. Yet, few studies have attempted to combine projections of changes in climate hazards and socio-economic trends and exposure (cf. Bouwer et al, 2010).

This chapter describes how socio-economic trends and developments can be combined effectively in flood risk studies. In various climate change impact assessments, efforts have been undertaken to combine socio-economic and climate scenarios (e.g. Lorenzoni et al, 2000a, 2000b; O'Brien and Leichenko, 2000; Strzepek et al, 2001; Alcamo et al, 2006; Verburg et al, 2006). However, methodological reflection on the process (choices, considerations, discussions, struggles, compromises, unproductive steps, flaws, practical adjustments, experiments, difficulties, challenges and local solutions) is generally lacking. If available, this normally consists of very short descriptions of some main steps or the provision of a simple scheme. In this chapter we build on the experience from two recent research projects to reconstruct different ways of combining socio-economic and climate scenarios in these flood risk studies.

This chapter is set up as follows. In sections 3.2 and 3.3 we explain what socio-economic scenarios are and why they are relevant for flood risk studies. In sections 3.4 and 3.5 the two case studies are introduced that combine socio-economic and climate scenarios: one qualitative and one quantitative approach. We conclude the chapter by discussing some lessons learned regarding the combination of socio-economic trends and developments in flood risk studies.

3.2 What Are (Socio-Economic) Scenarios?

A way of systematically exploring future trends and developments, and evaluating their potential impacts, is by developing scenarios, which can be understood as descriptions of hypothetical situations and stories about a possible future. Most scenarios take the present as the starting point and coherently describe various 'hypothetical sequences of events' (Kahn, cited in Aligica, 2004, p75). By identifying so-called drivers of change, different narrative storylines are constructed that are then fleshed out quantitatively (using models) or qualitatively. This mapping of a 'possibility space' (Berkhout and Hertin, 2002; Berkhout et al, 2002) is aimed towards systematic assessment of the implications of various evaluations of uncertainty, and ultimately the identification of robust strategies.

Socio-economic scenarios are scenarios in which future population and human development, economic conditions, changes in land cover, water supply and demand, agricultural developments, energy consumption and biodiversity (or a selection of these trends) are systematically explored. General objectives underlying these assessments are to characterize the sensitivity, adaptive capacity and vulnerability of environmental, social and economic systems (Carter et al, 2001).

Well-known and widely used examples of socio-economic scenarios in climate assessments are the *Special Report on Emission Scenarios* (SRES) developed by the Intergovernmental Panel on Climate Change (Nakicenovic and Swart, 2000), the *Global Environment Outlook 3* (GEO-3) by the United Nations Environment Programme (UNEP, 2002) and the *Foresight Futures* developed by UK Foresight Programme (Foresight Futures, 2002).

Despite many variations in goals, focus and approach, these socio-economic scenarios share common features. First, the standard tool to develop socio-economic (and other) scenarios is the two-dimensional scenario matrix (Ringland, 2002; Berkhout and Hertin, 2002; van 't Klooster, 2008; van Asselt et al, 2010). The scenario matrix is composed of two axes (A and B) and four quadrants (see Figure 3.1).

Most socio-economic scenarios consider more or less similar 'key uncertainties' (the drivers determining the four storylines) and are therefore based on similar assumptions. As a result, the scenarios generated by the scenario studies are quite similar, as shown in Table 3.1. The table, for example, indicates that the A1, markets first and world markets scenarios share many similarities that can be characterized as a conventional world where market forces dominate (first row).

One explanation for the overlap in scenario approaches and storylines is that existing scenario sets are often 'recycled' as a starting point for new scenario studies. For example, the Dutch WLO (*Welvaart en Leefomgeving*, translated as 'prosperity, well-being

Box 3.1 Different ways of assessing the future

In the scholarly literature on long-term assessment, a distinction is often made between three styles: the predictive, the exploratory and the normative style (cf. Godet and Roubelat, 1996). The main assumption employed in the predictive style is the idea of continuity. It is assumed that the future will be shaped by processes, mechanisms and factors that shaped the past. It is furthermore assumed that those processes, mechanisms and factors are sufficiently known and understood to support prediction. In the predictive style, the product is a forecast – namely, a clear picture of what the future will look like (see, for example, Armstrong, 2001, p2). This approach to the future is generally very data intensive and involves quantitative analysis and modelling as dominant means to predict the future.

The *predictive style* is usually associated with projection and baseline scenarios. Projections are aimed at producing very specific statements about the value of a particular indicator on a specified moment in the future, such as a country's gross domestic product (GDP) in five years' time. *Baseline scenarios* (also referred to as 'reference scenarios', 'business-as-usual scenarios' and 'best-guess scenarios') are generally more integrative and look 15 years ahead or less. These baseline scenarios are used for policy optimization (i.e. identifying the best way to reach a particular objective: the fastest, most cost-effective, fairest, most secure policy variant). Examples of such baseline scenarios from the environmental domain are the *GEO-2000 Alternative Policy Study for Europe and Central Asia* (van Vuuren and Bakkes, 1999) and *Clean Air for Europe (CAFE) Programme* (Pye and Watkiss, 2004).

While the predictive style aims at one future forecast, the *exploratory style* aims at surveying multiple futures. The basic assumption is that the future is inherently uncertain. Therefore, it makes much more sense to develop various scenarios of how the future may develop. The aim is explicitly not to predict the future, but to explore the 'possibility space' (Berkhout and Hertin, 2002) – namely, the wealth of thinkable options for the future, including the most radical outlooks. In an exploratory context, scenarios are often referred to as thought experiments and 'what-if' analysis. These scenarios consist of several (qualitative and/or quantitative) descriptions of hypothetical situations, contrasting stories about possible futures. Scenario practitioners generally work with sets of approximately four scenarios. This approach combines analysis/data with imagination/narratives (see also van Asselt et al, 2010). Such *exploratory scenarios* are aimed at supporting strategic orientation, addressing questions such as what alternative worlds do we need to prepare ourselves for; and what to do if our overall direction is too risky. Recent examples of such exploratory scenarios are the *Special Report on Emission Scenarios* (Nakicenovic and Swart, 2000) and the *Global Environment Outlook 3* scenarios (UNEP, 2002).

The *normative style* is formed by studies designed to support vision-building: what is the future that we want to fight for or avoid? Two common approaches are *backcasting* and so-called *critical futures*. In such normative futures, generally a long time horizon – beyond 25 years – is set, as these scenarios anticipate considerable change. Examples of such normative scenarios are Gallopín et al (1997), Raskin et al (2002) and Robinson et al (2006) (see also Chapter 11 in this volume).

Source: van Asselt et al (2010)

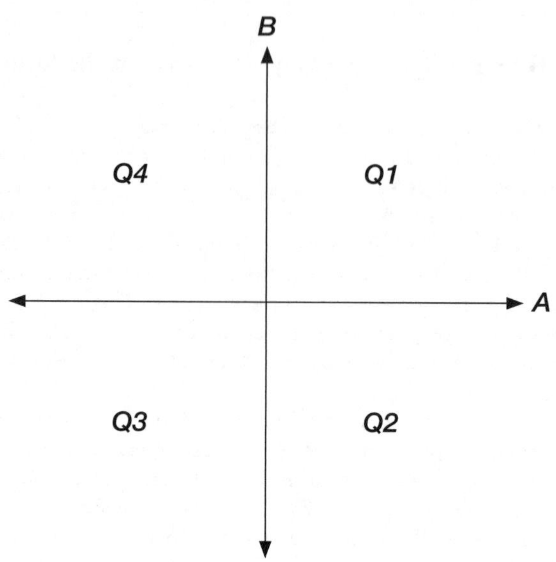

Figure 3.1 *The scenario matrix*

Source: van Asselt et al (2010)

Table 3.1 *Similarities between the* Special Report on Emission Scenarios *(SRES),* Global Environment Outlook 3 *(GEO-3),* Foresight Futures *and* Welvaart en Leefomgeving *(WLO) socio-economic scenarios*

Scenario	SRES	GEO-3	Foresight Futures	WLO
Conventional worlds				
Market forces	A1	Markets first	World markets	Global economy
Policy reform	B1	Policy first	Global sustainability	Strong Europe
Barbarization				
Breakdown	A2		National enterprise	Transatlantic markets
Fortress world		Security first		
Great transitions				
Eco-communalism	B2		Local stewardship	Regional communities
New sustainability paradigm		Sustainability first		

Source: adapted from Millennium Ecosystem Assessment (2005); for SRES scenarios, see Nakicenovic and Swart (2000); for GEO-3 scenarios, see UNEP (2002); for Foresight Futures scenarios, see Foresight Futures (2002); and for WLO scenarios, see Okker et al (2006)

and quality of the living environment') (Okker et al, 2006)[1] and the Sustainability Outlooks (RIVM-MNP, 2004)[2] scenario studies adopted the scenario matrix developed in the Netherlands Bureau for Economic Policy Analysis's (CPB's) *Four Futures of Europe* (de Mooij and Tang, 2003), which again resembles that of the Intergovernmental Panel on Climate Change's (IPCC's) SRES.[3] At the same time, WLO is again used in several other Dutch theme studies, such as for land use (MNP, 2008) and shipping (Province of Limburg, 2008), as well as in several regional scenario studies. Many scenario studies also reuse existing scenarios in climate assessments (see, for example, van Drunen et al, 2011).

3.3 Why are Socio-Economic Scenarios Relevant for Flood Risk Studies?

3.3.1 Interactions between climate and society

Awareness of the interconnectedness of human and environmental systems is nothing new. For example, in 1864, Marsh already provided indications of the impacts of human beings upon the environment (Marsh, 1864) and during the early 19th century the biologist Lamarck introduced the concept of the Earth's 'biosphere' (see Friedman, 1985, for a comprehensive historical overview).

Today, there is broad recognition that our activities have impacts upon the environment. For example, the IPCC concluded in its last assessment report that 'most of the observed increase in global average temperatures since the mid-20th century is very likely due to the observed increase in anthropogenic greenhouse gas concentrations' (IPCC, 2007, p39). The potential future developments in greenhouse gas emissions are assessed by the IPCC in the SRES scenarios (Nakicenovic and Swart, 2000). Human-induced changes are widely recognized as having the potential to significantly modify the structure and functioning of the Earth's system as a whole. At the same time, we have become aware that we are dependent upon the state of the environment and that human development may be constrained or even reversed by its future deterioration (Rotmans and de Vries, 1997).

Hence, it is very likely that societal developments affect our future climate. However, the reverse is also true. The IPCC's *Fourth Assessment Report* (Schneider et al, 2007) estimates the effects of climate change on the vulnerability of society. Even if the global mean temperature remains lower than 2°C above 1990 to 2000 levels in 2050, current climate impacts would be exacerbated. While in low-latitude countries agricultural productions would decrease, it would increase agricultural productivity in some other countries. The *Fourth Assessment Report* envisages serious impacts upon biodiversity, agricultural production and water security if the global mean temperature is 2°C to 4°C above 1990 to 2000 levels in 2050.

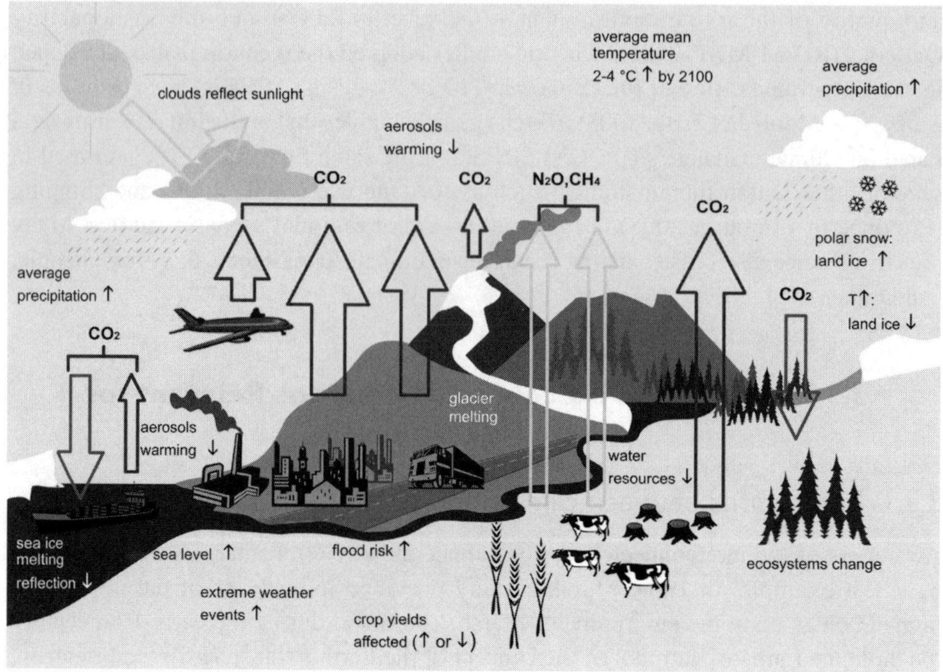

Figure 3.2 *Interactions between climate and society*

Source: Adapted from US Climate Change Science Program. See www.usgcrp.gov/usgcrp/images/ocp2004/
pages/OCP04-05_fig2.htm (*Our Changing Planet* FY 2004-05) for original.

Furthermore, socio-economic developments, such as changes in populations, institutions, economies and land use, will affect the impacts of, and adaptation to, climate change (Tol, 1998). For example, economic prosperity may imply that there is an increasing need for urbanization and industrialization, and may lead to an increase in asset values, which again may increase the vulnerability towards climate change impacts such as flooding. Prosperity may also imply that people have more funds available to adapt. A system's adaptive capacity is defined as its ability to adjust to climate change (including climate variability and extremes) to moderate potential damages, to take advantage of opportunities, or to cope with the consequences (McCarthy et al, 2001). The adaptive capacity of communities also depends upon peoples' ability and willingness to act collectively (Adger, 2006). Here, risk perceptions of people may be a dominant factor.

Notwithstanding the recognition of the mutual interactions between environmental and human systems, many flood risk assessments ignore the cumulative impact of human activities upon the environment or treat 'the environment' and 'the human' as two more or less separate domains. Most studies mainly focus on climate scenarios. In order to illustrate this, a previous analysis of different projects initiated by the Dutch Climate Changes Spatial Planning programme (CcSP)[4] indicated that about one third of the projects that would require the use of socio-economic scenarios based on their project description did not actually use them at all (van Drunen et al, 2011).

In the CcSP studies, the climate scenario outputs represent the *exogenous* variables. Exogenous variables are external to the considered process (i.e. they come from outside the system and are not hidden by being directly coupled to the system). Socio-economic variables are often coupled to the system under investigation (in climate assessments, this system is usually an economic sector, coastal defence or a specific region) and as such are considered *endogenous* (cf. Chermack, 2004). For example, the endogenous variable economic growth is assumed to lead to an increase of capital in areas that may be exposed to the exogenous variable flooding. Chermack (2004) explains this climate 'bias' as a result of the use of models in which the model structure only allows for the incorporation of exogenous variables. In addition, it is likely that most researchers involved in climate assessments are natural scientists and engineers who tend to focus on physical impacts and have less feeling for (future) human needs, goals and determinants of behaviour (Steg and Vlek, 2009).

In their simulation of future flood risk in The Netherlands, Maaskant et al (2009) and Bouwer et al (2010) combined climate and socio-economic scenarios. They used a combination of: a high-growth socio-economic scenario and an extreme climate change scenario; and a low-growth socio-economic scenario and a moderate climate change scenario. Both studies show that consideration of both the local conditions (such as local topography, projected locations of population growth and where people are located) and changes in flood conditions (sea-level rise and increased river discharges) are essential for arriving at reliable estimates of the future risk of flooding. Maaskant et al (2009) show for a major flood-prone area in The Netherlands (South Holland) that the combined impacts of sea-level rise and population growth leads to an estimated doubling in the potential number of fatalities. Bouwer et al (2010) show for their study area (Land van Heusden) that a combination of climate and socio-economic change may increase expected losses by between 97 and 790 per cent when no preventive measures are taken, mostly as a result of asset value increase, which leads to a doubling of losses. Studies such as those by Maaskant et al (2009) and Bouwer et al (2010) teach us that land-use change and increasing exposure affects flood risks at least as much as climate change.

3.4 Combining Socio-Economic and Climate Scenarios

This section introduces two related projects that combine socio-economic and climatic scenarios. These projects exemplify how such combinations can be achieved. The objectives and main components of the projects are described in this section, while the two subsequent sections highlight specific components of the two projects. Section

3.5 discusses the quantitative land-use modelling approach that was applied in both projects to generate future spatial patterns, and section 3.6 describes the qualitative approach taken in the Safety First project to further enhance the selected socio-economic scenarios.

3.4.1 LANDS

LANDS is a CcSP project aimed at identifying climate-driven spatial changes in land use and land development. It integrates changes in agriculture, industry, housing and environmental sectors within balanced national visions and regional solutions by applying the Land-Use Scanner. This model simulates future land use by integrating sector-specific inputs from dedicated models (see section 3.5). The LANDS project proposes a combination of socio-economic and climatic scenarios, and produces a set of related land-use simulations that provide input for Safety First[5] and several other CcSP projects, such as those focusing on flood risk, adaptation measures in the Rhine Basin and the spatial distribution of vegetation, optimizing the nature conservation potential. Subsequently, the results of these projects are fed into the Land-Use Scanner to simulate adjusted land-use patterns that take the possible impact of climate change into account.

3.4.2 Safety First

The Safety First project ran from 2006 until 2008 and was funded by CcSP, Living with Water and the Dutch Ministry of Transport, Public Works and Water Management. It investigated how long-term changes in climate, land use, governance and socio-economic trends will affect flood safety in The Netherlands. The project delivered a decision support system (DSS) that uses maps and images to show how spatial adaptation responses can make The Netherlands climate proof in the long term. The prototype of the DSS is described in Aerts et al (2008, Chapter 10). 'Future awareness' among the users of the DSS is increased by systematically evaluating water-safety policy options against different combinations of climate and socio-economic scenarios. The proposed users' session involves five steps:

1   The Netherlands in the long term: a combination of socio-economic and climate scenarios;
2   the effects in the 'do nothing' option, shown in maps;
3   solutions: the user selects possible sets of measures;
4   robustness of solution: an effects table and maps show the robustness of the sets of measures;
5   moments of investments: here it can be decided where turning points are to be expected (i.e. when it needs to be decided to invest or not).

The DSS challenges the user to 'play' with the available information. Hence, he or she will develop some sensitivity for the key parameters in the system and their implications for water safety in The Netherlands.

Safety First provided inputs for the Deltacommissie (2008) that advised the Dutch government about flood protection in the coming century. Socio-economic (and climate) scenarios played a crucial role in this project. The scenarios were outlined by stakeholder consultations in several workshops. An important theme was the elaboration of more extreme scenarios, taking uncertainties and discontinuities into account.

3.4.3 Combining climate and socio-economic scenarios

For the construction of the scenarios in LANDS and Safety First, it was felt useful to take advantage of existing scenarios. Less than a year before the start of both projects, two major and authoritative Dutch scenario studies were published: the WLO socio-economic scenario study, a co-production of three Dutch planning bureaus (Okker et al, 2006), and an updated set of climate scenarios by The Royal Netherlands Meteorological Institute (KNMI, 2006). Both scenario studies covered a wide spectrum of drivers of change and impacts, and of the directions that developments could take over the coming decades that are relevant for LANDS (see Boxes 3.2 and 3.3 for descriptions of the scenarios).

An important advantage of both scenario studies was that the background material was accessible for the LANDS project team as both research groups were willing to share their expertise and to assist the project team with further modification of the scenarios.

In order to determine the usefulness of the existing scenario studies and to explore to what extent the scenarios needed to be modified, the Safety First project team organized two workshops in which the WLO and KNMI scenarios were discussed. The WLO workshop was attended by a broad group of water experts and stakeholders, such as representatives from the ministry water department, the provinces, the municipalities, water-related research institutes, universities and consultancy firms. Three WLO project members from the three planning bureaus were present. They introduced the WLO scenarios to the workshop participants, provided clarification during the discussions and reflected on the workshop outcomes. The workshop participants set up three 'pluses, minuses and interesting issues' (PMI) matrices about WLO. The main conclusions were that WLO provided a good basis for the scenarios to be used in Safety First; but they wanted to look further into the future (2100) and users wanted to consider more extreme variants of the scenarios (van Drunen et al, 2007).

Box 3.2 The The Royal Netherlands Meteorological Institute (KNMI)
climate scenarios used in LANDS and Safety First

The climate scenarios used in LANDS and Safety First were developed by The Royal
Netherlands Meteorological Institute (KNMI). The G scenario is a moderate scenario
that involves an average global temperature increase of 1°C in 2050 compared to 1990. In
this scenario, the air circulation patterns remain unchanged. In the W scenario, the global
temperature will increase by 2°C in 2050. The G+ and W+ scenarios involve temperature
increases of 1°C and 2°C *and* changes in the air circulation patterns. Specifically, in the '+'
scenarios there will be more easterly winds during the summer and more westerly winds
during the winter, causing warmer and drier summers and milder and wetter winters.
The anticipated temperature increase depends upon greenhouse gas emissions, while the
anticipated change in circulation patterns (or not) depend upon physical uncertainties.
The key features of the four scenarios are summarized in Table 3.2.

Table 3.2 *Dutch climate change scenarios for 2050 relative to 1990*

Climate variable	G	G+	W	W+
Absolute sea-level rise (cm)	15–25	15–25	20–35	20–35
Winter				
Mean temperature	+0.9°C	+1.1°C	+1.8°C	+2.3°C
Mean precipitation	+4%	+7%	+7%	+14%
Yearly maximum daily mean wind speed	0%	+2%	–1%	+4%
Summer				
Mean temperature	+0.9°C	+1.4°C	+1.7°C	+2.8°C
Mean precipitation	+3%	–10%	6%	–19%
Potential evaporation	+3%	+8%	+7%	+15%

Source: KNMI (2006)

Note that the climate scenarios for The Netherlands are more extreme than the
Intergovernmental Panel on Climate Change (IPCC) scenarios because of its geophysical
characteristics. By 2050, the scenarios already show markedly different results.

Box 3.3 The *Welvaart en Leefomgeving* (WLO) socio-economic scenarios used in LANDS and Safety First

The Dutch WLO scenarios have a time horizon until 2040 and are set around two key uncertainties (see Figure 3.3). The vertical axis ranges from successful international cooperation at the top, to an emphasis on national sovereignty at the bottom; the horizontal axis ranges from a strong role for the public sector at the left, to private responsibility at the right. The combination of the two key uncertainties yields four scenarios for Europe and its countries. Each of the four scenarios is described by a specific storyline.

Figure 3.3 *Schematic overview of the WLO scenarios*

The scenarios differ greatly in their socio-economic assumptions, such as expected population growth and gross domestic product (GDP). The regional communities scenario, for example, foresees a population of about 16 million inhabitants in 2040, while close to 20 million are expected in the global economy scenario. Economic growth, as expressed in the yearly increase in GDP, also differs considerably between 0.7 per cent (regional communities) and 2.6 per cent (global economy) per scenario.

As a formal scenario exercise, WLO coupled approximately 40 quantitative models. These models include a global model that assesses economic developments, trade and energy supply; national and regional demographic models; a labour market model; transport models for people and freight; an agricultural model; energy models; and environmental models (Okker et al, 2006, pp205–209).

WLO did not use the KNMI 2006 scenarios, but the central scenario published in 2000 that indicated an average temperature increase of 1°C in 2050. This central scenario can be compared to the KNMI 2006 G-scenario (Riedijk et al, 2007).

Source: Okker et al (2006)

3.4.4 Consistent storylines for LANDS and Safety First

Only a limited number of scenarios were selected to provide a workable set of case studies for the various research projects related to adaptation and mitigation in the CcSP programme. To simply couple all four available socio-economic scenarios to the four climate scenarios would result in 16 different combinations. This was an overflow of information for the various adaptation and mitigation projects and, furthermore, contained the risk that each project would select its own specific combination to start from. The latter would imply that the proposed adaptation or mitigation measures cannot be readily integrated within a coherent set of spatial strategies as they all start from different assumptions regarding climate and society. Therefore it was decided to reduce the set of socio-economic scenarios to the two scenarios that describe the broadest range of possible futures. In order to further reduce the set of scenarios, climate and socio-economic scenarios were linked in a logical way.

Based on their common roots in the SRES storylines, the global economy (GE) scenario was associated with the SRES A1 scenario family, as were the high temperature rise (W or W+) climate change scenarios. The regional communities (RC) scenario was related to the SRES B2 family, as can be done with the lower rise in temperature (G or G+) scenarios. Table 3.3 summarizes the resulting set of scenarios. The advantage of selecting the extremes on both sides of the bandwidth in terms of socio-economic developments is that the full extent of possible changes in land use is preserved, while at the same time the full variability in climate scenarios is preserved. The Safety First project also selected the GE and RC scenarios to reflect the potential variation in socio-economic conditions and related land-use patterns.

In LANDS and Safety First, the KNMI climate scenarios and WLO socio-economic scenarios were used to develop (new) consistent storylines. In practice, existing scenarios are rarely exactly what is needed for another exercise in terms of focus, geographical area and time horizon. The two cases show very different methodologies for expanding upon these storylines and tailoring the scenarios. LANDS takes a quantitative approach and the scenarios are elaborated upon by adding socio-economic components to the climate scenarios (see the following section). Safety First is a more qualitative approach.

Table 3.3 *Combined scenarios in LANDS*

	Regional communities (RC) scenario	**Global economy (GE) scenario**
Circulation change	Moderate rise in temperature (G+)	Strong increase in temperature (W+)
No circulation change	Moderate rise in temperature (G)	Strong increase in temperature (W)

Source: Riedijk et al (2007, p23)

Box 3.4 Characterization of the socio-economic scenarios

The *global economy scenario* is characterized by rapid economic and population growth, liberalization, private initiatives, market-based solutions and efficient technologies. National governments concentrate on their core tasks, such as the provision of pure public goods, protection of property rights and setting the 'rules of the game'. Priority domains are national security, communication and trade (Hilderink, 2004). Government involvement in the functioning of the agricultural market and spatial policy is limited (Aerts et al, 2008).

   The *regional communities scenario* is characterized by modest multilateral cooperation, and collective arrangements to maintain an equitable distribution of welfare and to control local environmental problems. The public sector expands and local and regional governmental institutions play an important role. The population stabilizes and the economy grows at a slow pace. Priority areas are social welfare, equity and environmental protection (Aerts et al, 2008).

Its approach is somewhat similar to the UK Climate Impacts Programme (UKCIP) study of East Anglia (Lorenzoni et al, 2000a, 2000b).

3.5 Expanding the Storylines: The LANDS Approach

The LANDS project outlines integrated socio-economic and climate change scenarios with a time horizon until 2040. In order to expand upon the storylines, two subsequent steps were taken:

1   Existing socio-economic and climate scenarios were combined.
2   Regional projections of anticipated land-use change were made.

The combination of socio-economic and climate scenarios was described in the previous section. Building upon the existing socio-economic scenarios and their assumptions related to, for example, demography and economic growth, regional projections of land use were established. The general scenario descriptions were made regionally explicit with the help of several sector-specific models and a number of additional assumptions. These calculations have been performed by various specialized institutes that, for example, provided the expected amount of residential development, the demand for industrial and commercial land use and office space, and projections for agricultural land-use changes. A concise description of the basic characteristics of the underlying regional models and a short discussion on the related quality issues is provided elsewhere (Dekkers and Koomen, 2006), while a more detailed account of the regional land-use demand and underlying assumptions is provided by Riedijk et al (2007).

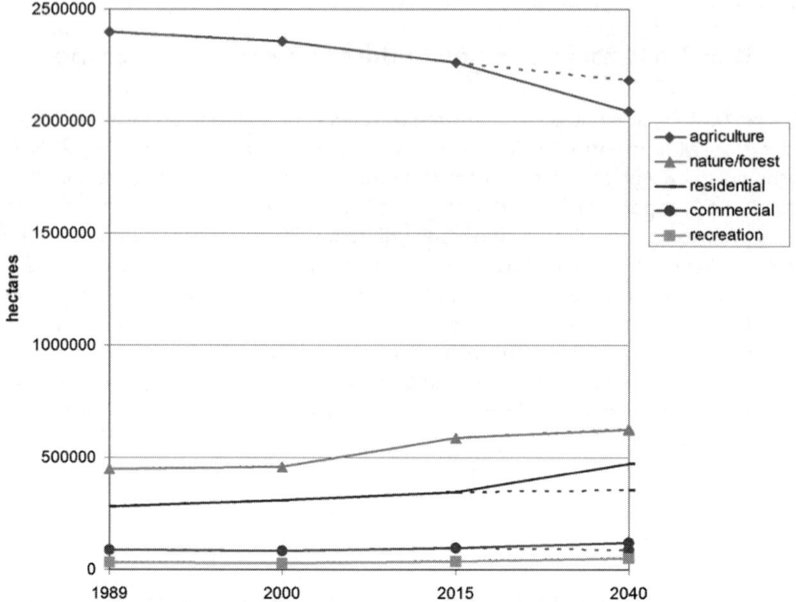

Figure 3.4 *Land use in The Netherlands, 1975–2040*

Source: CBS (2009) for historical trend; Riedijk et al (2007) for scenario projections

3.5.1 Regional projections of anticipated land-use change

The impact of the socio-economic scenarios upon future land use was simulated and visualized using the Land-Use Scanner model. These simulations were subsequently used as the basis for developing adaptation and mitigation strategies.

The Land-Use Scanner is a GIS-based model that simulates future land use based on the integration of sector-specific inputs from dedicated models (Hilferink and Rietveld, 1999; Koomen et al, 2008). The model is based on demand–supply interaction for land, with sectors competing for allocation within suitability and policy constraints. Land-use simulations are generally scenario driven, with a series of coherent assumptions regarding variables such as economic growth or level of government intervention, determining the way in which the supply and demand for land unfolds (Koomen et al, 2005). The renewed model configuration used for this project applies a 100m grid offering a very detailed view on possible spatial patterns in the future. It distinguishes 17 land-use types, out of which the model allocates 11. The remaining 6 types, mainly related to infrastructure and water, have a predefined location that is not influenced by model simulation. Their location is either a continuation of current land use or consists of predefined approved plans, as is the case with, for example, long-planned railway links (for a more detailed description of the most recent model version and its calibration and validation, see Loonen and Koomen, 2009). The model has been applied extensively in planning-related applications ranging from the regional to the supra-national catchment scale (e.g. Borsboom-van Beurden et al, 2007; Dekkers and Koomen, 2007; Koomen et al, 2010). The model has recently been incorporated in the European Union's ClueScanner: the pan-European model developed for the Directorate-General for the Environment to simulate, amongst other aspects, the potential spatial impact of new biofuel policies in Europe (Perez-Soba et al, 2010).

The simulations resulting from the two scenarios show two clearly diverging patterns. The regional communities scenario shows a modest increase in residential areas despite the fairly limited population growth. This growth can be largely ascribed to the minor increase in households and residential preferences for a rural living environment. Urban growth is most notable in the central and western part of The Netherlands (see Figure 3.5). Arable farming diminishes strongly in this scenario. Greenhouse horticulture disappears in many areas, especially from its current stronghold south of The Hague. From the land-use simulation it can be concluded that existing semi-natural areas are enlarged in a number of cases. New conservation areas are developed along the major rivers in the country. Clusters of outdoor recreation arise in attractive landscapes, particularly in the northern and western part of The Netherlands.

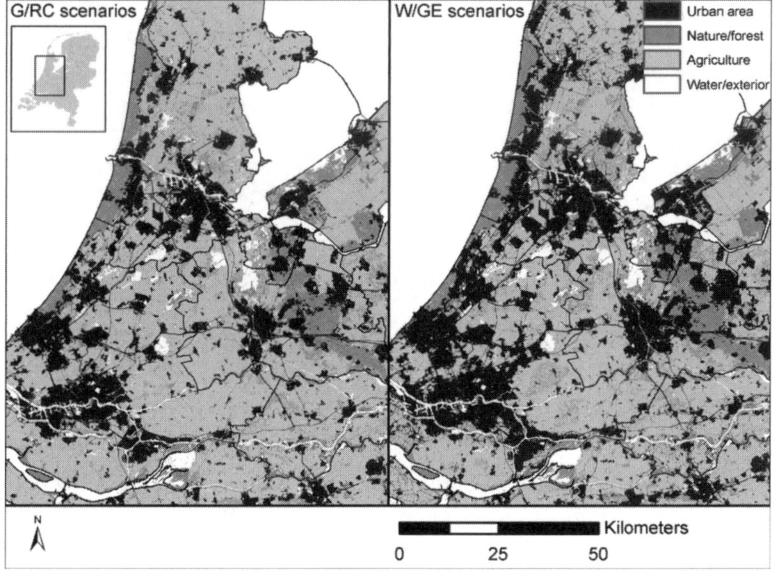

Figure 3.5 *Simulated land use for the low temperature rise/regional communities (G/RC) and high temperature rise/global economy (W/GE) scenarios in 2040*

Source: Koomen et al (2008)

The most striking land-use change in the global economy scenario is the strong increase in urban land use (see Figure 3.5, right). Residential land use expands substantially around the larger cities in the Randstad, as well as around many smaller villages in the rural areas. Commercial land use also increases strongly. This increase takes place in the Randstad and bordering parts of the intermediate zone. The urbanization causes a serious deterioration of the quality and openness of the landscape, as well as in designated national landscapes.

The appearance of the rural areas will also change. Arable farming will, to a large extent, disappear and be replaced by grassland for dairy farming. Capital-intensive forms of farming will also demand more land. Greenhouse horticulture expands around the main ports of Rotterdam and Amsterdam Airport Schiphol. New semi-natural areas will mainly be developed along the major rivers, where strong constraints are imposed on the expansion of urban functions and capital-intensive forms of farming. Recreation also claims more space, especially in the attractive small-scale landscapes of, for example, the Achterhoek. A limited description of the land-use simulations for the W scenarios has also been published in the *Sixth National Environmental Outlook* (MNP, 2006).

3.6 Expanding the Storylines: The Safety First Approach

This section explains how the storylines were expanded to analyse flood risk in The Netherlands and draft adaptation measures to minimize this risk.

In order to fit the specific purposes of the project, the socio-economic and climate scenarios used in LANDS were further modified by:

* extending the time horizon from 2040 to 2100;
* establishing more variation between the scenarios (i.e. more discontinuous scenario plots) by stretching the WLO scenarios in such a way that they fit better to the Dutch (institutional) water context;
* including non-linear events and developments (i.e. more discontinuous storylines).

3.6.1 Extending the time horizon from 2040 to 2100

As was already shown in Boxes 3.2 and 3.3, the time horizon of the socio-economic and climate scenarios varied: the socio-economic scenarios of WLO ran until the year 2040, whereas the climate scenarios provided an outlook until 2100 (and beyond). Whereas in LANDS the time horizon was set to 2040, in Safety First a longer time horizon was needed. For example, investments in coastal defence are supposed to last for many decades and therefore assessments of their efficiency and effectiveness require long-term scenarios.

In order to extend the time horizon until 2100, a number of scenario characteristics and assumptions had to be simplified, especially with regard to technological, governmental and behavioural patterns (van der Hoeven et al, 2007). To do this consistently, the choice was made to go back to the IPCC SRES scenario storylines (Nakicenovic and Swart, 2000) on which the WLO scenarios were indirectly based (see also Table 3.1). In addition, several follow-up studies were consulted that explored the long-term population and demographic developments, such as fertility, mortality and migration (e.g. de Jong and Hilderink, 2004; Hilderink, 2004).

3.6.2 Discontinuous scenario plots

In order to estimate the future vulnerability against flooding under uncertainty, we aimed to build more extreme assumptions into the scenarios. Additional scenarios were created in order to analyse the potential impact of a more extreme sea-level rise, with a rise in sea level of 60cm and 85cm, and including a third extreme variant of 150cm in 2100 in which the effects of melting of the Greenland and West Antarctic ice sheets are taken into account. These should be viewed as worst-case scenarios and aim to provoke policy-makers to rethink the applicability of current safety strategies for the long-term future.

Table 3.4 *Additional discontinuous scenarios in Safety First*

Year	Q_{1250} Rhine (m³/second)	Q_{1250} Meuse (m³/second)	Sea-level rise (cm)
2040	16.7	4.2	25
2100	18.0	4.6	60 85 150
Far future	18.0	4.6	500

Note: Q_{1250} is the estimation of the river discharge that (statistically) occurs once per 1250 years.
Source: Aerts et al (2008)

3.6.3 Non-linear events and developments

Due to uncertainty and complexity in the long term (especially with a time horizon of 2100 and beyond), the occurrence of disruptive events and developments was considered not just a possibility, but a probability.

In order to systematically explore such discontinuities and ways of dealing with them in our scenario exercise, we organized four additional workshops. In two backcasting workshops (van de Kerkhof et al, 2007) the activities that are required to reach a climate-proof Netherlands in 2100 were identified. The fundamental question underlying backcasting exercises is: 'If we want to attain (or avoid) a certain future, what

Figure 3.6 *Determining future images with a map and coloured clay*

Source: Photo provided by Tjeerd Stam

actions must be taken?' Backcasting exercises generally have their starting point in the distant future. Working backwards from a normative long-term future is assumed not only to facilitate breaking through the evolutionary paradigm – the gradual incremental unfolding of the future – but also to treat surprise random events and changed conditions. The participants used maps, clay, paper sheets, post-it memos and marker pens to visualize their insights (see Figure 3.6; see also Chapter 11).

During a Discontinuity Workshop, we systematically identified and further explored various climate-related and more socio-economic discontinuities. A discontinuity was defined as an abrupt or gradual break in a dominant development in society (trend break) as a result of crossing a certain threshold (e.g. Valkering et al, 2010). By means of so-called 'what-if' thought experiments and the use of various brainstorm techniques to stimulate participants to think 'out-of-the-box', various abrupt or more gradual discontinuities were identified. Next, these potential discontinuities were classified in terms of 'impact' and 'probability' (see Figure 3.7).

The high-impact discontinuities were further explored in terms of underlying principles and (systemic) mechanisms, and potential first-order (effects on water safety) and second-order effects (more general socio-economic, environmental and institutional effects).

A 'governance' workshop, attended by policy-makers and researchers, started with two extreme future perspectives and subsequent water management options to prevent flooding. A key question that was addressed was how to identify the necessary policies, institutional changes and new roles for stakeholders (van 't Klooster et al, 2007).

The workshops generated a long list of possible discontinuities (Aerts et al, 2008, p50) and possible implications for water safety in The Netherlands. Based on the

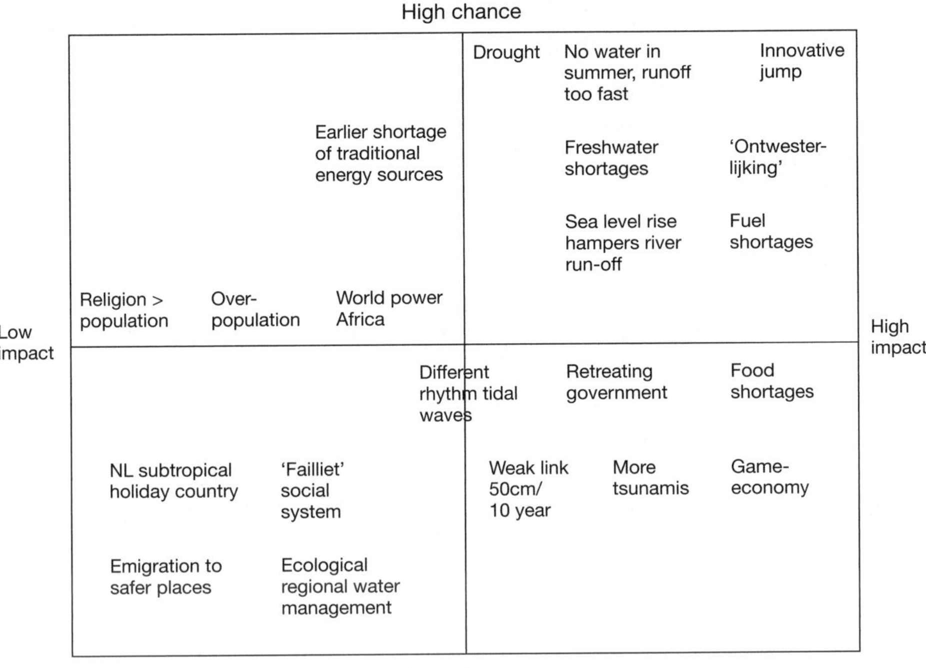

Figure 3.7 *Overview of possible discontinuities, clustered according to level of impact and chance*

evaluation of these discontinuities, the Safety First team included the policy option 'elevation' in its analysis. This option and three other policy options were evaluated in the extended regional communities and global economy WLO scenarios (Aerts et al, 2008, pp128–134).

Furthermore, based on these workshop outcomes, some low-possibility–high-impact climate events were taken into consideration, such as the possibility of severe economic decline, the shutdown or slowdown of the thermohaline circulation, the possibility of super-storms, extreme precipitation and extreme river discharges, as postulated effects of global warming.

3.7 Discussion and Conclusions

In climate assessments, socio-economic scenarios are as important as climate scenarios. However, in many flood risk studies, this appears to be not fully acknowledged. Therefore the adaptation options formulated in these projects are not likely to be 'future proof': they may work well under current circumstances, but not any more in 2050 or 2100.

LANDS and Safety First provided examples of how the combination of socio-economic and climate scenarios could be operationalized. LANDS followed a more quantitative approach, whereas Safety First is an example of a more qualitative scenario approach. The most salient lessons are summarized as follows:

- In general, the choice for including or excluding socio-economic scenarios must be substantiated well, as was done in the LANDS and Safety First projects.
- Reusing existing scenarios can be very time efficient; but such scenarios rarely offer precisely what is needed for another exercise in terms of focus, geographical area and time horizon. This, therefore, calls for finding ways of tailoring existing scenario descriptions to case study-specific needs.
- There is no standard approach for combining socio-economic and climate scenarios in flood risk studies. Which approach works depends – among other things – upon project aims, required output and the availability of time and data.
- The application of the scenario axis technique – the most popular framework for scenario-building – has many advantages: it is easy to understand, it generates a reasonable number of scenarios and it draws a clear picture of the 'uncertainty space'. Its main drawback is that it is not very suitable for taking into account discontinuities: temporary or permanent, sometimes unexpected, breaks in dominant conditions in society (van Notten et al, 2005). These changes in direction can evolve slowly or happen suddenly. It is possible to test if the developed scenarios are resilient towards discontinuities.
- It is essential that 'policy audiences' not only *apply* scenarios, but also *engage in the process* of developing scenarios. Projects that are most likely to actually influence decision-making processes involve policy-makers and other stakeholders in the scenario development (Berkhout and Hertin, 2002). When policy- and decision-makers are engaged in the process, 'future thinking' may be anchored in their organizations and the actual decision-making processes, which will hopefully lead to a decrease in misgovernment and more effective adaptation policies. Safety First is an example of how stakeholders can be involved.

The choice within LANDS and Safety First to build on existing scenario studies, rather than constructing new scenarios from scratch, proved to be very effective as it enabled the researchers to jump-start the project. At the same time, this choice also has an important limitation. By combining existing socio-economic and climate scenarios without re-running the models with a different set of parameters, feedback mechanisms were only qualitatively taken into account. There are not many examples of studies that take into account feedback mechanisms. The methodology for doing this is still premature and the exercise very time consuming (Haasnoot et al, 2009, and Valkering et al, 2010, provide examples of such an integrated scenario approach).

Notes

1 WLO was a co-production of three Dutch planning bureaus: The Netherlands Bureau for Economic Policy Analysis (CPB), The Netherlands Environmental Assessment Agency (MNP) and The Netherlands Institute for Spatial Research (RPB).
2 The *Sustainability Outlooks* were produced by the Dutch National Institute for Public Health and the Environment (RIVM).
3 One of the CPB dimensions – successful internal cooperation versus a preponderance of national sovereignty – resembles that of the IPCC SRES, whereas the other dimension chosen by CPB is a strong public sector versus a preponderance of private initiatives. The resulting storylines focus on Europe with European Union decision-making and solidarity as central themes.
4 This is a 90 million Euros research programme that covers more than 60 projects. The programme promotes joint learning between researchers and spatial planning practitioners, with themes such as climate scenarios, mitigation, adaptation and integration. Together with its successor Knowledge for Climate, it is the most important climate research programme in The Netherlands (CcSP, 2009).
5 Flood risk is a local phenomenon that depends upon characteristics such as topography and accumulated assets that vary greatly over space. In order to estimate potential future flood risk, it is thus necessary to downscale national scenarios to the local level. The Safety First project therefore applied the Land-Use Scanner model to create spatially explicit outlooks on the future.

References

Adger, W. N. (2006) 'Vulnerability', *Global Environmental Change*, vol 16, no 3, pp268–281

Aerts, J. C. J. H., Sprong, T. and Banning, B. (eds) (2008) *Aandacht voor Veiligheid*, Leven met Water, Klimaat voor Ruimte, DG Water, The Netherlands

Alcamo, J., Kok, K., Busch, G., Priess, J. A., Eickhout, E., Rounsevell, M., Rothman, D. S. and Heistermann, M. (2006) 'Searching for the future of land: Scenarios from the local to global scale', in J. E. F. Lambin and H. J. Geist (eds) *Land-Use and Land-Cover Change: Local Processes and Global Impacts*, The IGBP Series, Springer-Verlag, Berlin and Heidelberg, Germany

Aligica, P. D. (2004) 'Prediction, explanation and the epistemology of future studies', *Futures*, vol 35, no 10, pp1027–1040

Armstrong, S. J. (ed) (2001) *Principles of Forecasting: A Handbook for Researchers and Practitioners*, Kluwer Academics Publishers, Boston/Dordrecht/London

Berkhout, F. and Hertin, J. (2002) 'Socio-economic futures scenarios for climate impact assessment', *Global Environmental Change*, vol 10, no, pp165–168

Berkhout, F., Hertin, J. and Jordan, A. (2002) 'Socio-economic futures in climate change impact assessment: Using scenarios as "learning machines"', *Global Environmental Change*, vol 12, no2, pp83–95

Borsboom-van Beurden, J. A. M., Bakema, A. and Tijbosch, H. (2007) 'A land-use modelling system for environmental impact assessment', in E. Koomen, J. Stillwell, A. Bakema and

H. J. Scholten (eds) *Modelling Land-Use Change: Progress and Applications*, Springer, Dordrecht, The Netherlands, pp281–296

Bouwer, L. M., Bubeck, P. and Aerts, J. C. J. H. (2010) 'Changes in future flood risk due to climate and development in a Dutch polder area', *Global Environmental Change*, vol 20, pp463–471

Carter, T. R., Hulme, M. and Lal, M. (2001) *Guidelines on the Use of Scenario Data for Climate Impact and Adaptation Assessment, Version 1*, Intergovernmental Panel on Climate Change, Task Group on Scenarios for Climate Impact Assessment, Geneva, Switzerland

CBS (Centraal Bureau voor de Statistiek) (2009) *Statline*, Centraal Bureau voor de Statistiek, http://statline.cbs.nl/statweb

CcSP (2009) *Introduction Research: Climate Changes Spatial Planning*, www.klimaatvoorruimte. nl/pro3/general/start.asp?i=1&j=1&k=0&p=0&itemid=113, accessed 13 February 2011

Chermack, T. J. (2004) 'Improving decision-making with scenario planning', *Futures*, vol 36, pp295–309

de Jong, A. H. and Hilderink, H. B. M. (2004) *Lange-termijn bevolkingsscenarios voor Nederland*, RIVM, Bilthoven, The Netherlands

de Mooij, R. and Tang, P. (2003) *Four Futures of Europe*, Centraal Planbureau, The Hague, The Netherlands

Dekkers, J. E. C. and Koomen, E. (2006) *De rol van sectorale inputmodellen in ruimtegebruik-simulatie: Onderzoek naar de modellenketen voor de LUMOS toolbox*, Spinlab Research Memorandum SL-05, Vrije Universiteit, Amsterdam, The Netherlands

Dekkers, J. E. C. and Koomen, E. (2007) 'Land-use simulation for water management: Application of the Land Use Scanner model in two large-scale scenario-studies', in E. Koomen, J. Stillwell, A. Bakema and H. J. Scholten (eds) *Modelling Land-Use Change: Progress and Applications*, Springer, Dordrecht, The Netherlands, pp355–373

Deltacommissie (2008) *Achtergrondrapporten bij het advies*, www.deltacommissie.com/advies/ achtergrondrapporten

Foresight Futures (2002) *Foresight Futures 2020: Revised Scenarios and Guidance*, UK Department of Trade and Industry, London

Friedman, H. (1985) 'The science of global change: An overview', in T. F. Malone and J. G. Roederer (eds) *Global Change*, Cambridge University Press, Cambridge, UK

Gallopín, G., Hammond, A., Raskin, P. and Swart, R. (1997) *Branch Points: Global Scenarios and Human Choice: A Resource Paper of the Global Scenario Group*, Polestar Series Report no 7, Stockholm Environment Institute, Sweden

Godet, M. and Roubelat, F. (1996) 'Creating the future: The use and misuse of scenarios', *Long Range Planning*, vol 29, no 2, pp164–171

Haasnoot, M., Middelkoop, H., van Beek, E. and van Deursen, W. P. A. (2009) 'A method to develop sustainable water management strategies for an uncertain future', *Sustainable Development*, doi:10.1002/sd438

Hilderink, H. B. M. (2004) *Populations and Scenarios: Worlds to Win?* RIVM report 550012001/2004, RIVM, Bilthoven, The Netherlands

Hilferink, M. and Rietveld, P. (1999) 'Land Use Scanner: An integrated GIS based model for long term projections of land use in urban and rural areas', *Journal of Geographical Systems*, vol 1, no 2, pp155–177

IPCC (Intergovernmental Panel on Climate Change) (2007) *Climate Change 2007: Synthesis Report Summary for Policymakers*, IPCC, Cambridge University Press, Cambridge, UK

KNMI (2006) *KNMI '06 Klimaatscenario's*, www.knmi.nl/klimaatscenarios/knmi06

Koomen, E., Kuhlman, T., Groen, J. and Bouwman, A. A. (2005) 'Simulating the future of agricultural land use in the Netherlands', *Tijdschrift voor Economische en Sociale Geografie* [*Journal of Economic and Social Geography*], vol 96, no 2, pp218–224

Koomen, E., Loonen, W. and Hilferink, M. (2008) 'Climate-change adaptations in land-use planning: A scenario-based approach', in L. Bernard, A. Friis-Christensen and H. Pundt (eds) *The European Information Society: Taking Geoinformation Science One Step Further*, Springer, Berlin, Germany, pp261–282

Koomen, E., Koekoek, A. and Dijk, E. (2010) 'Simulating land-use change in a regional planning context', *Applied Spatial Analysis and Policy*, doi:10.1007/s12061-010-9053-5

Loonen, W. and Koomen, E. (2009) *Calibration and Validation of the Land Use Scanner Allocation Algorithms*, PBL Report 550026002, The Netherlands Environmental Assessment Agency, Bilthoven, The Netherlands

Lorenzoni, A., Jordan, M., Hulme, M., Turner, R. K. and O'Riordan, T. (2000a) 'A co-evolutionary approach to climate change impact assessment: Part I. Integrating socio-economic and climate change scenarios', *Global Environmental Change*, vol 10, pp57–68

Lorenzoni, A., Jordan, M., O'Riordan, T., Turner, R. A. and Hulme, M. (2000b) 'A co-evolutionary approach to climate change impact assessment: Part II. A scenario-based case study in East Anglia (UK)', *Global Environmental Change*, vol 10, pp145–155

Maaskant, B., Jonkman, S. N. and Bouwer, L. M. (2009) 'Future risk of flooding: An analysis of changes in potential loss of life in South Holland (The Netherlands)', *Environmental Science & Policy*, vol 12, pp157–169

Marsh, G. P. (1864) *Man and Nature or Physical Geography as Modified by Human Action*, Scribner, New York, NY

McCarthy, J. J., Canziani, O. F., Leary, N. A., Dokken, D. J. and White, K. S. (eds) (2001) *Climate Change 2001: Impacts, Adaptation, and Vulnerability: Contribution of Working Group II to the Third Assessment Report of the Intergovernmental Panel on Climate Change*, Cambridge University Press, Cambridge

Meehl, G. A., Stocker, T. F., Collins, W. D., Friedlingstein, P., Gaye, A. T., Gregory, J. M., Kitoh, A., Knutti, R. R., Murphy, J. M., Noda, A., Raper, S. C. B., Watterson, I. G., Weaver, A. J. and Zhao, Z. C. (2007) 'Global climate projections', in S. Solomon, D. Qin, M. Manning, Z. Chen, M. Marquis, K. B. Averyt, M. Tignor and H. L. Miller (eds) *Climate Change 2007: The Physical Science Basis. Contribution of Working Group I to the Fourth Assessment Report of the Intergovernmental Panel on Climate Change*, Cambridge University Press, Cambridge, UK, and New York, NY

Millennium Ecosystem Assessment (2005) *Scenarios Assessment*, www.millenniumassessment. org/en/Scenarios.aspx

MNP (2006) *Nationale Milieuverkenning 6 2006 – 2040*, Milieu en Natuurplanbureau, Bilthoven, The Netherlands

MNP (2008) *Nederland Later: Tweede Duurzaamheidsverkenning, deel Fysieke Leefomgeving Nederland, RIVM*, Bilthoven, The Netherlands

Nakicenovic, N. and Swart, R. (2000) *Special Report on Emission Scenarios*, Cambridge University Press, Cambridge, UK

O'Brien, K. L. and Leichenko, R. M. (2000) 'Double exposure: Assessing the impacts of climate change within the context of economic globalisation', *Global Environmental Change*, vol 10, no 3, pp221–232

Okker, R., Janssen, L. and Schuur, J. (2006) *Welvaart en Leefomgeving*, CPB, MNP and RPB, The Hague, The Netherlands

Perez-Soba, M., Verburg, P. H., Koomen, E., Hilferink, M., Benito, P., Lesschen, J. P., Banse, M., Woltjer, G., Eickhout, B., Prins, A.-G. and Staritsky, I. (2010) *Land Use Modelling: Implementation – Preserving and Enhancing the Environmental Benefits of 'Land-Use Services'*, Final report to the European Commission, DG Environment, Alterra, Wageningen UR/Geodan Next/ Object Vision/ BIOS/ LEI and PBL, Wageningen, The Netherlands

Province of Limburg (2008) *Netwerkanalyse vaarwegen en binnenhavens*, Province of Limburg, Maastricht, The Netherlands

Pye, S. and Watkiss, P. (2004) *Baseline and MTFR Scenarios for Cost-Benefit Analysis of Air Quality Related Issues in Particular in the Clean Air for Europe (CAFE) Programme*, AEA Technology Environment, Didcot, UK

Raskin, P., Banuri, T., Gallopin, G., Gutman, P., Hammond, A., Kates, R. and Swart, R. (2002) *Great Transition: The Promise and Lure of the Times Ahead*, Report of the Global Scenario Group, Stockholm Environmental Institute, Boston, MA

Riedijk, A., van Wilgenburg, R., Koomen E. and Borsboom-van Beurden, J. (2007) *Integrated Scenarios of Socio-Economic and Climate Change: A Framework for the 'Climate Changes Spatial Planning' Programme*, VU/MNP, Spinlab Research Memorandum, Amsterdam/ Bilthoven, The Netherlands

Ringland, G. (2002) *Scenarios in Public Policy*, Wiley, Chichester, UK

RIVM–MNP (2004) *Quality and Future: A Exploration of Sustainable Development* (in Dutch), SDU, Bilthoven, The Netherlands

Robinson, J., Carmichael, J., van Wynsberghe, R., Tansey, J., Journeay, M. and Rogers, L. (2006) 'Sustainability as a problem of design: Interactive science in the Georgia Basin. Special issue on interactive sustainability', *The Integrated Assessment Journal*, vol 6, no 4, pp165–192

Rotmans, J. and de Vries, B. (eds) (1997) *Perspectives on Global Change: The TARGETS Approach*, Cambridge University Press, Cambridge, UK

Schneider, S. H., Semenov, S., Patwardhan, A., Burton, I., Magadza, C. H. D., Oppenheimer, M., Pittock, A. B., Rahman, A., Smith, J. B., Suarez, A. and Yamin, F. (2007) 'Assessing key vulnerabilities and the risk from climate change', in M. L. Parry, O. F. Canziani, J. P. Palutikof, P. J. van der Linden and C. E. Hanson (eds) *Climate Change 2007: Impacts, Adaptation and Vulnerability. Contribution of Working Group II to the Fourth Assessment Report of the Intergovernmental Panel on Climate Change*, Cambridge University Press, Cambridge, UK, pp779–810

Steg, L. and Vlek, C. (2009) 'Social science and environmental behaviour', in J. J. Boersema and L. Reijnders (eds) *Principles of Environmental Sciences*, Springer, Dordrecht/Boston, MA, pp97–142

Strzepek, K., Yates, D., Yohe, G., Tol, R. and Mader, N. (2001) 'Constructing "not implausible" climate and economic scenarios for Egypt', *Integrated Assessment*, vol 2, pp139–157

Tol, R. S. J. (1998) 'Socio-economic scenarios', in J. F. Feenstra, I. Burton, J. B. Smith and R. S. J. Tol (eds) *Handbook on Methods for Climate Change Impact Assessment and Adaptation Strategies*, IVM, UNEP, Nairobi, and VU University, Amsterdam

UNEP (United Nations Environment Programme) (2002) *Global Environment Outlook 3*, EarthPrint, www.unep.org/GEO/geo3

Valkering, P., Offermans, A., van der Brugge, R., van Lieshout, M. and Rijkens-Klomp, N. (2010) 'Scenario analysis of perspective change to support climate adaptation: Lessons from a pilot study on Dutch river management', *Regional Environmental Change*, doi:10.1007/s10113-010-0146

van Asselt, M. B. A., van 't Klooster, S. A., van Notten, P. W. F. and Smits, L. A. (2010) *Foresight in Action: Developing Policy-Oriented Scenarios*, Earthscan, London

van de Kerkhof, M., Stam, T., Aerts, J. C. J. H., van 't Klooster, S. A. and Walraven, A. (2007) *Een backcasting analyse van een klimaatbestendig en waterveilig Nederland*, Working paper of the Safety First project, W-07/20, IVM–VU University, Amsterdam, The Netherlands

van der Hoeven, N., Jacobs, C., Koomen, E. and Aerts, J. C. J. H. (2007) *Beknopte beschrijving van sociaaleconomische scenario's voor het jaar 2100*, VU University Amsterdam, The Netherlands

van Drunen, M., van 't Klooster, S. A., Offermans, A., and Aerts, J. C. J. H. (2007) *Socio-economische scenario's voor klimaatbestendigheidstudies*, Report W-07/06, Institute for Environmental Studies, VU University, Amsterdam, The Netherlands

van Drunen, M., van 't Klooster, S. and Berkhout, F. (2011) 'Bounding the future: The use of scenarios in assessing climate change impacts', *Futures*, vol 4, no 4, pp488–496

van Notten, P. W. F., Sleegers, A. M. and van Asselt, M. B. A. (2005) 'The future shocks: On discontinuity and scenario development', *Technological Forecasting & Social Change*, vol 72, pp175–194

van 't Klooster, S. A. (2008) *Toekomstverkenning: ambities en de praktijk. Een etnografische studie naar de productie van toekomstkennis bij het Ruimtelijk Planbureau (RPB)*, PhD thesis, Eburon, Delft, The Netherlands

van 't Klooster, S. A., Cornelisse, C., Aerts, J. and Huitema, D. (2007) *Verslag van de workshop 'Veerkracht van Waterinstituties onder Klimaatverandering'*, www.adaptation.nl

van Vuuren, D. and Bakkes, J. (1999) *GEO-2000 Alternative Policy Study for Europe and Central Asia: Energy-Related Environmental Impacts of Policy Scenarios, 1990–2010*, UNEP/DEIA&EW/TR.99-4 and RIVM 4002001019, Nairobi and Bilthoven

Verburg, P. H., Kok, K., Pontius, Jr., R. G. and Veldkamp, A. (2006) 'Modeling land-use and land-cover change', in E. F. Lambin and H. J. Geist (eds) *Land-Use and Land-Cover Change: Local Processes and Global Impacts*, IGBP Series, Springer-Verlag, Berlin/Heidelberg, Germany

4

Vulnerability of Port Infrastructure for the Port of Rotterdam

A. J. (Joost) Lansen and S. N. (Bas) Jonkman

4.1 Introduction

The port of Rotterdam is the largest port in Europe and functions as a trans-shipment hub as well as a gateway of goods to a large hinterland, including The Netherlands, Germany and Belgium. The port area developed over time on the riverbanks of the Meuse River, which flows into the North Sea at Hoek van Holland, about 30km west of Rotterdam. The greater part of the port area is occupied by liquid bulk storage, chemical industry of all sorts and container terminals. A wide range of other commodities can also be found in the port of Rotterdam.

Although much of the inhabited polders of the western part of The Netherlands are protected by a system of primary flood defences, large parts of the port of Rotterdam are located outside this system ('un-embanked areas'). The Rotterdam harbour hosts a significant amount of (petro) chemical industries in un-embanked areas. In order to reduce the frequency of flooding, port facilities and infrastructure have been constructed at elevated terrains on riverbanks or reclaimed land. In addition, the Maeslant Storm Surge Barrier has been constructed, which protects parts of the port area (see also Chapter 10). For newly developed areas, a design ground elevation of + 5.0m above mean sea level (amsl) is used and this corresponds to a flooding frequency of 1/10,000 per year. The most important infrastructure is located at higher elevations.

The expected sea-level rise associated with climate change will result in an increase in the probability of flooding of the port infrastructure and industrial estates. As flood risk is generally defined as the probability of flooding multiplied by the damage due to flooding, the level of flood risk will rise accordingly. The potential consequences of flooding of typical port infrastructure were exemplified during Hurricane Katrina in 2005, which caused large-scale wind and flood damage in New Orleans and other areas of Louisiana and the Gulf Coast. Several industrial facilities flooded, causing spills and releases of industrial goods (Kok et al, 2006; Pine, 2006). The spills of a flood event cause environmental damage, economic damage and substantial societal disruption to the surrounding areas after the event.

Figure 4.1 *Elevation of the Rotterdam Rijnmond area*

Note: N.A.P. is the Amsterdam Ordnance Datum, the vertical datum used in the Netherlands.
Source: CGIAR (http://srtm.csi.cgiar.org/, 2008); more information is provided in Jarvis et al (2008)

The impact of sea-level rise on the vulnerability and actual risk of the un-embanked areas has not been studied extensively and a reliable method to assess the actual flood risk of port infrastructure in these un-embanked areas does not exist. In addition, the vulnerability of port infrastructure is case specific and is highly dependent upon the types of infrastructure and land use that are present, local topography (e.g. elevation) and the types of hazards that can lead to damage (e.g. flooding and windstorms such as hurricanes). This assessment was therefore based on expert knowledge resulting in a qualitative evaluation of the flood vulnerabilities of industries and port infrastructure.

High water levels in the port of Rotterdam occur as a result of North Sea storm surges in combination with high (spring) tides. During high water events, the water level will rise gradually. Inundation will also happen gradually, and can be predicted relatively well in advance. Flow velocities are low – generally below 0.5m per second – and the inundation depth of the un-embanked areas can be assumed not to exceed 1m as the elevation of these areas is relatively high (see also Figure 4.1). Because of this, flooding of port facilities could lead to damage to the facilities and release of hazardous materials, which could result in damage and loss of life in the exposed population. These domino effects are generally not taken into account in the existing risk assessments for the chemical industry in The Netherlands. Earlier research (Cruz et al, 2001) showed that oil refineries are susceptible to windstorm and flood events, and recommended that further risk quantification and expert elicitation should be undertaken.

This chapter is based on a research report (Lansen and Jonkman, 2010) in which further details can be found. The study assesses the flood vulnerability of chemical plants within the port area of Rotterdam. The significance of the additional flood risk to those facilities was compared to the existing risk profile of the area. Only un-embanked areas have been taken into account. Overall, the following questions have been addressed in this study:

- What is the vulnerability of port infrastructure in un-embanked areas to flooding?
- How do we evaluate this vulnerability to flooding in comparison to other (flood) risks?

This chapter is structured as follows. Section 4.2 provides an overview of the vulnerability of different types of functions and infrastructure in the harbour region. Section 4.3 focuses on a further analysis of the risks to chemical facilities in un-embanked areas. Section 4.4 discusses the additional risks to these facilities due to flooding in the context of several perspectives for risk evaluation. Closing remarks and some recommendations are included in section 4.5.

4.2 A Qualitative Assessment of Flood Vulnerability of Port Infrastructure: Methodology and Results

4.2.1 Approach

The study started with a qualitative assessment of the vulnerability to flooding of port infrastructure by means of expert elicitation. Experts from various fields of expertise (container terminal, external safety, chemical, flood risk, all from consultancy firms, central government and the port authority itself) were invited to discuss and perform an assessment using the methodology presented below. The main objective of this qualitative assessment was to identify the most vulnerable components in the port area. Vulnerable within this context was defined as susceptible to large consequences if flooded.

The land use and infrastructure in the port area was divided into several categories, of which it is assumed that the type of damage and consequences differ significantly. The following main categories were used:

- dry bulk;
- liquid bulk;
- containers;
- RoRo (roll on/roll off);
- business areas and other land use;
- public/logistic infrastructure (rails, roads, electricity cables, water supply, etc.).

In addition, several subcategories were given to provide a more detailed overview of the types of infrastructure and land use for the category.

The experts were asked to give a qualitative judgement of these land-use categories for a hypothetical flood event in which all the facilities would be exposed to a water depth of 1m. The vulnerability of these categories is expressed using the following consequential damage categories (see also Chapter 6):

- casualties;
- societal disruption;
- environmental risk/ecological damages;
- economic damage.

In order to harmonize the scoring of the experts, descriptions of the effects of different categories were given before the assessment (see Table 4.1). In addition, the likelihood of damage should a flood occur was ranked qualitatively using qualitative descriptions that were related to a range of probabilities (1, 10 and 100 per cent). The overall scoring methodology that was presented to the experts is shown in Table 4.1.

Table 4.1 *Scoring of the probability of damage given a flooding event and the potential consequences*

	Low	**Middle**	**High**
Probability of damage given a flooding event	~ 1%	~ 10%	~ 100%
Consequences			
Casualties	None to a few (< 5), especially individuals at the facility/location	A few (~ 10), also externally	Many (> 100)
Societal disruption	Effects are brief and limited to the facility/location	Hindrance and damage within 10km of the location of flooding	The complete region and/or nation experiences sustained hindrance and damage as a result of flooding
Environmental risk/ ecological damages	The range and severity of effects are limited and short term	The range of effects is in the order of kilometres and limited to a number of weeks	Large area affected and negative effects long term
Economic damage	< 1 million Euros in damage	A few dozen million Euros in damage	Hundreds of millions to billions Euros in economic damage

The vulnerability assessment includes so-called chain or domino effects. For instance, if a container terminal cannot be operated for a considerable time after a flood event, disruption to society as a result of the interruption is likely to occur.

4.2.2 Results of the qualitative assessment

For every category of port infrastructure, experts were asked to rank the probability of damage given flooding, and to rank the possible consequences if flooded. The ranking was performed after discussion and sharing of ideas on the topic. An example of the scores is provided in Table 4.2, for the category 'dry bulk'. The consequences of flooding of dry bulk areas are primarily the runoff of material into surrounding waters and the interruption of business. Effects include delayed transport of goods and minor environmental damage due to goods dissolving into the river.

Experts were asked to assess the vulnerability to flooding according to terrain use, based on the information provided and the discussions. Port infrastructure categories most vulnerable to flooding were concluded as follows:

- public infrastructure;
- liquid bulk;
- containers;
- other categories:
    - business areas and other land use;
    - RoRo;
    - dry bulk;
    - general harbour facilities.

The qualitative assessment of port infrastructure in the port of Rotterdam shows that infrastructure and liquid bulk (e.g. oil, gas and liquid petroleum gas), in particular, are vulnerable if flooded. Large quantities of casualties are not expected, but large societal disruption is likely to occur.

For liquid bulk, due to the amount of chemical installations and the variety of goods stored in industrial port areas, it is expected that flooding of some plants could potentially release hazardous goods. These goods will result in health effects in addition to environmental damage. Storage of liquid bulk and the transportation of liquid bulk to the hinterland might also be severely interrupted during a flooding event.

The category 'public logistic infrastructure' is also very vulnerable if heavily flooded. This category includes power supply, information and communications technology (ICT), and roads and tunnels. Power failure will affect many other operations. These are not only local, but extend to regional and, potentially, to national level. ICT is very important for crisis management during flood events. Roads, tunnels and pipelines provide transport of people and goods. This is severely interrupted during floods.

Table 4.2 *Example of scoring the vulnerability of port infrastructure for dry bulk*

	Potential effect of flooding	Probability of damage given a flooding event of 1m	Casualties	Societal disruption	Environmental damage	Economic damage
Facilities: cranes, conveyor belts, silos, sheds	• Corrosion • Failure and damage of electricity and information and communications technology (ICT) • Instability of buildings on footings	8 (high) 4 (medium) 0 (low)	0 (high) 0 (medium) 12 (low)	0 (high) 6 (medium) 6 (low)	0 (high) 0 (medium) 12 (low)	0 (high) 10 (medium) 2 (low)
Storage of goods: zinc, iron, coal, etc.	• Washout of material (loss of goods) • Instability of toxic goods in surface water	3 (high) 9 (medium) 0 (low)	0 (high) 0 (medium) 12 (low)	1 (high) 1 (medium) 10 (low)	2 (high) 5 (medium) 7 (low)	0 (high) 9 (medium) 3 (low)

Given the terrain height of most of the port areas, the probability of flooding is rather low (in the order of 1/4000 to 1/10,000 per year). In general, however, the chain effects of all categories are very vulnerable, mainly affecting the provision of goods to the hinterland. Operations might have to be terminated for a period of time if the port terrains are flooded. The port of Rotterdam serves large parts of Northwest Europe and the system effects of flooding of the port can be considerable.

The number of casualties, given a flood scenario with a depth of 1m, is expected to be very small. Dozens of casualties, at most, are expected if chemical facilities are damaged and hazardous materials released. This is due to the fact that highly populated areas are generally not within immediate range of hazardous facilities. Furthermore, it is expected by most of the attendees that effective measures are taken during installation processes to prevent explosions or the release of hazardous substances, as well as during the design and maintenance of storage facilities. In addition, if flooding is forecasted, plant operators will probably put operational procedures into action, which also reduces the vulnerability of industrial areas.

Given the complexity and interrelatedness of numerous functions in the port area, a site- and scenario-specific assessment of the risks was deemed preferable. The qualitative assessment revealed interesting thoughts and insights into the categories and mechanisms which should be considered when determining the actual flood risk of these (un-embanked) port areas. In the following section, a case study determines the vulnerability to flooding of a chemical plant, and the potential consequences to human life, using a more quantitative approach.

4.3 A Quantitative Assessment of Flood Vulnerability of Port Infrastructure

4.3.1 Approach

Although some research has been conducted on the vulnerability of chemical installations to flooding (Campedel et al, 2008; Cazzoni et al, 2010) no generally accepted method is available. In addition, there is limited insight into this particular vulnerability for the Rotterdam port area. An approach similar to that used in external safety risk assessment methodologies for the (chemical) industry was adopted in this study (e.g. the Hazard and Effects Management Process, or HEMP). HEMP is a structured and systematic analysis methodology involving the identification, assessment and control of hazards and the recovery from effects caused by a release of the hazards. The following steps are usually made:

1 *Hazard identification.* A multidisciplinary team reviews a hazard checklist to identify those hazards which are relevant to the business throughout its total life cycle. This list is the start of the hazard and effects analysis.

2   *Identification of hazard scenarios.* The experience of the team is used to identify the failure trees or hazard scenarios (top events).

3   *Risk assessment.* For each hazard scenario, the risk is assessed and hazards are prioritized according to their potential consequences and effects (on the basis of realistic worst-case assumptions).

During the risk assessment process, most undesirable events, circumstances and consequences are identified. These can be presented in a so-called bow-tie model (see Figure 4.2). In this study into the vulnerability of port infrastructure to flooding, it is assumed that flooding will be an additional hazard scenario of the chemical plant. Campedel et al (2008) provide a similar approach for estimating the flood risk of chemical installations.

It is clear that estimating the flood risk of a chemical plant requires case-specific information. The potential effects of flooding of a chemical installation depend to a large extent upon the type of goods which are processed, the storage conditions, the transportation methods on the terrain, the safety procedures in place, etc. In an initial attempt to draw some general conclusions about the actual risk for the port of Rotterdam, an approach was proposed which provides insight into the actual flood risk of chemical installations in the port area. A case study was developed using known information of an existing chemical plant (the study is not site specific). This information was used to quantify the flood risk with the following steps:

- A worst-case scenario under 'normal' plant operations was selected (i.e. a worst case without flooding). The effects and consequences are known.
- A characterization of flood events for the port of Rotterdam was made.
- It was assumed that the worst-case scenario without flooding is also the worst-case scenario with flooding. The consequences and effects of the worst-case scenario

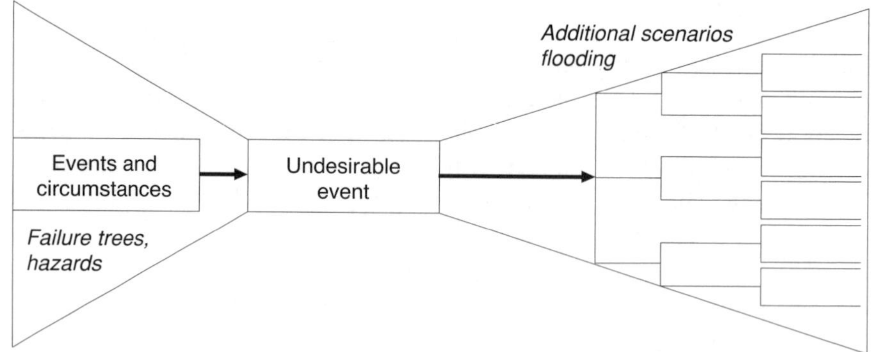

Figure 4.2 *Bow-tie model which schematically indicates the analysis of risks for a chemical plant and the additional scenarios due to flooding*

Source: Lansen and Jonkman (2010)

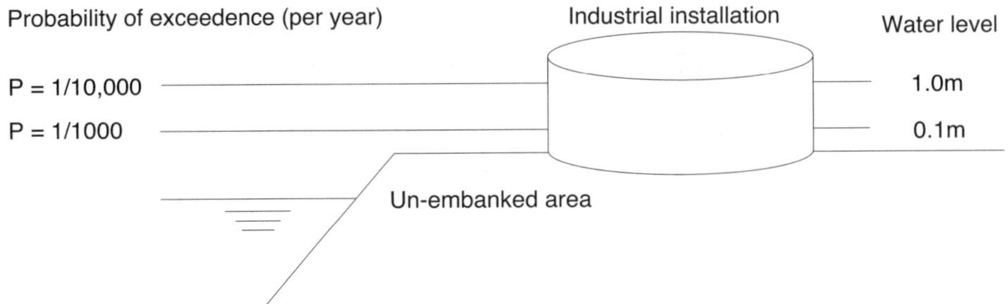

Figure 4.3 *Flooding of areas not defended by primary flood defences*

Source: Lansen and Jonkman (2010)

with flooding were estimated by using the flood characteristics, and the impact under these circumstances was estimated by comparing this to the worst-case scenario without flooding.

The probability of the worst-case scenario to occur, given flooding, is assumed to depend upon the flood depth. Flood velocity is not separately taken into account in this study because we assume that for the typical application of un-embanked areas, the velocities are expected to be relatively low (see Asselman, 2010). The fact that higher flood depths will also lead to higher velocities during flooding and runoff is included in the higher probability of failure estimate.

4.3.2 Results

The following causes can be identified for a chemical plant, leading to an occurrence probability of a worst-case scenario given the inundation depth onsite. The depth is determined by an extreme water level with a low return frequency in combination with the terrain elevation of the un-embanked area. The inundation depths and corresponding return frequency are typical for the port of Rotterdam. In addition, it can be assumed that during high water conditions, wind velocities are high due to the large correlation of extreme water levels and high wind velocities.

Low water levels (~ 0.1m) can cause buoyancy effects of empty or gas-filled pipelines. Destabilization due to erosion by local high velocities may also result in local scour holes underneath pipelines supported at ground level. The consequences include multiple pipeline rupture, resulting in airborne emissions and water pollution. Water pollution may spread by water (provided that the density is lower than the density of water). Toxic goods dissolved in water may react. In both cases, toxic substances will evaporate. As a result, airborne emissions may reach high concentrations over large areas due to the fact that the water distributes the material over large areas.

Table 4.3 *Characteristics of the worst-case scenario (without) flooding and the effects of the additional flood scenarios*

	Worst-case scenario (without flooding)	Effects of additional flood scenarios	
Scenario	Hexafluoropropene release		
		Flood depth: 0.1m	Flood depth: 1m
Probability (per year)	10^{-5}	10^{-4}	10^{-5}
Conditional probability of damage given flooding	–	1%	10%
Total probability (per year)	10^{-5}	10^{-6}	10^{-6}
Primary cause	3 inch nozzle rupture	Same	Same
Wind	1.5m/second	Beaufort/force 12 from sea (> 32m/second)	Beaufort/force 12 from sea (> 32 m/second)
Emission period	12 minutes	10 hours (total storm duration up to 50 hours)	10 hours (total storm duration up to 50 hours)
Warning (days in advance)	None	2–5	2–5

Higher water levels (~ 1m): additional damage of pipelines, tanks and vessels due to floating objects and debris may occur. Waves and wind may result in the collision of floating objects, resulting in serious damage. The amount of leakage increases due to multiple pipe rupture and because elevated storage facilities and process installations are subject to flooding.

Table 4.3 presents specific circumstances for a worst-case scenario from an existing plant. For this scenario, the consequences are known from previous studies on the potential effects of the release of hazardous goods. The specific circumstances and flood characteristics are also given, which are used to determine the effects and consequences of flooding. Flood frequencies are derived from the water-level statistics and ground elevations of the port of Rotterdam.

The effects and consequences of the regular worst-case scenario and the scenario with flooding are compared in Table 4.3. Health effects are expressed relative to the *Emergency Response Planning Guidelines* (ERPG) threshold values. The ERPG is a system of guidelines that is intended to provide estimates of concentration ranges where one might reasonably anticipate observing adverse effects as a consequence of exposure to a specific toxic substance.

We have made a distinction between the following values:

- *Health effects:* due to exceeding ERPG level 2. This is the maximum airborne concentration below which it is believed that nearly all individuals could be exposed for up to one hour without experiencing or developing irreversible or other serious health effects, or symptoms that could impair their abilities to take protective action.
- *Serious health effects:* due to exceeding ERPG level 3. This is the maximum airborne concentration below which it is believed that nearly all individuals could be exposed for up to one hour without experiencing or developing life-threatening health effects.

For this case study, it is assumed that strong winds will reduce the effects of the release of hazardous products into the air by dispersion. On the other hand, floodwater will run off the industrial site, eventually leading to a larger distribution of hazardous substances. The estimate of the number of casualties is used to evaluate the flood risk of a chemical installation in terms of individual risk and group risk.

The case study was intended to estimate the order of magnitude of the flood risk of a chemical installation due to flooding. The methodology proposed for this study follows standard procedures in the chemical industry. Using emission calculations from a known worst-case scenario in combination with knowledge about flood circumstances, a judgement on the effects and consequences was made. Based on this hypothetical but realistic case study, the following conclusions are drawn:

- The number of expected immediate casualties as a result of the installation's flooding, as presented in the case study, is not expected to increase due to flooding. This means that the number of offsite casualties for this case study is limited to nil. This also signifies that flooding does not increase the effects of the release of hazardous substances with respect to casualties.
- The health effects of the worst-case scenario, including flooding, might, however, be worse than the health effects without flooding (~100 people with health effects, compared to ~10 health effects): the toxic material can be distributed by water to distant locations. In addition, the scale of simultaneous damages is expected to be larger due to flooding of the whole industrial site. These factors are fundamentally different from a worst-case situation in which the hazardous goods are contained on the dry soil in a situation without flooding.

Table 4.4 presents the results of a quantitative estimate of casualties, affected persons, economic damage and cultural damage. These estimates are based on the considerations above.

Table 4.4 *Overview of effects on casualties, affected persons, economic damage, environmental damage and cultural damage based on the case study*

Consequence category	Subcategory	Worst-case scenario: No flooding	Scenarios including flooding	
			Flood depth: 0.1m	Flood depth: 1m
Casualties	–	None/limited	None/limited	
Affected persons (health effects)	Onsite health effects	Serious effects for (operators)	Health effects are smaller as it is expected that the site will be evacuated due to warning	
	Offsite health effects	1000	100	2000
	Serious offsite health effects	10	n/a	n/a
Economic damage		Two months of downtime; expected damage in the range of 10–100 million Euros	Comparable	Five months of downtime; similar damage to plant, but larger amounts of claims
Environmental damage		Significant contamination of air; limited water pollution	Comparable	Larger contamination of air; limited water pollution
Cultural damage		Minor	Minor	Minor

Note: n/a = not available.

The following additional remarks are made:

- The area which is affected by releases and emission of hazardous goods during a flooding event increases due to distribution by runoff from the industrial site: the toxic water can flow from the site towards urban areas. This results in more severe effects for a case including flooding.
- However, the effect of high wind speed is also to be taken into account. Wind has a dispersive effect on emissions; hence, the 'worst-case' scenario under normal operations mostly occurs under low wind speed conditions. Contrary to this, flooding will occur under high wind speeds. This has a positive effect on airborne emissions, reducing the consequences and effects of toxic release. This results in less severe effects for a case including flooding.

- An industrial site will probably be (partly) evacuated during flooding, as a flood event can be predicted between two to five days in advance. Mitigating measures are also taken when a flood is forecasted. This results in less severe effects for a case including flooding.

Further research will need to involve a specific installation using relevant information from the plant owner. In addition, the effect of simultaneous flooding, leading to an accumulation of risk, should be considered. The relevance of this study is found in the approach for assessing the flood risk of chemical installations. A multidisciplinary approach is necessary to incorporate flood risk within the safety philosophy of the industry. Knowledge-sharing between disciplines seems vital in this respect.

4.4 Discussion in the Context of Risk Evaluation

4.4.1 General

The previous section provided insights into the additional risk due to flooding of a chemical installation in un-embanked areas. This information is now discussed in the context of risk evaluation approaches. In the risk evaluation phase, results from the risk quantification are used as inputs in order to determine whether a risk level is acceptable or not. In other words, the question 'How safe is safe enough?' is posed (Starr, 1967).

No direct regulations or policies exist to assess the tolerability of (additional) flood risks to chemical installations. Therefore, three perspectives are used that are associated with relevant policies in The Netherlands:

1 comparison with flood risk policies for areas protected by flood defences (embanked areas);
2 external safety standards for chemical installations;
3 safety regulations used by the chemical industry.

4.4.2 Flood risk policies for embanked areas in The Netherlands

In The Netherlands, no policies exist that directly regulate the flood risks of un-embanked areas. For the flood defences that protect the dike rings (or embanked areas), safety standards reflect the return period of the water level that flood defences are designed to safely withstand. For most of the flood defences near Rotterdam, the safety standard is 1/10,000 per year.

Currently, an extensive nationwide study is examining the risks (probabilities and consequences) faced by protected areas. It is therefore interesting to compare the (additional) flood risk for the chemical installation to risk estimates for the dike rings.

During high water events in the port of Rotterdam (with a return probability of flooding of 1/10,000 per year), it is very likely that the dike of South Holland will also face a considerable risk of flooding (see Jonkman et al, 2008, for an extensive risk study on the flood risks to people in this dike ring). South Holland is one of the largest flood-prone areas in The Netherlands, with 3.6 million inhabitants; it is also the most densely populated area in the country and includes major cities. The severity of flooding and its consequences depend upon the breach location. In the most extreme coastal flood scenario, breaches occur at two locations (Den Haag and Ter Heijde). An area of approximately 230 square kilometres with more than 700,000 inhabitants could be flooded. It is expected that the probability of fully evacuating this densely populated area is limited because the time available (approximately one day) is insufficient. Calculations also suggest that this flood scenario could lead to more than 3000 fatalities. For other scenarios, the estimated numbers of fatalities are smaller. By combining the probability and consequence estimates for these scenarios, a so-called FN curve (frequency vs. number) can be constructed (also indicated as societal risk). This plot of the frequency distribution of multiple-casualty events shows the probability of flooding resulting in a large number of fatalities (see Figure 4.4 for the estimated societal risk for South Holland).

In Figure 4.4 a conservative estimate of the additional societal risk due to flooding of un-embanked areas is shown. Based on the findings in section 4.3.2, the probability of such an event is estimated to be in the order of magnitude of 10^{-6} per year. The number

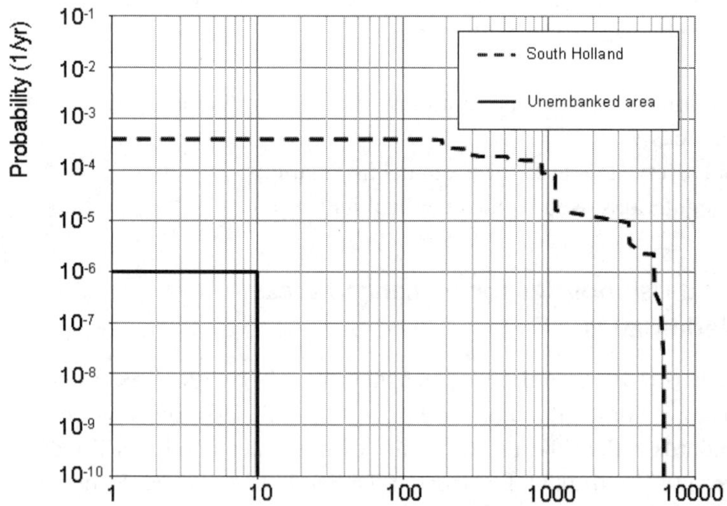

Figure 4.4 *Societal risk for South Holland and additional societal risk for un-embanked areas*

Source: Jonkman et al (2008)

of fatalities that could occur is estimated to be approximately ten. It is clear that the additional risk due to flooding of un-embanked areas and consequent damage to chemical installations is much smaller than the estimated societal risk level for embanked areas in South Holland. This is no surprise as the number of potentially affected people in the dike ring is large, but much smaller for the individual installation.

A similar comparison can be made for individual risk. The probability of death for a person in South Holland due to flooding, the so-called individual risk, is small and in most areas is lower than 10^{-6} or 10^{-7} per year (Jonkman et al, 2008). The additional individual risk for flooding of the un-embanked area can be found by multiplying the probability of such an event by the conditional probability of death of an individual near the facility. The probability of an accident at a chemical installation due to flooding is in the order of magnitude of 10^{-6} per year. The conditional probability of death could be in the range of 0.1, so the total addition to the individual risk for those most affected could be 10^{-7} per year. Thus, the (additional) individual risk for the case study involving flooding of a chemical plant is expected to be of the same order of magnitude as the individual risk in the dike ring. To date, no policy has been defined with regards to an accepted standard for the individual flood risk or a standard for group risk. These standards are being developed in the coming years by the province of South Holland.

4.4.3 External safety: The Dutch major hazards policy

Dutch major hazards policy deals with the risks to those living in the vicinity of major industrial hazards, such as chemical plants and liquid petroleum gas (LPG) fuelling stations. The cornerstones of the Dutch major hazards policy are (Bottelberghs, 2000; Ale, 2002):

- quantitative risk analysis;
- individual and societal risk as risk-determining parameters; and
- quantitative acceptability criteria for evaluating levels of individual and societal risk.

Individual risk is defined as the probability of death of an average unprotected person that is constantly present at a given location.

Individual risk criteria are reference levels for evaluating individual risks. The individual risk criteria were given legal status in 2004 by the External Safety Decree. These limits to individual risk prevent disproportional individual exposure. Permits for property developments or plant modifications are denied if vulnerable objects would then be located within the 10^{-6} contour (see Figure 4.5).

An individual risk criterion alone cannot prevent the too frequent occurrence of multi-fatality accidents. As shown in Figure 4.5, the area affected by an accident can differ considerably from the area that is defined by an iso (individual)-risk contour. When individual exposures are low, there could still be a chance that a single accident

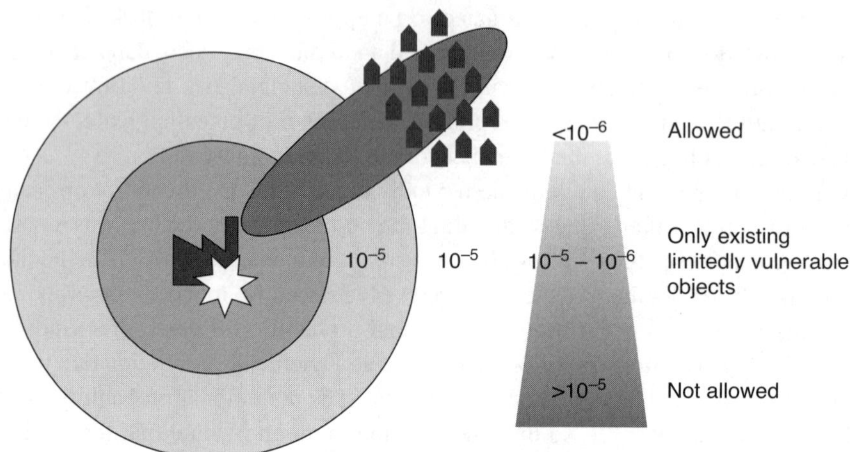

Figure 4.5 *Individual risk contours around a hazardous establishment and the area affected by an individual accident scenario*

Source: Jonkman et al (2011)

kills a large number of people. While a vast number of small accidents can go by hardly being noticed, multi-fatality accidents can shock a nation. Psychometric studies have indeed shown that 'dread', or catastrophic potential, is an important factor in explaining risk perceptions (Slovic, 1987). In order to prevent the too frequent occurrence of large-scale accidents, societal risk criteria were implemented in The Netherlands. Societal risk is graphically represented by an FN curve that shows the exceedance frequency of the number of fatalities ($P(N = n)$) on a double log scale (see Figure 4.6).

The Dutch societal risk criterion of $10^{-3}/n^2$ per installation per year, where n^2 is the the square of the number of fatalities, was initially developed for LPG fuelling stations. It was later applied to all Seveso establishments. Similar societal risk criteria thus apply to hazardous establishments of different character and size despite considerable differences between the marginal costs of risk reduction in different cases (see Jongejan, 2008, for further discussion).

The estimates of the individual and societal flood risk levels are compared with the limits for external safety. Figure 4.7 compares the (additional) risk for the case study involving flooding of a chemical plant in the port of Rotterdam and the societal risk limit for installations (indicated in the grey box). It is clear that the risk is smaller than the limit for installations; in fact, the risk due to flooding should be added to the normal risk. In this context it is noted that both the axes of the FN curve are logarithmic: the more to the right on the horizontal axis, the smaller the visual contribution of an additional event with ten fatalities. Regarding the individual risk, it is estimated that the additional contribution from flooding (order of magnitude 10^{-7} per year or smaller) is below the individual risk limit of 10^{-7} per year.

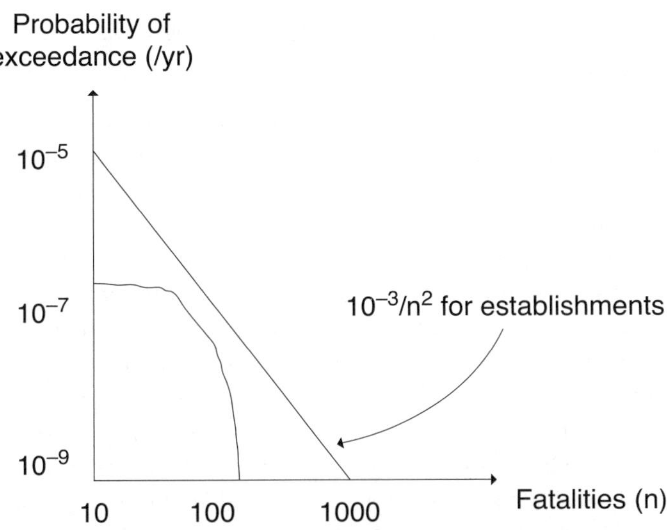

Figure 4.6 *The Dutch societal risk criterion for hazardous establishments and a fictitious FN curve*

Source: Jonkman et al (2011)

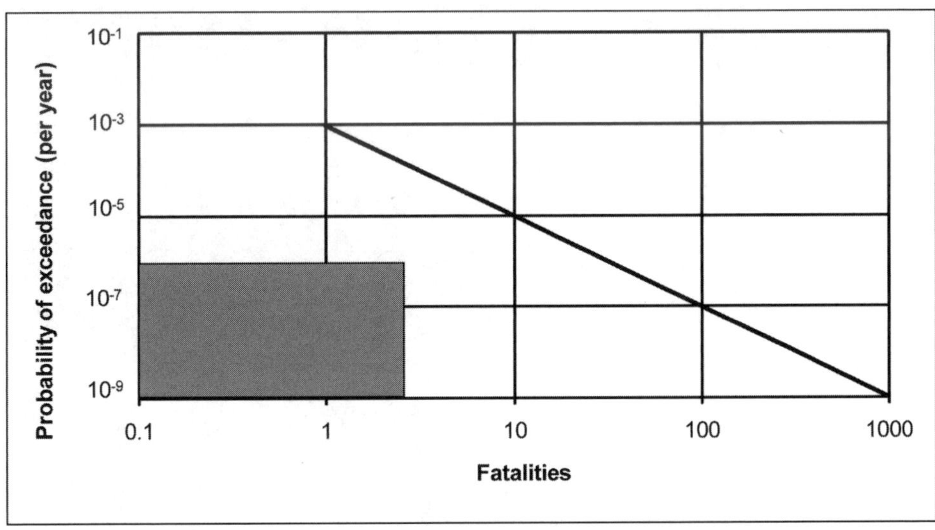

Figure 4.7 *Comparison between the (additional) risk for the case study involving flooding of a chemical plant in the port of Rotterdam and the societal risk limit for installations (indicated by the grey box)*

Source: Lansen and Jonkman (2010)

4.4.4 Discussion in the context of safety frameworks by the chemical industry

Companies within the chemical industry have adopted their own safety and risk policies to ensure the safety of their companies, installations, employees and the surroundings of the company. The risk of the installation is evaluated by considering the probability/frequency of accidents and their consequences.

An example of such a 'risk assessment matrix' is given in Figure 4.8, which considers a qualitative description of the accident frequency (horizontal axis) and several consequence categories on the vertical axis. By combining the two, the overall risk can be positioned in the matrix. Depending upon the position of the quantified risk, it is possible to judge how the risk should be treated, and this is a basis for further analysis on the need for risk reduction measures. Figure 4.8 also indicates the (additional) risk of the chemical installation. Based on section 4.3.2, the additional flood risk can be classified as an event with severe effects and a probability that can be described as 'heard of in the industry'. This would imply that this additional risk is in the middle region of the matrix. The additional risk would therefore require further attention and further evaluation of the need of risk reduction measures. Companies should therefore be involved in evaluations of this type of risk and in recommendations for further risk reduction measures.

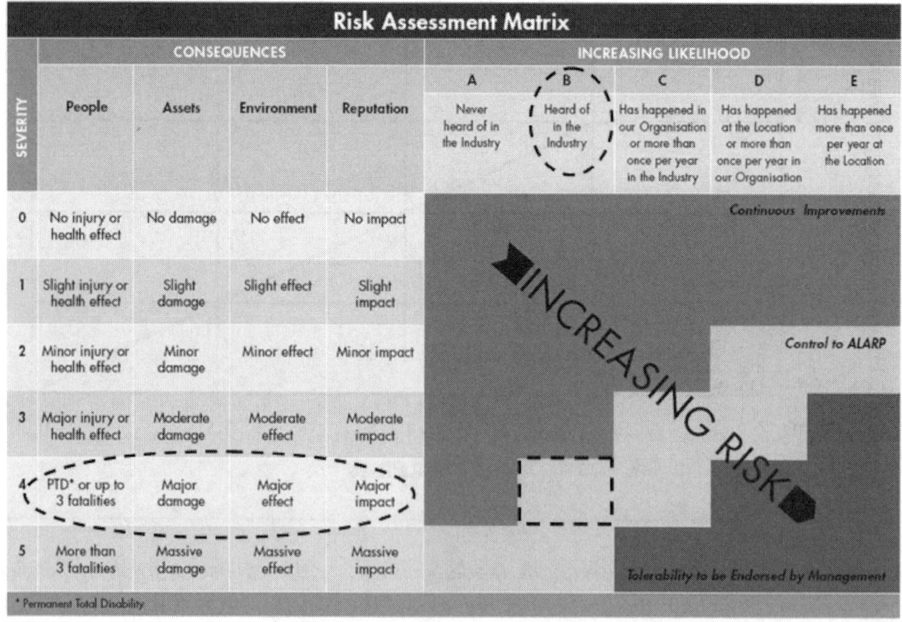

Figure 4.8 *Risk assessment matrix*

Source: www.eimicrosites.org

4.5 Conclusions

In general, it is expected that the additional risk of flooding of chemical facilities is relatively small for sufficiently highly elevated areas that are un-embanked by flood defences (> +5.0m amsl). Flooding and consequent frequency of accidents are relatively small. The (additional) risks in these areas are low relative to the risks of flooding of embanked areas, such as South Holland. Although they are expected to be acceptable for most cases within the frameworks of external safety policy and chemical companies, further consideration on the additional level of risk due to flooding is recommended. Overall, it is expected that societal disruption and economic damage (for the company itself) will be the most relevant consequence types for the case study.

Eventually, if it is concluded in the risk evaluation that current risks are unacceptable, additional risk reduction measures should be taken. These could consist of the following:

- structural measures, such as raising the terrain to reduce the probability of flooding or a (more) flood-resistant design for the facility;
- measures related to the management of the facility (e.g. a shutdown of the plant or the movement of dangerous goods in a threatening situation, or a timely warning and evacuation of people living in the vicinity).

The following issues need to be considered in further investigations:

- A simplified method has been used to estimate whether the additional level of risk due to flooding is expected to be significant. Further development of more advanced methods for a risk assessment of these interconnected events and systems is recommended.
- During a high water event or storm surge, extensive parts of the un-embanked areas (the so-called Maasvlakte) could be flooded and multiple facilities may be simultaneously affected. In this study, the risk for a single (hypothetical) facility has been assessed, and assessment of the cumulated risk for multiple installations is recommended. An assessment should be conducted for the whole region to analyse the presence, elevation and risks of existing installations in un-embanked areas.
- The (additional) risks of chemical facilities due to flooding will depend upon the characteristics of the substance that is affected, the interaction between the substance and water, and the role of weather conditions (wind, temperature, etc.). Note that this is different from 'normal' flood risk assessments where location-specific risk assessments are not made.
- The case referred to one type of scenario and facility. Other substances and facilities could lead to different scenarios and risks, which, in some cases, could be more severe. It is therefore necessary to consider the additional risks of flooding of chemical installations on a case-by-case basis.

- Currently, it is not clear how responsibilities are defined in flood risk management of chemical installations in un-embanked areas. Is this a public task as part of the flood risk policy for un-embanked areas? Which government entity (municipal, provincial or national) is responsible? Do flood risks have to be included in external safety regulations for chemical installations in these specific cases? These questions should be explored in more detail with the relevant stakeholders.

Further consideration of these questions necessitates cooperation between people and organizations from different fields (e.g. external safety, flood risk management) and backgrounds (policy and experts). As part of the research project, a workshop was organized with people from various organizations and backgrounds, providing an important network. One significant conclusion from the workshop was that the knowledge derived from this study should be consolidated and made available to different parties.

Acknowledgements

This research project was carried out within the framework of the Dutch national Knowledge for Climate research programme (www.knowledgeforclimate.org), Project HSRR02: Flood Risk in Un-embanked Areas. This research programme is co-financed by the Ministry of Housing, Spatial Planning and the Environment (since 2011, the Ministry of Infrastructure and Environment), the Municipality of Rotterdam and the Port Authority of Rotterdam. Rinske van der Meer (Port of Rotterdam), Nick van Barneveld (Municipality of Rotterdam) and Henk de Visser (Royal Haskoning) are gratefully acknowledged for their contributions to this work.

References

Ale, B. J. M. (2002) 'Risk assessment practices in The Netherlands', *Safety Science*, vol 40, no 1, pp105–126

Asselman, N. E. M. (2010) 'Flooding characteristics: Flow velocities in the downstream reaches of the river Rhine and Meuse', *Knowledge for Climate Report*, no 022B/2010

Bottelberghs, P. H. (2000) 'Risk analysis and safety policy developments in the Netherlands', *Journal of Hazardous Materials*, vol 71, pp117–123

Campedel, A. et al (2008) 'A framework for the assessment of the industrial risk caused by floods', in S. Martorell et al (eds) *Safety, Reliability and Risk Analysis: Theory, Methods and Applications*, Taylor & Francis Group, London

Cazzoni, V. et al (2010) 'Industrial accidents triggered by flood events: Analysis of past accidents', *Journal of Hazardous Materials*, no 175, pp501–509

Cruz, A. M., Steinberg, L. J. and Luna, R. (2001) 'Identifying hurricane-induced hazardous material release scenarios in a petroleum refinery', *Natural Hazards Review*, vol 2, no 4, pp203–210

Jarvis A., Reuter, H. I., Nelson, A. and Guevara, E. (2008) *Hole-Filled Seamless SRTM Data, Volume 4*, International Centre for Tropical Agriculture, Colombia

Jongejan, R. B. (2008) PhD thesis, Civil Engineering and Geosciences, Delft University of Technology, Delft, The Netherlands

Jonkman, S. N., Vrijling, J. K. and Kok, M. (2008) 'Flood risk assessment in the Netherlands: A case study for dike ring South Holland', *Risk Analysis*, vol 28, no 5, pp1357–1373

Jonkman, S. N., Jongejan, R. B. and Maaskant, B. (2011) 'The use of individual and societal risk criteria within the Dutch flood safety policy: Nationwide estimates of societal risk and policy applications', *Risk Analysis*, vol 31, no 2, pp282–300

Kok, M., Theunissen R., Jonkman S. N. and Vrijling, J. K. (2006) *Schade Door Overstroming: Ervaringen uit New Orleans*, TU Delft/HKV Publications, The Netherlands

Lansen, A. J. and Jonkman, S. N. (2010) 'Vulnerability of port infrastructure in unembanked areas', *Knowledge for Climate Report*, no 022D/2010

Pine, J. C. (2006) 'Hurricane Katrina and oil spills: Impact on coastal and ocean environments', *Oceanography*, vol 19, no 2, pp37–39

Slovic, P. (1987) 'Perception of risk', *Science*, vol 236, pp280–285

Starr, C. (1967) 'Social benefit versus technological risk', *Science*, vol 165, pp1232–1238

5

Storm Surge Modelling for the New York City Region

*Malcolm J. Bowman, Brian A. Colle
and Hamish Bowman*

5.1 Storm Surges and Flood Risk in New York City

The New York City (NYC) metropolitan area has a history of tropical cyclone landfalls near NYC, such as the hurricanes of 1821 (category 3), 1893 (category 2; Scileppi and Donnelly, 2007) and the category 3 Long Island Express hurricane in 1938 (Brooks, 1939). The most recent hurricane to make landfall on Long Island was Hurricane Gloria, which crossed central Long Island, New York, at 16:00 UTC on 27 September 1985 with 75 knot (kt) (~37m/second) sustained winds and a central pressure of 961 hectopascals (hPa) (Case 1986). More recently, tropical storm Floyd, on 16 September 1999, created a maximum surge of ~1m around NYC (Colle et al, 2008); but luckily the peak surges of Floyd and Gloria occurred near low tide, so only minor flooding was observed. Winter nor'easters, although not as violent as hurricanes, can inflict significant damage as they tend to persist much longer (e.g. the December 1992 storm). The consequences of some of these extreme weather events are discussed in Hill (1996), Bloomfield et al (1999), Bowman et al (2008) and Chapter 13 in this volume.

There have been no moderate flooding events at The Battery from 1997 to 2010 (only a minor event on 13 March 2010, which is the quietest period during the last 50 years). The elevation of the surrounding lip of the excavated World Trade Center site ('Ground Zero') would be less than 3.7m (12 feet) of water above the estimated storm surge of a worst-case scenario category 3 direct-hit hurricane (winds of 50m/second to 58m/second, or 112mph to 130mph) with a surge elevation of 7.5m (24 feet) at The Battery.

Most of the lower Manhattan financial district would be flooded and the city hall would sit on an island separate from the rest of Manhattan (see Plate 2). The ambitious new commercial development now taking place in Jersey City, New Jersey (west of the Hudson River opposite lower Manhattan), could readily be flooded even with a more modest direct-hit category 1 hurricane. Secaucus, New Jersey, would be an island in the completely flooded Hackensack Meadowlands (see Figure 5.1).

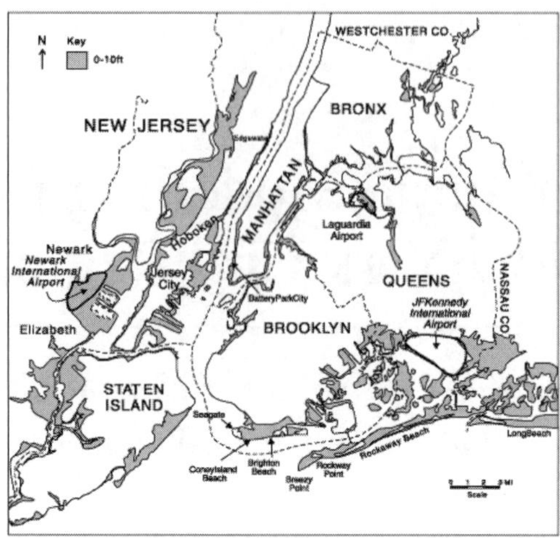

Figure 5.1 *The 100-year floodplain (grey tone) at current sea levels for metropolitan New York*

Source: Gornitz (2001)

What is now regarded as the 100-year flood of 3m (9.7 feet) above National Geodetic Vertical Datum (NGVD)[1] would inundate large segments of the metropolitan region: not just lower Manhattan and nearby coastal northern New Jersey State (see Plate 3), but extensive low-lying sections of the outer boroughs of southern Brooklyn and Queens. Also at risk are the La Guardia, John F. Kennedy and Newark international airports, Hackensack Meadowlands, New Jersey (Figure 5.1), and the entire southern coast of Long Island (see darkened terrain below the 3m (10 feet) contour line shown in Figure 5.1).[2]

The waterways surrounding New York City are particularly prone to flooding because of the topography and bathymetry of the region. The orientation of Long Island Sound makes it a natural funnel for strong north-easterly winds to drive surface waters into the western Sound, through the East River and into the harbour. North-easterly winds blowing parallel to the south shore of Long Island pile up surges against the coast due to Ekman transport directed to the right of the wind (the Coriolis effect). This coastal surge then flows hydraulically through the Narrows into New York Harbor. As a result, harbour surges are fed from two sources simultaneously: indirectly, from Long Island Sound, and directly, from the Atlantic Ocean.

The onset, duration and signatures of storm surges differ markedly between hurricanes and nor'easters. As illustrated in Figure 5.2, the surge from a hurricane lasts only a few hours, but rises and falls very rapidly. The extent of flooding therefore depends critically upon the state of the astronomical tide at the time. For nor'easters, on the other hand, the surge typically rises more slowly but may last a few days, running the

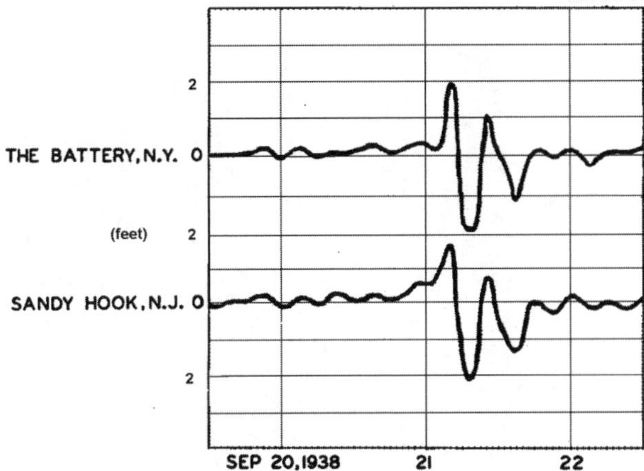

Figure 5.2 *Storm surge signature of the 21–22 September 1938 hurricane, known locally as the Long Island Express*

Note: Typically, a hurricane storm surge rises rapidly but lasts only a few hours. Dates are shown at noon. The hurricane was estimated to have killed between 682 and 800 people in eastern Long Island and southern New England, damaged or destroyed over 57,000 homes, and caused property losses estimated at US$306 million (US$4.7 billion in 2010).3
Source: Pore and Barrientos (1976)

risk of coastal flooding with a number of successive high tides, as illustrated in Figure 5.3 (Pore and Barrientos, 1976).

Colle et al (2008) described the climatology of minor and moderate regional storm surges and actual flooding events for New York City for almost a half century (1959 to 2007). Their analysis revealed that the number of minor flooding ('storm-tide') events has been steadily increasing since 1990 even though the number of minor surge events has been decreasing. This apparent paradox was explained by the fact that sea level has been rising by about 2.8mm per year at The Battery during the past 50 years. If this trend of sea-level rise at The Battery is removed, the number of minor flooding events no longer increases. Thus, sea-level rise over the last few decades may already be enhancing the number of nuisance flooding events around metropolitan New York (which trigger National Weather Service (NWS) coastal flood advisories), as described in the previous section.

Hurricanes are less frequent; but the risk of a catastrophic surge is high for metropolitan New York in the next 100 years. Lin et al (2010) used a risk-based assessment methodology to investigate hurricane storm risks for New York City. They coupled a statistical/deterministic hurricane model with a simple hydrodynamic model called Sea, Lake and Overland Surges from Hurricanes (SLOSH)[4] and then used the Advanced Coastal Circulation and Storm Surge (ADCIRC)[5] two-dimensional circulation model to

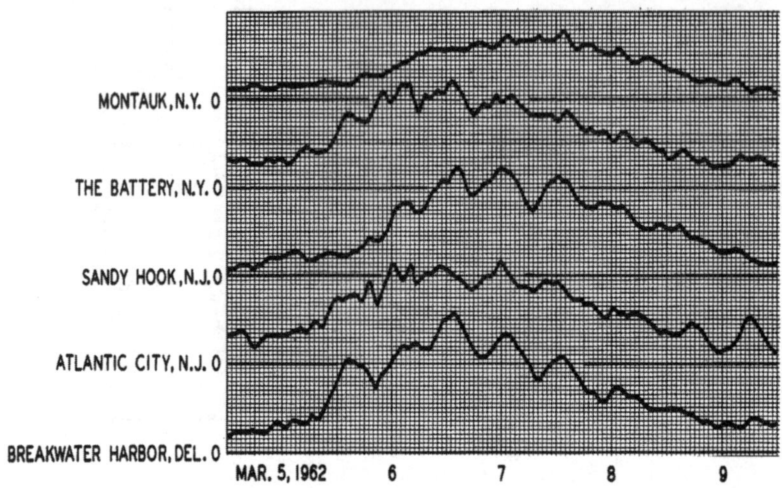

Figure 5.3 *Storm surge signature of the 5–8 March 1962 nor'easter*

Note: Typically, the storm surge signature of a nor'easter rises slowly, but may last several days. Dates are shown at noon.
Source: Pore and Barrientos (1976)

generate a large number (~ 1000) of synthetic surge events. The SLOSH model simulations compared favourably with the ADCIRC runs; but both suffered from inadequate resolution in critical areas of the domain (e.g. harbours, narrow rivers, bays and inlets). Their estimated return periods of surge heights is consistent with our predictions; but the ability to predict catastrophic storms signatures (for all groups) is still very limited.

Sea level continues to rise inexorably in the metropolitan New York region at the rate of about 30cm (1 foot) per century due to isostatic adjustment of the North American continent over the last 10,000 years (see Chapter 13 in this volume). Stated another way, the east coast is tilting and sinking at this rate; but the effect on coastal erosion and flooding is the same.

On top of this, global warming will increase the rate at which sea level rises by as much as 1m (3 feet) by 2100 (Horton et al, 2010), but which may rapidly accelerate (up to 2m) if the Greenland and West Antarctic ice sheets melt more rapidly than expected over the next few hundred years (Bowman et al, 2008). Recently accelerated ice melt in Greenland and West Antarctic indicates the potential for high levels of sea-level rise over multiple centuries. Neither the Greenland nor West Antarctic ice sheets have yet to significantly contribute to global and regional sea-level rise (SLR); but because potential SLR is large and the consequences dire for coastal societies, current melt patterns and accelerations should continue to be monitored very carefully.

As a result, the probability in one year of coastal flooding to a level of about 3m (10 feet) above mean sea level (amsl), now considered to be 1/100, may be as likely as 1/20

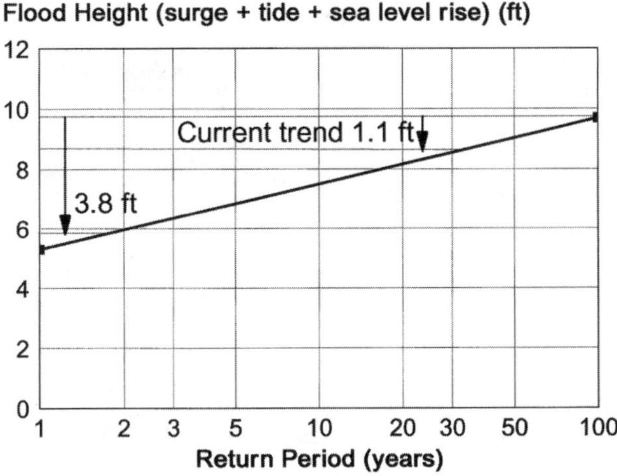

Figure 5.4 *The reduced return period due to rising sea level of the present 100-year flood in the New York City metropolitan region by the 2090s, showing the extremes from 30 years at the current rate of sea-level rise to as little as every second year*

Source: Gornitz (2001)

by the 2050s and 1/4 by the 2080s (see Figures 5.4 and 5.5). Colle et al (2010) found that as sea level rises by ~ 0.5m several decades from now, there are likely to be several moderate flooding events each year, even from relatively weak cyclones. With changing climate and weather, more severe and more frequent storms could make such flooding even more likely and severe.

Should increased freshwater discharge from the melting Greenland ice sheet slow or shut down the winter formation of North Atlantic Deep Water, this could affect the so-called 'conveyor belt' and consequentially alter the strength and location of the Gulf Stream. This, in turn, will further increase sea-level rise along the eastern seaboard as a consequence of the diminishment of the geostrophic balance across the Gulf Stream, which currently maintains a sea-level difference of about 1m between (lower) inshore and (higher) Sargasso Sea waters far offshore.[6]

5.2 Modelling Storm Surges in the Metropolitan New York Region

In order to study the nature and physics of storm surges near and around metropolitan New York, the Stony Brook Storm Surge Model (SBSS) system was developed (Bowman et al, 2008; Colle et al, 2008). It promises to have applications for short-term (two-day) emergency management protocols for coastal flooding/evacuation decision-making. It is useful in delineating coastal locations vulnerable to violent storm damage

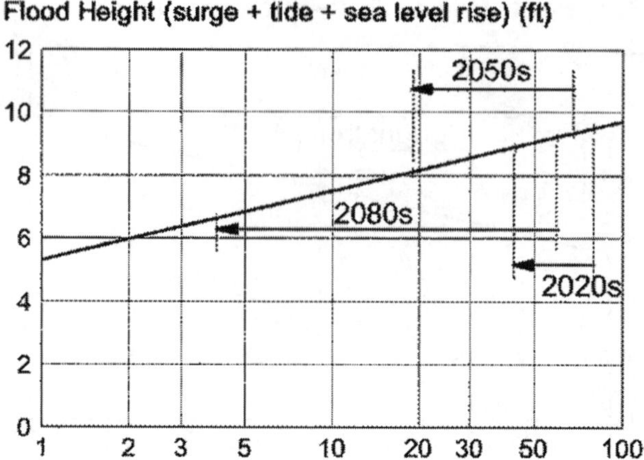

Figure 5.5 *The reduced return period due to rising sea level of the present 100-year flood in the New York City metropolitan region, showing the estimated range for the 2020s, 2050s and 2080s*

Source: Gornitz (2001)

and helps to determine indices of priority for adaptation to rising sea levels from future storms. It can be, and has been, used to help determine the efficacy of regional storm surge barrier systems (Morang, 2007; Bowman et al, 2008; Hill, 2011), such as are currently installed in New England, US, and Europe (e.g. the River Thames, UK; the Delta Project in The Netherlands; Venice Lagoon, Italy; and St Petersburg, Russian Federation).

Crucial to predicting accurate coastal flooding and warnings is not only the peak surge amplitude, but also the timing of surges with respect to local tidal conditions. If the peak surge coincides with local high tide, significant flooding may occur. If the peak surge occurs, say, six hours later (or earlier) at low tide, flooding is unlikely. This phasing is less important for nor'easters which may persist for several days, guaranteeing at least one coincident high tide/significant surge event. However, hurricane-related surges are short lived, so timing is crucial to understanding flooding risks. The Long Island Express hurricane surge of 1938 occurred at local high tide, as well as a coincident extra high spring tide[7] (see Chapter 13 in this volume). Huge damage and loss of life resulted.

As well as timing issues, the complex shallow-water fluid dynamics of the various bays, inlets and channels of the southern coasts of Long Island and eastern shores of northern New Jersey present difficult modelling challenges. The archipelago of islands strung along the southern shores of Long Island, Brooklyn and Queens are difficult to model accurately with sufficient computer speed. Obviously, rapidly produced (and readily updated) surge predictions are needed for effective emergency planning and execution strategies.

The SBSS model couples two research meso-scale weather-prediction models (MM5[8] and WRF[9]) to the Advanced Coastal Circulation and Storm Surge Model (ADCIRC)[10] along the eastern seaboard from the states of North Carolina in the south to Maine in the north. The hourly surface winds and pressures are used to derive the frictional drag stresses and elevations of the sea surface at each model node. These forcing functions force the so-called wind-driven currents which modify the tidal currents and, depending upon the wind direction, intensity and duration, pile up surges along the coast.

Major surges can be difficult to predict, especially a day or two in advance, as storms can rapidly change characteristics. As a result, we use an ensemble approach in which the atmospheric model is run several times using slightly different initial data and model parameterization of key physical properties of the atmosphere and ocean. This helps to span the range of uncertainty given that we never have sufficient observations to know about all the important structures – models are imperfect!

The WRF and MM5 models are run nine times each day to a time horizon of 48 hours; each run is used to force its own ADCIRC simulation. The models are programmed to run automatically each night, with idle time built in to allow for uncertainties in the arrival of key information from the NWS and National Ocean Service (NOS), which provide initial and boundary conditions from their operational global and regional forecast models. These two-day forecasts of surface (10m, or 33 feet) winds, sea-level pressure, tides and surges are then presented and updated daily on the Stony Brook Research Group website.[11]

Geographically, the outputs of the SBSS model extend along both coasts of Long Island Sound, adjacent areas in northern New Jersey (see Figure 5.6) and to the

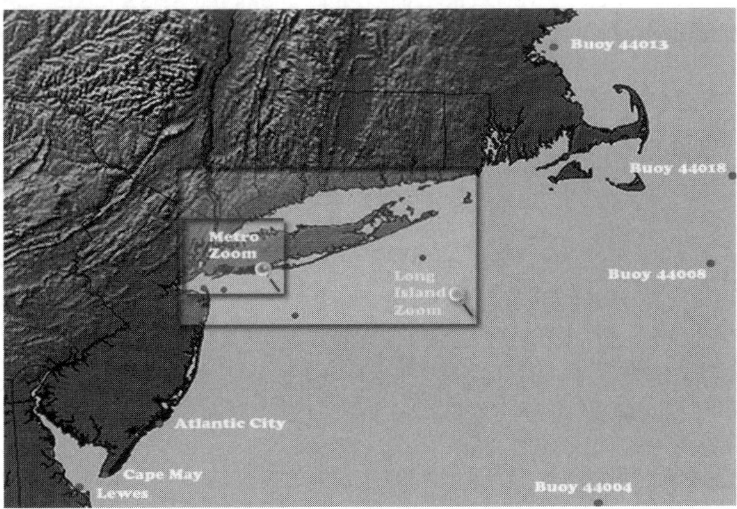

Figure 5.6 *The domain of the Stony Brook University Storm Surge System and regional Coastal Storm Surge/Coastal Warning System over the eastern seaboard*

extremely vulnerable Long Island's south shore. Current emphases are on refining very high-resolution grids (< 5m), access to more accurate bathymetry databases, plus identification and eradication of model instabilities that can be a nuisance in the archipelago of channels, islands and entrance channels of Long Island's Jamaica Bay, Great South Bay, Moriches and Shinnecock Bays.

Predicted tides and surges are compared to NOS observed water levels, distributed via the internet in real time at primary reference stations spaced along the coast and located in key locations within the various harbours, inlets and bays. Results are presented in graphical time-series plots, as well as animations of coastal ocean surges at three different resolutions. Results are posted on the Stony Brook Storm Surge Research Group website[12] for four stations in New York Harbor (The Battery, Bergen Point, Kings Point and Sandy Hook), four northern stations (Montauk, New York; Bridgeport, Connecticut; New London, Connecticut; and New Haven, Connecticut), four southern stations (Atlantic City, New Jersey; Cape May, New Jersey; Ocean City, Maryland; and Lewes, Delaware) and seven National Oceanic and Atmospheric Administration (NOAA) offshore data buoys (see Figures 5.6 to 5.8).

5.3 Presenting the Storm Surge Model Output

Figures 5.9 and 5.10 are representative water-level and storm surge prediction/observation plots for the 12 December 2010 nor'easter storm that struck the eastern seaboard, as experienced at the New Haven, Connecticut, NOS primary tidal station (see Figure 5.7 for station location; the storm surge is derived by subtracting the astronomical tide from the observed water level). The dashed line line is the SBSS model surge prediction, with the grey shading surrounding it being +/−1 standard deviation uncertainty (as

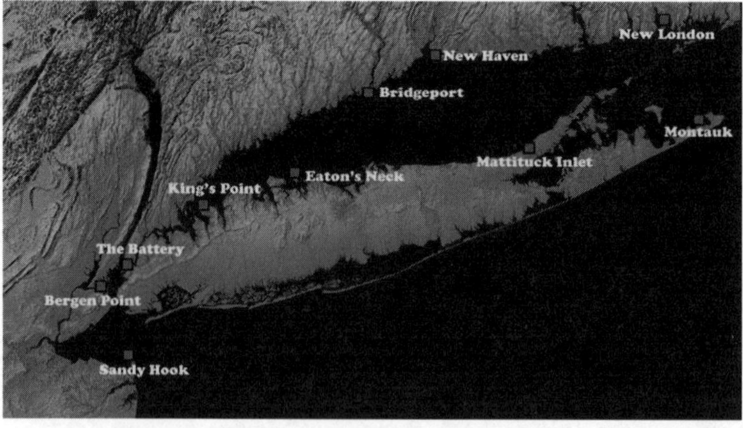

Figure 5.7 *Long Island regional locator map from the Stony Brook Storm Surge Model (SBSS)/Coastal Warning System*

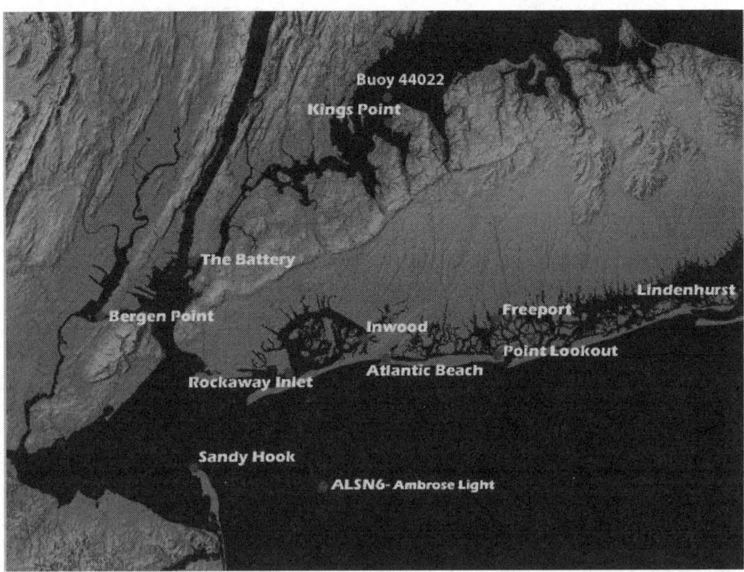

Figure 5.8 *Metropolitan New York locator map of the SBSS/Coastal Warning System*

generated by the ensemble of weather models running simultaneously). The grey line is the observed tidal height at the NOS tide gauge. The grey shaded area in Figure 5.10 fluctuating around a surge height of 0m is the error between the predicted and the observed surge. The vertical dotted line is the current time when the snapshot was taken. As time passes, the plot shifts right to left and new sections of the record are revealed. The predictions are proving useful for short-term predictions (up to 48 hours) for transmission to the NWS for possible coastal alerts and warnings, and for emergency response activities of community coastal emergency managers.

In addition to coastal tidal predictions, the SBSS model compares predictions of surface (10m) winds and sea-level pressure to real-time observations from a field of offshore National Data Buoy Center (NDBC) weather buoys on the continental shelf of the eastern seaboard (see Plate 4). This provides us with information on how accurately the model is able to track regional weather systems, particularly the passage of storm centres over the oceanic portion of the model domain.

Plate 5 is a snapshot of the passage of the low pressure centre of the 13 March 2010 winter nor'easter that produced a significant surge (~ 1m) in New York Harbor and Long Island Sound. The graph illustrates that the model was able to accurately predict the characteristics and timing of the low pressure centre as it passed by Buoy 44004 about 03:00 EST on 16 March (Buoy 44004 is located 200 nautical miles east of Cape May, New Jersey, and is not shown in Plate 4).

The model also displays an animation in areal form of the predicted surge characteristics at three resolutions (metro, Long Island and regional). The maximum predicted

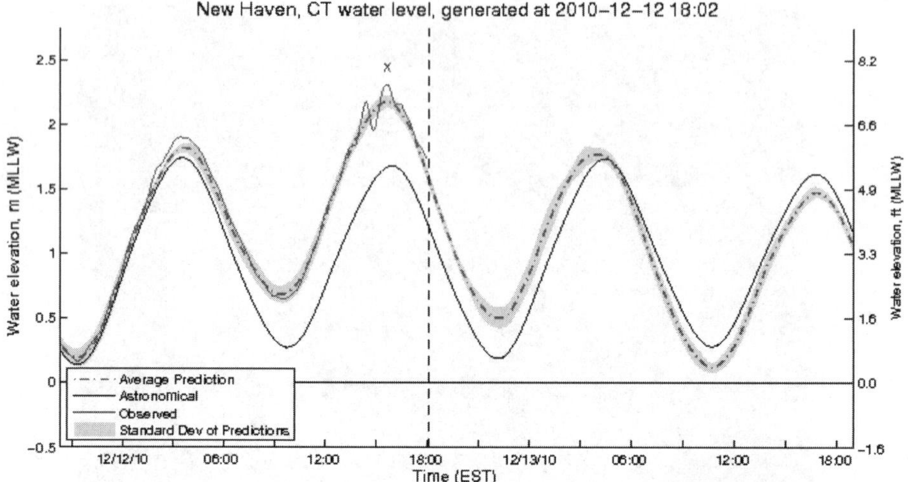

Figure 5.9 *Snapshot of water-level predictions and observations (combined astronomical tide and storm surge) for the 12 December 2010 nor'easter storm for New Haven, Connecticut (north shore of Long Island Sound; see Figure 5.7 for location)*

Note: The black line is the astronomical tide, and the dashed line is the water elevation prediction (astronomical tide plus the storm surge), with the grey shading surrounding it representing one standard deviation prediction uncertainty. The grey line is the observed tidal height at the National Ocean Service (NOS) tide gauge and the vertical dotted line is the current time when the snapshot was taken.

surge (regardless of peak time) within the run-time horizon (two days) is also presented (see Plate 6). These are useful as a summary in map form of what magnitude surges are expected and where the regions experiencing the greatest surges are likely to be located.

5.4 Towards a Coastal Storm Surge Early-Warning System

A coastal alerts early-warning system for metropolitan New York based on the SBSS model is being developed to broadcast impending storm surges from developing hurricanes and nor'easters with a 48-hour time horizon. If sea level is expected to rise more than 30cm (1 foot) above normal, an email *advisory* is automatically sent to subscribers. If sea level is expected to rise more than 60cm (2 feet) above normal, an email *warning* is sent to the local National Weather Service and other subscribers. If a station button is clicked on the Coastal Alerts webpage, the viewer is taken to a table of predicted surges, their magnitudes, timing and the level of uncertainty in the prediction.

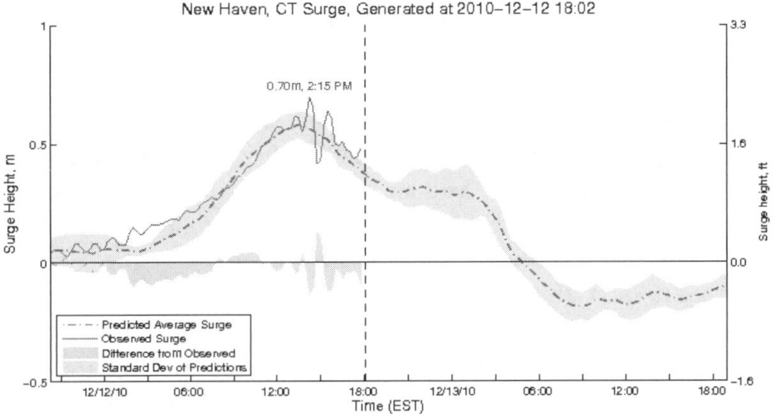

Figure 5.10 *Snapshot of storm surge predictions and observations (astronomical tides removed) for the 12 December 2010 nor'easter storm for New Haven, Connecticut (north shore of Long Island Sound; see Figure 5.7 for location)*

Note: The dashed line is the surge prediction, with the grey shading surrounding it representing one standard deviation prediction uncertainty. The grey line is the observed surge at the NOS tide gauge and the vertical dotted line is the current time when the snapshot was taken.

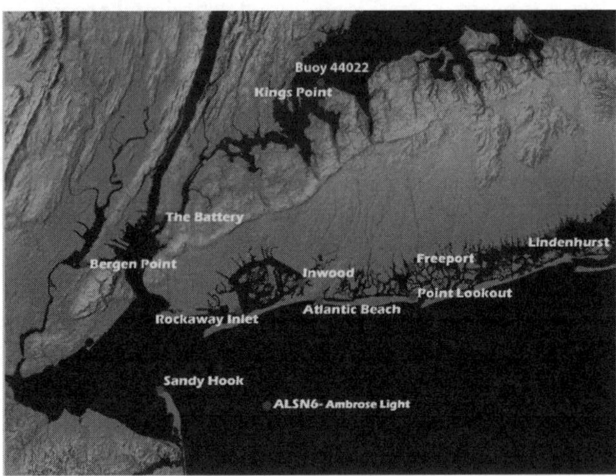

Figure 5.11 *The domain of the Stony Brook University Metro New York Coastal Warning System*

Note: When sea level is predicted to rise more than 30cm (1 foot) above normal, an email advisory is sent out to a predetermined email list, including the National Weather Service (NWS). When sea level is expected to rise more than 60cm (2 feet) above normal, an email warning is sent out. The NWS alone is responsible for making official coastal storm advisories and warnings based on several sources of information.

5.5 Inland Flooding Issues

In addition to ocean-derived storm surges, heavy precipitation events accompanying a major storm may lead to local flooding from inland sources. New York Harbor is built on a river delta system at the mouth of the Hudson River. The river and its tributaries drain a significant fraction of the land area of New York State.[13] In addition, several rivers located entirely in New Jersey State drain directly into New York Harbor and its approaches (Hackensack, Passaic and Raritan rivers).

The phasing of the various contributions from storm-swollen river flows from the Hudson's numerous tributaries define the height, timing and duration of the inland surge as the Hudson River discharges through New York Harbor and out to sea. As an example, Plate 7 illustrates the gauged river flows, their timing and amplitudes (units are in cubic feet per second during and after extra-tropical storm Floyd, in September 1999, which turned out to be a 1/100-year precipitation event in northern New Jersey (Colle, 2003), causing several drownings.[14]

As it turns out, the surface slope of the Hudson River is exceedingly small ($\sim 10^{-5}$) and the river drops only $\sim$ 0.5m (1.5 feet) from Troy to the ocean, a distance of about 240km (150 miles). This means that the river can store a large volume of water for limited periods (one or two days) and contain an inland surge that is typically quite small.

As the SBSS model was being developed, we examined the effects on freshwater discharge from extra-tropical storm Floyd. Floyd set rainfall records up and down the Atlantic Coast, including the Raritan River watershed in nearby New Jersey, Windsor Locks in western Connecticut, and Albany, New York (Colle, 2003). Most of the counties bordering the Hudson River below Albany became eligible for disaster assistance.[15] About 15cm (6 inches) of rain fell in New York City over a few hours.[16]

The surge that Floyd actually produced in New York Harbor is shown in Figure 5.12, which illustrates the deviations from mean sea level at The Battery over a period of 16 days, beginning on 15 September 1999. The ocean-derived storm surge peaked at about 1.1m (3.8 feet) just before midnight on 17 September. As the water level dropped after the peak of the storm surge, a secondary peak, about 0.36m (14 inches) in amplitude, occurred a few hours later. This secondary peak was identified as primarily the effect of rainfall runoff from nearby New Jersey rivers, although local runoff and rainfall directly into the waterways would have contributed a small additional amount. The third anomaly is a prolonged crest lasting about 2.5 days, beginning about 2.5 days after the New Jersey river peak, averaging about 0.14m (6 inches), which was identified as the combined runoff from the Hudson River watershed.

Various datasets are available to analyse the rainfall and river runoff in the region. In New Jersey, there are gauging stations measuring stream flows at regular time intervals on the major rivers emptying into the harbour; but the gauges are often located upstream, far from the river mouths. Likewise, there are gauging stations on the tributaries of the Hudson River, almost all of which are far upstream from metropolitan New York.

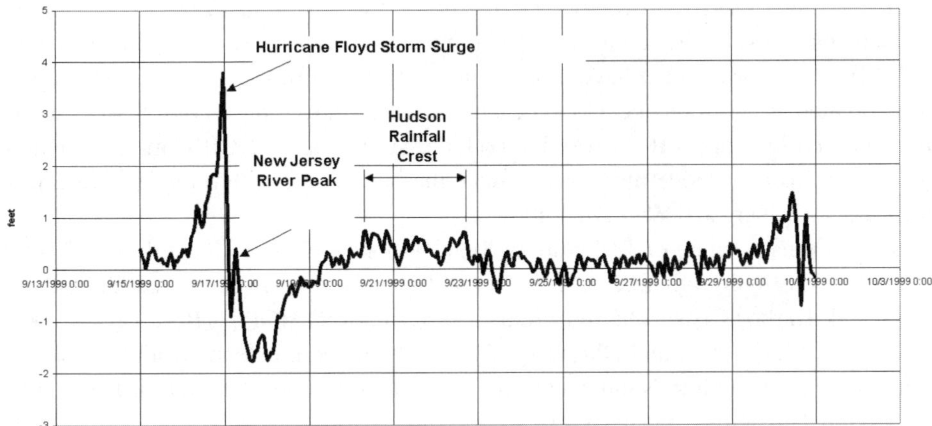

Figure 5.12 *Deviations from normal water level observed at The Battery during extra-tropical storm Floyd, September 1999*

Source: Bowman et al (2005)

Gauging stations at several points along the lower Hudson River measure only water levels, not stream flows.

From this example of Floyd-related precipitation, it was inferred that two possible flooding modes exist and need to be considered. One is a brief but large spike in runoff from the nearby rivers in New Jersey that immediately follows torrential rains. The other is the following accumulation of river flow and rainfall runoff from the Hudson River watershed, which covers a significant fraction of New York State.[17] For a single storm, these two pulses are observed typically days apart.

With the large water surface area in the metropolitan region and the slight slope ($\sim 10^{-5}$) of the navigable Hudson River, neither of these possibilities would threaten dangerous flooding levels *inside* storm surge barriers closed for several hours during the worst surge conditions. For this reason, inland surge is not currently computed or presented in the current version of the SBSS model.

5.6 On-going Efforts and Inter-comparison Studies

In addition to the Stony Brook Storm Surge Research Group's work, there are several teams investigating storm surge science in the New York metropolitan region.

Lin et al (2010) generated a model-based hurricane storm surge risk model (Emanuel 2006; Emanuel et al, 2006) by injecting a multitude (~ 1000) of synthetic eastern seaboard storms in a statistical/deterministic model based on SLOSH (Jelesnianski et al, 1992) and ADCIRC (Luettich et al, 1992; Westerink et al, 1992). They obtained some realistic results; but their system is limited to hurricanes with a symmetric wind field. Tropical storms often have more of an elongated north–south wind field as they develop more extra-tropical cyclone structures as they move up the coast (Colle, 2003). In addition, Lin et al (2010) report that both the SLOSH and ADCIRC models running in two-dimensional mode may overestimate the bottom stress and thus underestimate the surge (Weisberg and Zheng, 2008).

The Stevens Institute of Technology (SIT), Hoboken, New Jersey, is developing the SIT-NYHOPS hydrodynamic model. This is based on the forecasting module of the New York Harbor Observing and Prediction System (NYHOPS) (Bruno et al, 2006; Georgas 2010; Georgas and Blumberg, 2010). The domain extends from the coast of Maryland in the south to Nantucket in the north, and includes the Hudson River up to the federal dam across the Hudson River, located at Troy, New York.

The NOAA ET-SURGE ('extra-tropical') model[18] is based on the same two-dimensional shallow water dynamical equations as the widely used, but necessarily simplified, SLOSH model[19] (Jelesnianski et al, 1992; Burroughs and Shaffer, 1997) and the US National Centers for Environmental Prediction (NCEP) Global Forecast System (GFS) model for hourly wind and air pressure forcing. Unlike the SBSS and SIT-NYHOPS modelling systems, the ET-SURGE model predicts storm surge alone and does not include tides (Burroughs and Shaffer, 1997). Predicted storm surge levels are then added to the National Ocean Service (NOS)-predicted astronomical tides to obtain the total water levels.

DiLiberto (2009) and DiLiberto et al (submitted) compared these three storm surge models (SBSS, SIT-NYHOPS and NOAA ET-SURGE) for 75 available days during the 2007 to 2008 and 2008 to 2009 cool seasons for five stations located around the New York City/Long Island region. They showed that the SIT-NYHOPS model had the lowest root mean square (RMS) errors on average, while ET-SURGE had the largest RMS errors after hour 24. SBSS model errors were intermediate. None of the models incorporate the additional set-up arising from storm waves and swell as they break against the shore (the swell often persists days after the storm centre has passed).

As computer power increases, ensemble storm surge modelling will become more routine. This is important, since a single surge simulation will probably not verify correctly and thus possibly produce false alarms and missed events. DiLiberto et al (submitted) completed the first verification of the SBSS ensemble surge system, as well as describing the potential benefits of combining our predictions with the NOAA ET-SURGE and SIT-NYHOPS systems. A three-member ensemble (ENS3) of one of the SBSS members, NOAA ET-SURGE and SIT-NYHOPS improves the probabilistic score relative to the best deterministic member (SIT-NYHOPS), and the ENS3 has more skill and better reliability than the Stony Brook ensemble, thus illustrating the benefit of a multi-model storm surge ensemble for forecasting in this region.

5.7 Conclusions

The understanding and accurate simulation of storm surges for the New York metropolitan region is a challenging undertaking. Each storm has its own unique characteristics, such as type (hurricane, extra-tropical storm, winter nor'easter), intensity, duration, storm track (especially the propagation velocity, location and angle of attack at landfall for hurricanes), phasing of the surges *vis-à-vis* the tides, and the accurate enumeration of the complex bathymetry of the region. This includes the coastal archipelagos of low-lying islands, shallow inlet channels and branching rivers which characterize the very complex nature of the metropolitan New York coastal zone. Sophisticated surge-calculating algorithms need to be further developed, refined and then tested, verified and maintained for better regional operational weather prediction, wave refraction studies, coastal set-up from breaking ocean waves and swell, and river responses to extreme inland precipitation events. Given these challenges, it is paramount that storm surge forecasting and flood planning management use ensemble forecasting to improve the accuracy and stability of forecasts for emergency planning and decision-making.

Acknowledgements

The work upon which this chapter is based was supported by various grants from the New York Sea Grant Institute, the Eppley Foundation and the New York City Department of Environmental Protection. Douglas Hill is thanked for many stimulating discussions regarding the need for long-term planning studies of the consequences of climate change and sea-level rise by New York City government (Hill, 1996), and the need to consider *regional* coastal protection for the metropolitan region, in addition to specific local adaptations to the threats of rising sea level and increasing frequency of damaging storms (Hill, 2011). Roger Flood, Frank Buonaiuto and Charles Flagg are thanked for many useful suggestions on developing the technology, scope and intent of the modelling system. Robert Hunter, Christian Mirchel, Kate Rojowsky and Thomas Diliberto each made significant contributions to the development of the code and the interpretation of the results.

Notes

1 National Geodetic Vertical Datum of 1928.
2 The minimum elevation of Newark Airport is elsewhere reported at 10.3 feet, a discrepancy of 0.3 feet.
3 See http://en.wikipedia.org/wiki/New_England_Hurricane_of_1938.
4 See www.nhc.noaa.gov/ssurge/ssurge_slosh.shtml.

5   See www.unc.edu/ims/adcirc.
6   See www.psmsl.org/train_and_info/faqs.
7   See http://en.wikipedia.org/wiki/New_England_Hurricane_of_1938.
8   See www.mmm.ucar.edu/mm5.
9   See www.wrf-model.org/index.php.
10 See www.unc.edu/ims/adcirc.
11 See http://stormy.msrc.sunysb.edu.
12 See http://stormy.msrc.sunysb.edu.
13 See www.hudsonwatershed.org/watershed.html.
14 See http://en.wikipedia.org/wiki/Hurricane_Floyd.
15 See www.fema.gov/news/eventcounties.fema?id=387.
16 See http://en.wikipedia.org/wiki/Effects_of_Hurricane_Floyd_in_New_York.
17 See www.hudsonwatershed.org/sustainablewatershed.
18 See www.opc.ncep.noaa.gov/et_surge/et_surge_info.shtml.
19 See https://docs.google.com/viewer?url=http://slosh.nws.noaa.gov/sloshPub/pubs/Vol-33-Nu1-Glahn.pdf.

References

Bloomfield, J., Smith, M. and Thompson, N. (1999) *Hot Nights in the City*, Environmental Defense Fund, Washington, DC, p36

Bowman, M., Colle, B., Flood, R., Hill, D., Wilson, R. E., Buonaiuto, F., Cheng, P. and Zheng, Y. (2005) *Hydrologic Feasibility of Storm Surge Barriers to Protect the Metropolitan New York–New Jersey Region, Final Report*, Marine Sciences Research Center Technical Report, Stony Brook University, New York, p106

Bowman, M., Hill, D., Buonaiuto, F., Colle, B., Flood, R., Wilson, R., Hunter, R. and Wang, J. (2008) 'Threats and responses associated with rapid climate change in Metropolitan New York', in M. MacCracken, F. Moore and J. C. Topping (eds) *Sudden and Disruptive Climate Change*, Earthscan, London, p327

Brooks, C. F. (1939) 'Hurricanes into New England: Meteorology of the storm of September 21, 1938', *Geo. Rev.*, vol 29, pp119–127

Bruno, M. S., Blumberg, A. F. and Herrington, T. O. (2006) 'The urban ocean observatory: Coastal ocean observations and forecasting in the New York Bight', *J. Mar. Sci. Environ.*, vol C4, pp31–39

Burroughs, L. B. and Shaffer, W. A. (1997) 'East coast extra tropical storm surge and beach erosion guidance', *NWS Technical Procedures Bulletin*, no 436, National Oceanic and Atmospheric Administration, US Department of Commerce, US

Case, R. (1986) 'Annual summary: Atlantic hurricane season of 1985', *Monthly Weather Review*, vol 114, pp1390–1405

Colle, B. A. (2003) 'Numerical simulations of the extra tropical transition of Floyd (1999): Structural evolution and responsible mechanisms for the heavy rainfall over the northeast United States', *Monthly Weather Review*, vol 131, pp2905–2926

Colle, B. A., Buonaiuto, F., Bowman, M. J., Wilson, R. E., Flood, R., Hunter, R., Mintz, A. and Hill, D. (2008) 'New York City's vulnerability to coastal flooding', *Bulletin of the American Meteorological Society*, vol 89, pp829–841, doi:10.1175/2007BAMS2401.1

Colle, B. A., Rojowsky, K. and Buonaiuto, F. (2010) 'New York City storm surges: Climatology and analysis of the wind and cyclone evolution', *Journal of Applied Meteorology and Climatology*, vol 49, pp85–100

DiLiberto, T. (2009) *Verification of a Storm Surge Modeling System for the New York City–Long Island Region*, MSc thesis, School of Marine and Atmospheric Sciences, Stony Brook University, New York, NY

DiLiberto, T., Colle, B. A., Georgas, N., Blumberg, A. F. and Taylor, A. A. (submitted) 'Verification of multi-model storm surge ensemble around New York City and Long Island during the cool season', *Weather Forecasting*

Emanuel, K. (2006) 'Climate and tropical cyclone activity: A new model downscaling approach', *Journal of Climate,* vol 19, pp4797-4802, doi:10.1175/JCLI3908.1

Emanuel, K., Ravela, S., Vivant, E. and Risi, C. (2006) 'A statistical deterministic approach to hurricane risk assessment', *Bulletin of the American Meteorological Society*, vol 87, pp299–314, doi:10.1175/BAMS-87-3-299

Georgas, N. (2010) *Establishing Confidence in Marine Forecast Systems: The Design of a High Fidelity Marine Forecast Model for the NY/NJ Harbor Estuary and Its Adjoining Waters*, PhD thesis, Stevens Institute of Technology, Hoboken, NJ, p272

Georgas, N. and Blumberg, A. F. (2010) 'Establishing confidence in marine forecast systems: The design and skill assessment of the New York Harbor Observation and Prediction System, ver. 3 (NYHOPS v3)', in M. L. Spalding (ed) *11th International Conference in Estuarine and Coastal Modeling, 4–6 November 2009*, American Society of Civil Engineers, Seattle, WA, pp660–668

Gornitz, V. (2001) 'Sea-level rise and coasts', in C. Rosenzweig and W. Solecki (eds) *Climate Change and a Global City: The Potential Consequences of Climate Variability and Change, Metro East Coast,* Report for the US Global Change Research Program, Columbia Earth Institute, Columbia University, New York, NY

Hill, D. (ed) (1996) 'The baked apple? Metropolitan New York in the Greenhouse', *Annals of the New York Academy of Sciences*, vol 790, p221

Hill, D. (ed) (2011) *Against the Deluge: Storm Surge Barriers to Protect New York City*, Conference Proceedings, Polytechnic Institute of New York University, Brooklyn, NY, 30–31 March 2009

Horton, R., Gornitz, V. and Bowman, M. (2010) 'Climate observations and projections', *Climate Change Adaptation in New York City: Building a Risk Management Response*, Annals of the New York Academy of Sciences, vol 1196, p354

Jelesnianski, C. P., Chen, J. and Shaffer, W. A. (1992) *SLOSH: Sea, Lake, and Overland Surges from Hurricanes*, NOAA Technical Report, NWS 48, NOAA/AOML Library, Miami, FL

Lin, N., Emanuel, K. A., Smith, J. A. and Vanmarcke, E. (2010) 'Risk assessment of hurricane storm surge for New York City', *Journal of Geophysical Research*, vol 115, pD18121, doi:10.1029/2009JD013630

Luettich, R. A., Westerink, J. J. and Scheffner, N. W. (1992) *ADCIRC: An Advanced Three-Dimensional Circulation Model for Shelves, Coasts, and Estuaries. Report 1: Theory and Methodology of ADCIRC-2DDI and ADCIRC-3DL, Dredging Research Program*, Technical report DRP-92-6, Coastal Engineering Research Center, Vicksburg, MS

Morang, A. (2007) *Hurricane Barriers in New England and New Jersey: History and Status after Four Decades*, US Army Corps of Engineers, Final report ERDC/CHL TR-07-11, New Orleans, LA, p103

Pore, N. A. and Barrientos, C. S. (1976) 'Storm surge', *Marine Ecosystem Analysis Program, MESA New York Bight Atlas Monograph 6*, New York Sea Grant Institute, Albany NY, p44

Rosenzweig, C. and Solecki, W. (eds) (2001) *Climate Change and a Global City: The Potential Consequences of Climate Variability and Change, Metro East Coast, Report for the US Global Change Research Program*, Columbia Earth Institute, New York, NY

Scileppi, E. and Donnelly, J. P. (2007) 'Sedimentary evidence of hurricane strikes in western Long Island, New York', *Geochemistry Geophysics Geosystems*, vol 8, pQ06011, doi:10.1029/2006GC001463

US Army Corps of Engineers (1995) *Metro New York Hurricane Transportation Study*, Interim Technical Data Report no 74, New York, NY

Weisberg, R. H. and Zheng, L. (2008) 'Hurricane storm surge simulations comparing three-dimensional with two-dimensional formulations based on an Ivan-like storm over the Tampa Bay, Florida region', *Journal of Geophysical Research*, vol 113, pC12001, doi:10.1029/2008JC005115

Westerink, J. J., Luettich, R. A., Blain, C. A. and Scheffner, N. W. (1992) *ADCIRC: An Advanced Three-Dimensional Circulation Model for Shelves, Coasts, and Estuaries. Report 2: User's Manual for ADCIRC-2DDI, Dredging Research Program*, Technical report DRP-92-6, Coastal Engineering Research Center, Vicksburg, MS

6

Flood Risk Modelling

Jennifer K. Poussin, Philip J. Ward, Philip Bubeck,
Lidia Gaslikova, Aurel Schwerzmann
and Christoph C. Raible

6.1 Introduction

Flooding poses serious threats to coastal cities, including both economic damage and loss of lives (Nicholls, 2004; Nicholls et al, 2008; Dasgupta et al, 2009). More than 50 per cent of the world's population live in cities, and worldwide more than 40 million people living in coastal cities are exposed to floods with a return period of 100 years (Nicholls et al, 2008). Within the next 30 years, more than two-thirds of the world's cities will be vulnerable to flooding due to factors including sea-level rise, climate change, subsidence and socio-economic changes (Rosenzweig and Solecki, 2001; Bouwer et al, 2007; IPCC, 2007b).

In the past, flood management has concentrated on providing protection against floods through technocratic measures such as storm surge barriers and dikes (Aerts and Droogers, 2004). These measures were aimed at maintaining clear borderlines between land and water. However, fuelled by the knowledge that climate change, and broader environmental and socio-economic changes, will make it increasingly expensive to provide the desired safety standards, and the recognition that the probability of flooding can never be reduced to zero, there is currently an international shift towards a more integrated system of flood risk management (Few, 2003; Merz et al, 2004; de Bruijn, 2005; Büchele et al, 2006). In this context, flood risk is defined as the probability of flooding multiplied by the potential consequences, such as economic damage or loss of lives (Smith, 1994). The level of flood risk therefore depends on:

- the hazard characteristics, such as flood depths and extent, flood duration or flow velocity (Milly et al, 2002; Kundzewicz and Schellnhuber, 2004);
- the exposure characteristics in flood-prone areas, such as number of people, land use and value of assets (Kundzewicz and Schellnhuber, 2004); and
- the vulnerability of the exposed assets and population to the hazard, which can vary largely depending upon the location (e.g. in developed or developing countries) (Kron, 2005).

This move from traditional flood management to flood risk management can be seen at several scales, ranging from international to local. For example, in Europe flood risk management has been given added impetus by the European Flood Directive (EFD) (Directive 2007/60/EC), which requires member states to assess whether watercourses and coastlines are at risk from flooding, to map the flood extent, and to take measures to reduce flood risk. At the city level, some cities are developing and implementing ambitious plans to become 'climate-proof' (Aerts et al, 2009), and one element of this may be to develop long-term visions, including flood risk reduction measures, such as is the case in the Rotterdam Climate Initiative (RCI) and PlaNYC 2030 (New York).

Nevertheless, the scientific field of flood risk assessment (and, as part of this, the related science of flood damage assessment) lags behind the better-developed fields of hydrology and hydraulics (Büchele et al, 2006). As a consequence, there has been more and more attention focused on flood risk and damage assessment in both the scientific and policy spheres during recent years. A risk-based approach is advantageous as it can allow us to gain a better understanding of the effects of physical and socio-economic change on both the flood hazard and its consequences. It therefore provides the opportunity to evaluate risk-reducing strategies for both now and in the future. It can provide data for different purposes, such as the support of decisions on the allocation of tax money or the implementation of insurance premium differentiation.

In this chapter we provide an overview of some of the main terminologies and approaches that are being used and developed in this field. In section 6.2 we begin with a definition of the key terms, before describing the main generic applications of flood risk modelling in section 6.3. The general methodological framework of flood risk assessment is presented in section 6.4. In section 6.5, there is a discussion on how scenarios can be used to evaluate flood risk developments over time. Some case study examples are presented in section 6.6, followed by a brief discussion of how several studies have addressed uncertainty in section 6.7. Section 6.8 gives an example of the assessment of the impact of climate change in the North Sea region. Finally, the chapter ends with conclusions on the main points in section 6.9.

6.2 Definitions

In this chapter, we define flood risk as the probability of flooding (hazard) multiplied by the consequences of flooding (damage) (Smith, 1994). Flood risk can therefore also be given as the average total expected flood damage per year. This is represented conceptually in Figure 6.1, whereby the total risk is represented by the integral of the area under the damage-exceedance probability curve. In practice, practical considerations of time and resource availability dictate the number of points that are used to develop such a damage-exceedance probability curve; for example, in Figure 6.1 the damage has been calculated for four flood probabilities, and the curve interpolated between these points. An important facet is to define the probability at which damage begins to occur,

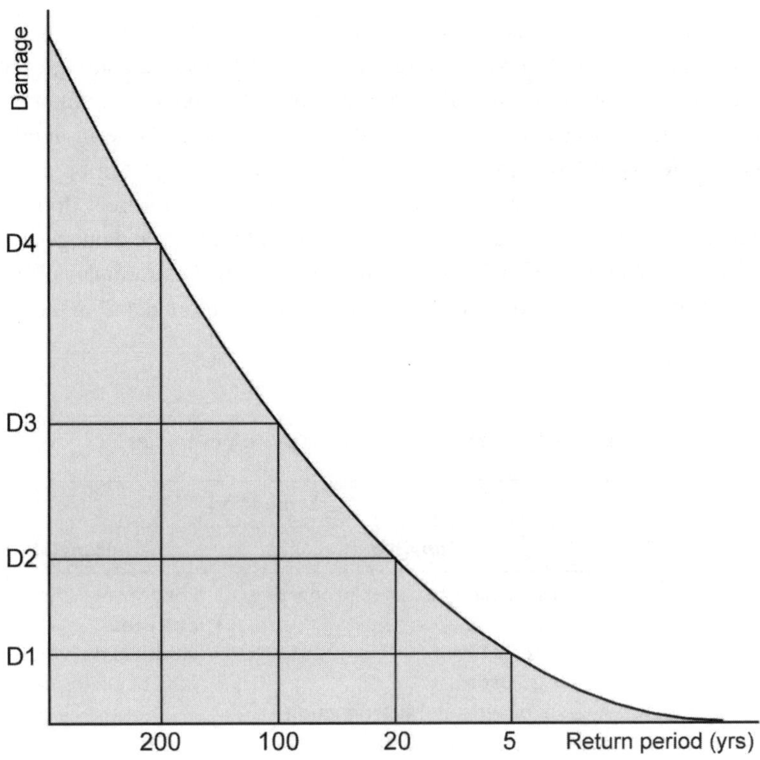

Figure 6.1 *Theoretical damage-exceedance probability curve: The area under the curve (in grey) represents the risk, expressed as the average total expected damage per year*

and also to estimate damage for flood events with low probabilities since these are the events where the highest damage occurs.

Flood hazard can be defined as the exceedance probability of a potentially damaging flood situation in a given area and within a specified period of time (Merz et al, 2007). An example for coastal flooding is the water depth associated with storm surges with varying probabilities of occurrence, expressed as once per a given number of years (i.e. return periods). There are several indicators of flood hazard; but by far the most commonly used in flood risk assessment is the inundation depth (e.g. Smith, 1994; Merz et al, 2007; Middelmann-Fernandes, 2010). Several studies have found that for many cases the water depth is the flood characteristic that has the largest influence on flood damage (e.g. Penning-Rowsell et al, 1994; Wind et al, 1999). Other factors that have been found to be important flood hazard parameters include flow velocity (e.g. Sangrey et al, 1975; USBR, 1988; Clausen and Clark, 1990; Marco, 1994; Jonkman et al, 2008; Middelmann-Fernandes, 2010), inundation duration (Parker et al, 1987; Lekuthai and Vongvisessomjai, 2001; FEMA, 2005) and sediment or contamination load (Haehnel and Daly, 2002; Penning-Rowsell et al, 2003; Thieken et al, 2005).

In terms of the consequences of flooding, most flood risk assessments are limited to the detrimental effects (damages), although there may be positive consequences, such as the replenishment of groundwater, or the delivery of water and sediments to high biological diversity areas such as wetlands (Merz et al, 2004). The term flood damage refers to all varieties of harm caused by flooding. Flood damage can be divided into several categories; usually they are separated into direct and indirect flood damage, both of which are often further divided into tangible and intangible damages (Parker et al, 1987; Smith and Ward, 1998; Penning-Rowsell et al, 2003). Examples of these four categories are shown in Table 6.1, and each is described in more detail in the following paragraphs.

Table 6.1 *Classes of flood damage with examples*

		Evaluation	
		Tangible	*Intangible*
Categories	Direct	Physical damage to assets/properties: • Buildings • Contents • Infrastructure • Loss of agricultural production	• Loss of life • Evacuees • Health and physical effects • Loss of ecological goods
	Indirect	• Loss of industrial production • Traffic disruption • Emergency costs	• Post-flood recovery • Migration • Psychological damages • Increased vulnerability of survivors

Source: adapted from Messner and Meyer (2005); Merz (2006); FLOODsite (2007)

Direct damages are those that occur due to the physical contact of floodwater with humans, property or any other objects (Smith and Ward, 1998; Merz et al, 2004; Büchele et al, 2006). These can include damage to buildings, economic assets, loss of crops and livestock, immediate health impacts, loss of lives and loss of ecological goods. They are often measured as damage to stock values. Indirect damage is a damage that is induced by the direct impact, but occurs outside of the space and/or time of the flood event. Examples of indirect damage include disruption of traffic, trade and public services (Büchele et al, 2006). Indirect damages are often measured as loss of flow values.

Tangible damages are those that can be relatively easily evaluated in monetary terms (e.g. damage to assets, loss of production, etc.). They can be subdivided into direct damages (such as physical damage to properties, contents and infrastructure), and indirect damages (which can be more difficult to estimate and include damages such as traffic or industrial disruption, and emergency costs).

Intangible damages are more difficult to evaluate in monetary terms (Lekuthai and Vongvisessomjai, 2001). There are fewer studies on their valuation than for tangible damages. However, some authors still provide information on their valuation (e.g. Mitchell and Carson, 1989; Bateman et al, 2002; Kahn, 2005). Intangible damages include social and environmental impacts of floods; again, they can be subdivided into direct and indirect intangible damages (see Table 6.1). Social impacts include health or psychological impacts (Smith and Ward, 1998), as well as loss of human lives. When describing flood damage, information on loss of lives is usually given separately from the calculated economic damage; examples of studies addressing loss of lives include Brown and Graham (1988), Funnemark et al (1998), DeKay and McClelland (1993) and Maaskant et al (2009).

It is clearly acknowledged in the flood damage assessment literature that direct intangible damage or indirect damage can play an important role in evaluating flood damage (Penning-Rowsell and Green, 2000). However, by far the largest part of the flood damage literature focuses on direct tangible damages, which are considered as a good indicator of the severity of flood disasters. Direct tangible damages will therefore form the main focus of this chapter.

6.3 Generic Applications

Flood risk modelling has become an important tool in flood risk assessment and management for three main reasons. First, it can help to gain a better understanding of current flood risk in vulnerable areas and provide data on how adaptation measures can reduce that risk. Second, flood risk modelling can be used to assess how the risk will change in the future as a result of environmental and socio-economic changes. Third, since the social purpose of flood risk management is to reduce flood damages, it can be used to determine the relative effectiveness of alternative intervention strategies both now and in the future (Zerger and Wealands, 2004).

Based on these notions, FLOODsite (2007) identifies the following three main spheres in which risk analysis can be used in public policy:

1   *Supporting decisions on financial allocation of tax money*: comparing flood risk (in terms of annual expected flood damage) with other societal risks (such as damage due to illness, earthquakes, terrorism, etc.) provides policy-relevant information about the significance and severity of different risk factors in society. This kind of information can provide decision support on the allocation of tax money to various policy fields all aimed at reducing societal risk.

2   *Project appraisal*: information on annual expected flood damage is vital for decision-making on flood risk management measures and investments. The risk-reducing effects of different flood management measures can be estimated and compared by means of flood risk analysis.

3   *Accountability*: flood risk analysis results can be used to justify public investments and to demonstrate the appropriateness of public spending. They also provide useful insights into why some flood protection measures may be highly efficient in one basin, but not necessarily in another.

Another key use of flood risk analysis is the production of flood risk maps. Merz et al (2007) summarize several of the most important uses of flood risk maps, as listed below:

- raising awareness amongst people at risk and decision-makers;
- providing information for land-use planning and urban development, investment planning and priority-setting;
- helping to assess the feasibility of structural and non-structural control measures;
- serving as a basis for deriving flood insurance premiums;
- allowing disaster managers to prepare for emergency situations;
- developing land-use regulations, building codes and insurance.

Examples of such flood risk mapping studies include those in Austria, Italy and the Czech Republic, where maps are created by, or in association with, insurance companies in order to differentiate premiums according to the flood risk (de Moel et al, 2009). In Finland, the UK, France, Poland and Germany, legislations enforce the use of maps. The degree to which these legislations are binding differs between the countries.

6.4 Methodological Framework

In this section we describe the general methodological framework that is used for flood risk assessment. As stated previously, this discussion focuses mainly on direct, tangible damages. Even for direct, tangible flood damage assessment, a plethora of methodologies exist depending upon factors such as scale and data availability (Meyer and Messner, 2005). However, any flood risk analysis involves three basic facets (adapted from FLOODsite, 2007) – namely:

1   project definition and selection of appropriate approach;
2   data gathering and damage modelling; and
3   calculation and presentation of risk.

The most time- and resource-intensive phase of a flood risk assessment is often the data gathering and damage modelling (i.e. step 2), which forms the main thrust of this section. The discussion is based mainly on FLOODsite (2007), which provides more detailed explanations and guidelines for each step.

6.4.1 Project definition and selection of appropriate approach

Clearly, the geographical scale at which a risk assessment is to be carried out will be a key factor in choosing the approach; there are clear methodological differences between macro-, meso- and micro-scale approaches (FLOODsite, 2007). Macro-scale approaches are mainly used in investigations of large international areas, and consider aggregated damage within administrative or physical units (e.g. municipalities, homogeneous stretches of coastline, river basins). Meso-scale analyses tend to be carried out for regional flood risk assessments, and consider damage for aggregated land-use units or classes, such as residential areas, industrial areas, etc. Finally, micro-scale methods are used in local assessments in which an object-oriented approach is often used (i.e. damages are calculated for single properties, such as buildings) (Hall et al, 2005; Nicholls et al, 2008; Veerbeek and Zevenbergen, 2009). However, according to Meyer and Messner (2005), this simple macro–meso–micro differentiation is not so clear cut, and object-oriented assessments of flood damage are already applied on the regional and national scale (Hall et al, 2003). Still, the level of detail generally decreases as the geographical scale of study increases.

Next to the geographical scale of the study, the approach to be used is also strongly dependent upon the objective of the study (e.g. quick and approximate overview of risk versus detailed calculations for project implementation) (Gewalt et al, 1996), the availability of resources (budget and time restrictions) and the availability of pre-existing datasets (which will affect the budget and time constraints).

Another important consideration during the project definition phase is the selection of damage categories to be assessed. There are many categories of direct, tangible damages, such as residential properties (buildings, household goods), non-residential properties (buildings, machinery and equipment), technical infrastructure (streets, railways, flood defences, watercourses, etc.), vehicles and agricultural products (livestock and crops). Ideally, a flood risk assessment would take all of these into account; but, again, due to constraints of scale, budget, time and data availability, it may be necessary to prioritize those categories most relevant for the study at hand. In any case, damage statistics from observed floods help to identify the most important damage categories in a given region.

6.4.2 Data-gathering and damage modelling

There are three main types of data that are required in the majority of flood risk assessments – namely, data on hazard characteristics, data on exposure characteristics and data on vulnerability characteristics. These are then combined to estimate the potential damage for a given flood event.

6.4.2.1 Hazard characteristics

A number of parameters can be used in damage and risk assessment to characterize the flood hazard of a given flood event, such as inundation depth (e.g. Smith, 1994; Merz et al, 2007; Middelmann-Fernandes, 2010); flow velocity (Sangrey et al, 1975; USBR, 1988; Clausen and Clark, 1990; Marco, 1994; Middelmann-Fernandes, 2010); inundation duration (Parker et al, 1987; Lekuthai and Vongvisessomjai, 2001; FEMA, 2005); and sediment or contamination load (Haehnel and Daly, 2002; Penning-Rowsell et al, 2003; Thieken et al, 2005). By far the most commonly used of these characteristics is the inundation depth (Merz et al, 2007; Apel et al, 2009). Several studies have identified inundation depth as the characteristic with the largest influence on flood damage (e.g. Penning-Rowsell et al, 1994; Wind et al, 1999), and nearly all damage functions are solely depth-damage functions (see section 6.4.2.3).

In addition to economic damage, floods can pose a serious threat to human lives and have caused large numbers of fatalities around the world. For example, Bangladesh was hit by a cyclone in 1991, causing storm surges that left between 67,000 and 139,000 victims (Chowdhury et al, 1993), and Hurricane Katrina, which hit New Orleans in 2005, caused the death of 1118 people (Jonkman et al, 2009). The inundation characteristic most commonly used as an indicator of flood hazard in assessments of loss of lives is the inundation depth (Jonkman et al, 2008). However, risk factors to life can vary with the types of flood. For large dam-break floods, the risk to life mainly originates from the resulting flood wave. Warning time is therefore a crucial parameter to be taken into account in approaches estimating mortality from dam-break floods or flash floods (Brown and Graham, 1988; Dekay and McClelland, 1993). For coastal flooding, local water depths and the pace of rising water levels are also important parameters to be taken into account when assessing the risk to life (Jonkman et al, 2008; Maaskant et al, 2009).

In risk analysis, it is not sufficient to examine the hazard characteristics for just one particular flood event. In order to develop a damage-exceedance probability curve (see Figure 6.1), flood events of several return periods should be considered (Hall et al, 2005). This can either be done by using maps and records of previous flood events in the studied area, and using these as proxies for future floods, or by using models to simulate the inundation depth for given coastal flood events (e.g. Thumerer et al, 2000; Bryan et al, 2001; Mastin and Olsen, 2002; Nicholls, 2002; Madsen and Jakobsen, 2004; Nicholls, 2004; Bates et al, 2005; Dawson et al, 2005; Nicholls et al, 2008; Purvis et al, 2008; Gaslikova et al, 2011). In essence, all of these methods attempt to predict inundation extents and depths based on some combination of process drivers (such as meteorology, tides and flood-defence design periods) (Dawson et al, 2005). There are two main approaches: planar models and hydrodynamic models. Planar models use as input the water level of the tide and distribute this over a Digital Elevation Model (DEM) by means of some kind of flow-connectivity algorithm. For the planning and implementation of protection measures, predictions are typically obtained from hydro-dynamic models; for coastal flows, two-dimensional horizontal solutions of the shallow water equations are state of the art (Madsen and Jakobsen, 2004; Bates et al, 2005). Such models require accurate topographic and bathymetric data at a high resolution and are computationally demanding. As a result, Bates et al (2005) developed a simplified two-dimensional hydraulic model with much less computational demand, following similar advances in fluvial inundation modelling (Horritt and Bates, 2002).

6.4.2.2 Exposure characteristics

The potential damage that may result from a flood event depends upon the flood exposure characteristics in the potentially inundated area. In flood risk assessment, these exposure characteristics can be represented by several different kinds of datasets, such as buildings and assets, land-use type, and number of people living or working in the inundated area.

For the evaluation of economic damage, the most widely used exposure characteristics are representations of the land use in the areas susceptible to flooding. In order to get an approximate economic valuation of the exposed assets, monetary values can be assigned to the different land-use types (FLOODsite, 2007). For detailed valuations of the potential damage, more precise knowledge of the assets at risk in the study area, and their values, is required. This could include, for example, the number of houses, businesses and the presence of facilities such as hospitals, industrial plants or petrol stations. The amount of information and detail obtained on the assets at risk will therefore depend upon the size of the study area, the availability of the data, and its quality and content. The values of the assets vary with their types (e.g. houses or industrial buildings), the period considered (market prices vary in time) and the geographical location (Merz et al, 2010). For a specific type of asset, such as a residential home, two types of values can usually be determined: the value of the building and the value of the contents. Concerning the specific problem of evaluating the exposure of human life, relevant data on exposure characteristics include the number of people living in the affected area; the availability of flood shelters; and the warning and evacuation time (Jonkman et al, 2008).

Official statistics usually provide useful data at the national, regional or local scale, such as gross and net value of fixed assets or population statistics (FLOODsite, 2007; Merz et al, 2010). In addition, depending upon the scale and the required precision of the flood risk assessment, other databases may be needed. At the macro-scale, the values of the assets can be aggregated and an equal spatial distribution of the assets in the study area can be considered. At the meso- and micro-scale, however, the required level of detail increases, and the spatial distribution of the assets needs to be closer to the real distribution (Merz et al, 2010).

6.4.2.3 Vulnerability characteristics

Vulnerability can be defined as the 'relationship between the severity of hazard and the degree of damage caused' (UN DHA, 1993). This can be denoted by damage functions (Füssel, 2007), which are typically used to assess direct, tangible flood damages. These functions provide the level of the damage that can be expected given the respective hazard and exposure characteristics described above if an area is actually flooded. The most common damage functions are stage-damage functions (Middelmann-Fernandes, 2010), which relate stage height (i.e. inundation depth) to damage.

Hypothetical stage-damage functions are shown in Figure 6.2, where the expected economic damage is shown on the y-axis, the inundation depth on the x-axis, and each line on the graph represents a relationship between inundation depth and damage for a given aggregated land-use type. The figure also shows the two main kinds of damage functions (Middelmann-Fernandes, 2010) – namely, absolute damage functions (see Figure 6.2a) and relative damage functions (Figure 6.2b).

Absolute damage functions estimate the actual expected damage in monetary terms for given inundation characteristics. Relative damage functions, on the other hand, show the expected damage as a proportion of the maximum potential damage that could possibly occur. Absolute damage functions are developed for specific regions using data on the market value of assets and their susceptibility to certain inundation characteristics (e.g. inundation depth); they estimate the absolute damage per property or per unit area of land depending upon the inundation and exposure characteristics (e.g. inundation depth and land use, respectively). However, since they are developed specifically for a particular region, it can be difficult to transfer these to other regions. In theory, relative damage functions can be transferred more easily between regions since they simply describe the proportion of the maximum damage that would occur for different inundation and exposure characteristics. Hence, by using market data to

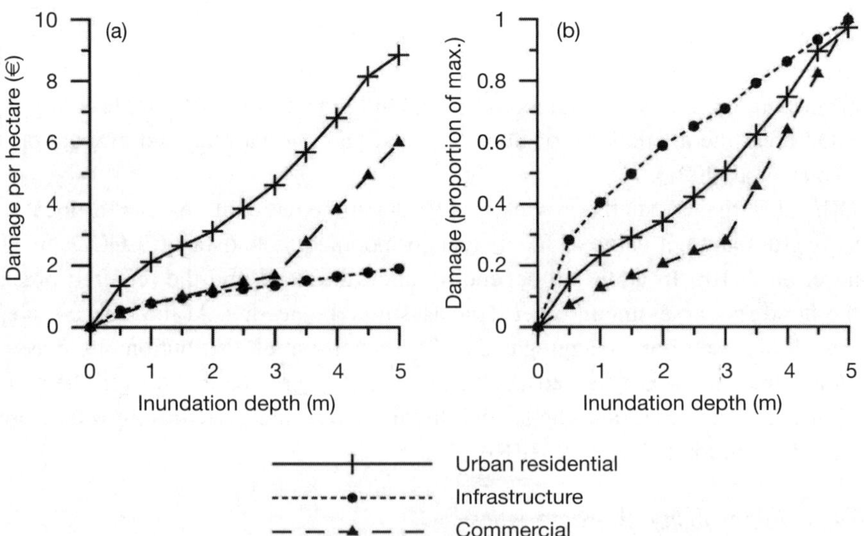

Figure 6.2 *Hypothetical absolute (a) and relative (b) stage-damage functions for use in flood damage modelling*

estimate the maximum potential damage for each exposure class, they can be applied to any other region. Of course, in practice this transferability is less simple since the shape of the relationships is not the same in different regions due to differences in building styles or different development stages.

Similar functions are also used in many assessments of loss of lives; a comprehensive overview of methods is provided by Jonkman et al (2008). A common feature of many of these methods is the use of a function that relates flood mortality to inundation characteristics. An example of such an approach is given in section 6.6.

6.4.3 Calculation and presentation of risk

As stated in section 6.2, flood risk is the product of hazard, exposure and vulnerability characteristics. It can be considered as the average total expected flood damage per year for all kinds of floods. Therefore, in theory, risk estimates should be based on a fully probabilistic analysis of the damages associated with floods of all return periods. However, due to time and resource constraints, flood risk assessments, in practice, simply assess damage for a small number of inundation scenarios and then interpolate the damage estimates for floods of different return periods. The average annual damage, or risk, can then be calculated as the integral of the area under a damage-exceedance probability curve, as shown in Figure 6.1.

Spatial representations of flood risk can also be portrayed in flood risk maps, which are useful for the reasons listed in section 6.3. Damage and risk mapping can be carried out relatively easily in modern geographic information systems (GIS) programmes, and can be used for communication and stakeholder participation (Fuchs et al, 2009, Van Alphen et al, 2009).

6.5 Scenarios

Flood risk is expected to increase in many coastal regions in the future due to on-going socio-economic development in risk-prone areas, as well as due to global warming (IPCC, 2007b; Bouwer et al, 2010), In order to develop sustainable flood management strategies, to design appropriate risk mitigation measures and to optimize spatial planning, it is important to gain insights into the pace and magnitude of possible changes in flood risk over time. This is especially true because many investments in flood protection or programmes for spatial planning often take 20 to 30 years to implement (Maaskant et al, 2009). However, estimating future flood risk is no easy task because both climate change and socio-economic developments involve considerable uncertainties, making it difficult, if not impossible, to accurately estimate future flood risk.

A possible way to deal with these uncertainties is to use climate and socio-economic scenarios in flood risk assessment. Scenarios are sets of assumptions reflecting alternative and preferably contrasting future developments that are coherent, internally consistent and plausible (Kuik et al, 2008). Usually, scenarios are not assigned probabilities, but show a range of plausible outcomes. Hence, contrasting scenarios can be used to visualize the possible bandwidth of future flood risk developments and its uncertainty (Bouwer et al, 2010; Te Linde et al, 2010). A widely used set of such scenarios derives from the *Special Report on Emission Scenarios* (SRES) provided by the Intergovernmental Panel on Climate Change (IPCC, 2000) (see Chapter 3 for an overview on scenarios).

In order to estimate the development of flood risk over time, information from both climate and non-climate scenarios needs to be taken into account. Climate change scenarios can be used to derive information on changes in hazard characteristics, such as sea level, waves and storm surges (IPCC, 2007a), river discharges (Belz et al, 2007) or flood probabilities (Te Linde et al, 2010). Socio-economic scenarios can be used to determine the change in exposure and vulnerability characteristics of social systems over time, such as land use (Verburg et al, 2008) or population (Maaskant et al, 2009).

The task of gathering detailed (object-based) information on future exposure is very difficult and time consuming. As a result, the majority of methods applied to assess future development of flood risk use information on land use as an input parameter to calculate damages, loss of life and, consequently, risk due to flooding (ICPR, 2001; Aerts et al, 2008; Maaskant et al, 2009; Bouwer et al, 2010), Thus, land-use change projections have been proven to be especially relevant for studies investigating future developments in flood risk. A modelling framework that enables the development of land-use change projections on the basis of scenarios is provided by the CLUE model (Verburg and Overmars, 2009). The CLUE model can be used to translate aggregate changes in demand for different land-use types, reflecting different development scenarios, into spatially explicit land-use patterns for a given area. The (future) demand of different economic sectors in terms of land use can be taken from existing projection or economic models. Spatially explicit land-use patterns are derived from the aggregated changes by applying dynamic spatial allocation rules that can be either based on empirical analyses, user-specified decision rules, neighbourhood characteristics or a combination of these methods (Verburg et al, 2008). The CLUE model has, for instance, been used to derive land-use change projections for four different scenarios for the EU-15 states (i.e. the member states of the European Union as of 1 January 1995) (Verburg et al, 2008). An example of how these land-use projections have been used in an analysis to estimate future development of flood risk in the Rhine Basin and its uncertainty is provided by Te Linde et al (2011) (see section 6.6).

6.6 Examples of Flood Risk Assessment Applications

In this section, we provide brief overviews of several studies that examine flood damages and risk at different scales, using the classifications of macro-, meso- and micro-scale as introduced in section 6.4.1. Where possible, examples of flood risk assessments in coastal areas are provided. Furthermore, an example is given of how the future risk of flooding has been examined in terms of loss of life in South Holland, The Netherlands.

6.6.1 Macro-scale: Global exposure in port cities

The most comprehensive study of flood damage exposure at the global level is that of Nicholls et al (2008), which examined exposure to flood damage due to storm surge in 136 global port cities (with populations of over 1 million). The assessment investigates how climate change is likely to affect each port city's exposure to coastal flooding by the 2070s, alongside subsidence, population growth and urbanization. The analysis focuses on the *exposure* of population and assets (economic assets in cities in the form of buildings, transport infrastructure, utility infrastructure and other long-lived assets) to a 100-year surge-induced flood event (assuming no defences), rather than the risk of coastal flooding. Flood protection is not included explicitly as it is difficult to ascertain accurate and comprehensive data on flood protection in many, if not most, of the cities under study.

The following scenarios were used to examine changes in exposure to a 1/100-year water-level event:

- *current city (C):* situation in 2005;
- *future city, no environmental change (FNC):* current environmental situation with the 2070s economy and population scenario;
- *current city, climate change (CCC):* current socio-economic situation with the 2070s climate change and natural subsidence/uplift;
- *current city, all changes (CAC):* current socio-economic situation with the 2070s climate change, natural subsidence/uplift and human-induced subsidence;
- *future city, climate change (FCC):* future socio-economic situation with 2070s climate change and natural subsidence/uplift;
- *future city, all changes (FAC):* future socio-economic situation with 2070s climate change and natural subsidence/uplift and human-induced subsidence.

In order to demonstrate the land area and population exposed to inundation, the investigation took the form of an elevation-based GIS analysis, after McGranahan et al (2007). Current extreme water levels were taken from the database of the Dynamic Interactive Vulnerability Assessment model (DIVA) (DINAS-COAST Consortium, 2006; available at www.civil.soton.ac.uk). The water levels for each future scenario and each city were calculated by combining the appropriate relative sea-level rise, the 100-year return period extreme water level, a storm enhancement factor, and, where appropriate, natural and anthropogenic subsidence. The calculated water levels were used with the population distributions to estimate the exposed population and the value of exposed infrastructure assets at an elevation below the 100-year extreme water level (i.e. those that would be affected by a 100-year event in the absence of any flood defences). Population data were taken from Landscan 2002 and constrained using city extents from postcode data. For each city, the population distribution data were mapped onto the relevant DEM, giving a horizontal map of geographical cells with defined population and elevation. From this, the total populations within 1m vertical bands were extracted. The population 'exposed' below the contour defined by the extreme water level was estimated, thus corresponding to the 100-year event for each scenario. The calculation of exposed assets was based on national per capita gross domestic product (GDP) purchasing power parity (PPP) for 2005, obtained from the International Monetary Fund (IMF) database (see www.imf.org). The assets exposed and at risk were calculated directly from the population measures using a simple relationship between exposed population, GDP per capita and exposed assets.

In total, 136 cities were ranked, with 38 per cent of the cities located in Asia; these cities represent 65 per cent of the global exposed population. The largest share of the economic exposure is in the cities of North America, closely followed by Asia. The total value of exposed assets in all of the cities studied was estimated at approximately US$3000 billion under current conditions. By 2070, the value of these exposed assets could increase up to US$35,000 billion as a result of both climate change and socio-economic growth.

6.6.2 Meso-scale

6.6.2.1 Flood risk in the Rhine Delta

An example of a flood risk analysis on a meso-scale is provided by Te Linde et al (2011) for the Rhine Basin, including a part of the Dutch delta area in The Netherlands. Current flood risk is assessed using a simple damage model based on the two input parameters of inundation depth and land use. Calculated damages are combined with estimations of flood return periods for different sections along the Rhine to evaluate flood risk.

The impact of projected climate change and socio-economic developments upon future flood risk is assessed by using two climate change and two socio-economic scenarios. The latter are represented by land-use change projections for 2030 and are derived from a land-use model called the Land-Use Scanner (Loonen and Koomen, 2009). The two (contrasting) socio-economic scenarios used as input for the land-use model are based on the scenarios developed within the EUruralis project (Verburg et al, 2008; Verburg and Overmars, 2009). By combining a low climate change scenario with a low socio-economic scenario, as well as a high climate change scenario with a high socio-economic scenario, the possible bandwidth of future flood risk developments and the associated uncertainties are exemplified. Results indicate that future basin-wide flood risk might increase between 54 and 230 per cent between 2000 and 2030 due to future climate change and on-going socio-economic development in risk-prone areas. The use of both climate and socio-economic scenarios also allows one to separately assess the contribution of each driving factor on future flood risk. Flood risk, for example, has been found to increase by 43 to 160 per cent due to the effect of climate change on flood probabilities, while socio-economic developments contribute to an increase in risk of between 6.5 and 27 per cent.

6.6.2.2 Flood risk in England and Wales

In order to assess future flood risk in England and Wales, Hall et al (2005) used data on the location, the Standard of Protection (SOP) and conditions of the 34,000km English and Welsh flood defences, as well as on floodplain extent, topography, occupancy and assets values. So-called impact zones were determined using land-use data and 'impact information' (i.e. depth–damage curves and population data). For each impact zone, the probability of failure of the defences, and the water depth at which it would occur (y), were calculated. The resulting economic damage, or expected annual damage (R), was defined as a function of the damage and the probability density function for flood depths y in the impact zone studied. The total expected annual damage (T) in England and Wales was evaluated by summing the expected annual damage R obtained for each impact zone. The social damage was evaluated using the Social Flood Vulnerability Indices, which were developed by Tapsell et al (2002) in order to measure the impact that floods can have upon communities at risk. To calculate the total flood risk, the probability of flooding was estimated as the probability of failure of the flood defences evaluated using the SOP (an assessment of the return period at which the defences will be overtopped) and was combined with the total expected annual damage, also called annual average damage. In order to estimate the future flood risk, emissions and socio-economic scenarios were used.

6.6.3 Micro-scale

6.6.3.1 Flood damage model for the city of Dordrecht, The Netherlands

The case study area of the flood damage model developed by Veerbeek and Zevenbergen (2009) for Dordrecht represents 2096ha and includes over 5600 individual buildings. The objective of this damage model was to estimate detailed damages of a potential flood event at the urban scale by taking the density and heterogeneity of the city into consideration. In order to do so, a high level of detail was needed. Specific stage-damage functions were developed using data from the Dutch building sector. To incorporate the influence of flow velocity and flood duration on expected damages, damage functions using flow velocity and flood duration as input parameters were used. The flood damage model resulted in a map representing, for a defined flood event, the number of flooded buildings and the expected aggregated damage per damage cluster. In order to provide usable data for the management of the city, flood damages were also aggregated per administrative neighbourhood and represented with damage curves per neighbourhood. These curves combined the damage and the likelihood of flooding. In order to evaluate and visualize the evolution of future flood damage that the city can expect, the authors used climate change scenarios to estimate changes in sea level and river discharges for flood events with different probabilities (return periods) in 2050 and 2100. The results predicted an increase of the flood damage in Dordrecht in the coming century.

6.6.3.2 Flood risk in the Perth Metropolitan Area, Australia

Middelmann-Fernandes (2010) carried out an assessment of flood damage on the Swan River system in Perth, and made several interesting findings regarding the use of stage-damage functions. Although the method examines fluvial flooding from the Swan River, the methodology and findings are relevant for both fluvial and coastal flooding in coastal cities. The study estimates the direct tangible damage for residential structures and contents using two different damage functions: one is based on inundation depth only; the second is based on both inundation depth and flow velocity. The study found that the residential damages were underestimated compared to actual insured losses, when either the stage-damage functions or the velocity stage-damage functions were used in isolation. The velocity stage-damage function only identifies those buildings that fail by moving off their foundations, and therefore buildings with only partial damage are not accounted for in the damage calculation. On the other hand, the stage-damage function does not consider that buildings may fail and therefore leads to an underestimation of damage. The study finds that the use of these two methods in combination may provide more accurate information on residential flood risk.

6.6.4 Case study on the loss of human lives

Maaskant et al (2009) assess the potential loss of life as a result of flooding for the province of South Holland, The Netherlands. This is a densely populated delta area comprising large cities such as Amsterdam, Rotterdam and The Hague, and is largely located below sea level. This study has been selected because it not only assesses potential loss of life at present, but also the spatial and temporal change in flood mortality projected due to future climate change and socio-economic development. Furthermore, the study also provides information on the so-called societal risk level by multiplying flood probabilities with the number of potential fatalities (i.e. average fatalities per year, or risk), and constructs a loss–probability curve. Thus, it provides a good example of how the risk-based approach (see section 6.2) and the use of climate and socio-economic scenarios (see section 6.5) can be applied in the context of loss of life due to flooding in coastal areas.

An estimation of potential flood casualties for a given flood event is provided by combining information on inundation characteristics, the population exposed and an estimate of the mortality amongst the exposed population. Flood mortality is thereby defined as the number of fatalities divided by the number of exposed people. A flood mortality function is applied that relates mortality to flood depth. The mortality function applied by Maaskant et al (2009), as well as the 5 to 95 per cent uncertainty bounds, is given in Figure 6.3.

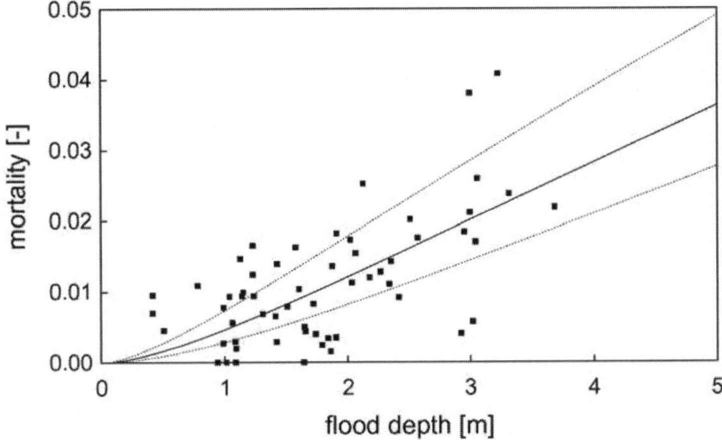

Figure 6.3 *Mortality function and 5 to 95 per cent uncertainty bounds*

Source: Maaskant et al (2009), with permission from Elsevier

In order to examine changes in flood mortality over time, information from both climate change and socio-economic scenarios are integrated within the modelling approach. Information on the future number, distribution and density of the population is derived from land-use projections for 2040. These projections of future land use for The Netherlands are available from a land-use model referred to as Land-Use Scanner (Schotten et al, 2001; Loonen and Koomen, 2009). The land-use change projections thereby reflect the *Welvaart en Leefomgeving* (WLO) scenarios discussed in Chapter 3. The effects of climate change on certain flood characteristics and, consequently, flood mortality are derived from Dutch climate change scenarios (Van den Hurk et al, 2006). Due to a projected rise in sea level and a change in river discharge, flood depths are assumed to increase and flood probabilities are projected to decrease. Taking these developments into account, Maaskant et al (2009) assess the effects of climate change and population growth on flood mortality, both separately and combined. Evaluating the effects of both of these changes on flood mortality is of interest because it allows for the estimation of the independent contribution of each driving factor.

Maaskant et al (2009) find that the share of flood victims per 1000 inhabitants increases by 29 per cent due to population growth in low-lying areas, by 15 per cent due to sea-level rise (and, consequently, higher water depths) and by 47 per cent if both effects are combined. They also show that the societal risk (i.e. the expected number of fatalities per year) increases by more than a factor 4 from 0.37 fatalities per year in 2000 to 1.46 fatalities per year in 2040. This considerable increase is attributed to the increase in the population exposed, as well as the decrease in flood probabilities and the rise in water depths.

6.7 Uncertainty Assessment

In section 6.5 we discussed how uncertainties regarding future developments can be addressed by the use of socio-economic and climate scenarios. In addition, uncertainties linked to flood risk assessments can originate from the data or the methods used. As such, uncertainties can arise from the damage functions, the generalization and categorization of land-use data, the use of aggregated data, or faults and inaccuracies in the basic data or in the methods used to estimate the data (USACE, 1996). In order to validate a flood risk analysis as part of a decision support tool, it is important to indicate the uncertainties of the analysis. The reliability of the results can be an important parameter for policy-makers to take into consideration before making a decision (Merz and Thieken, 2009). Ideally, a flood risk analysis should therefore contain the calculated risk results and their uncertainty bounds.

In scientific studies, uncertainty is commonly divided into two broad categories – namely, aleatory uncertainty and epistemic uncertainty. Oberkampf et al (2004) describe aleatory uncertainty as the 'inherent variation associated with the physical system or the environment under consideration'. It is sometimes referred to as variability, irreducible uncertainty and stochastic uncertainty. They state that epistemic uncertainty derives 'from some level of ignorance, or incomplete information, of the system or the surrounding environment'. This is also referred to as reducible uncertainty, subjective uncertainty and model form uncertainty. In their study on flood risk uncertainties in Cologne (Germany), for example, Merz and Thieken (2009) identify sources of aleatory uncertainty as factors, such as the variability of maximum runoff of a catchment in sequential years, while epistemic uncertainty is represented by factors such as a lack of runoff data when defining a probability density function of a river discharge. By increasing the amount and quality of available data, the epistemic uncertainties can be decreased; this is not the case for aleatory uncertainties. Therefore, Merz and Thieken (2009) decided for their study to analyse the two categories of uncertainties separately.

Different approaches can be used to evaluate the uncertainties of an analysis. USACE (1996) describes some approaches: the *Multi-Coloured Manual* gives examples of the integration of uncertainty parameters in damage valuation in England. Merz et al (2004) researched the uncertainties surrounding the relation between depth and damage. This analysis used empirical data and showed that only a small part of the flood depth data explained the resulting damage. De Moel and Aerts (2009) studied uncertainties related to inundation depths, land-use data and stage-damage functions for polder areas in the Dutch delta by varying the values of all of these parameters. They showed that a high proportion of the uncertainties are linked to the inundation depths and the stage-damage functions chosen. In addition, the authors also calculated absolute and proportional changes in flood damage for different land use and different stage-damage functions. The results showed that absolute changes in damage vary between a factor of 1.20 and 4.15, while proportional changes in damage vary between a factor of 1.04 and 1.20. Proportional estimations of the change in damage are therefore more robust than absolute estimations.

6.8 A Full Chain Example: Storm Surge Risks for the North Sea Region – Assessing Impacts upon Losses and Associated Uncertainties

In the study by Gaslikova et al (2011), the potential impact of changing climate upon coastal flood damage and the associated potential losses were assessed for a number of North Sea neighbouring countries. This was achieved by combining the climate and hydrodynamic models with the loss model from the Swiss Reinsurance Company (Swiss Re). The climate and hydrodynamic models provided the necessary climate change signal relevant for the inland flood assessments (i.e. water levels offshore). The inundation characteristics, in this case inland water depth, were determined by utilizing information about contemporary coastal protection and land elevation maps (90m resolution). To simulate the inland propagation, a planar approach was applied to the data (i.e. any water level above mean sea level in front of the coast could cause inland flooding). Coastal protection was parameterized by an empirical model employing a standard of protection and crest elevation for the coastal dikes, where protection failure probabilities were connected to the water levels near the coast. As the amount of water is limited, the inundated water level was assumed to linearly decrease with increasing distance from the flood source. Local flood depths were aggregated for postcode areas. The exposure characteristics were presented by the potentially insurable property distribution (according to today's values). Corresponding vulnerabilities were assigned based on Swiss Re's loss experience record. Flood-related losses were estimated for five countries in the North Sea Basin (Denmark, Germany, The Netherlands, Belgium and the UK).

According to the loss model methodology, a representative set of hazardous storm surge events should be considered in order to create a wide spectrum of flooding scenarios, which can further be aggregated to the statistical quantities like annual expected losses. Observations and historical records of storms and associated high water events do not provide a sufficient database for such a loss assessment. First, the spatial coverage is fragmentary and limited to the tide-gauge locations. Second, only events that actually happened in the past are recorded, whereas the equally potential but not realized events are skipped. Moreover, an additional method is necessary to project the historical climatology into the future in order to consider the potential effects of a changing climate. To omit these difficulties, a set of multi-decadal simulations based on historical (reanalysis of the past) as well as scenario (potential future) climate conditions (Weisse et al, 2009) was used for the loss assessments. They provided storm surge water levels, which were obtained by combining an atmospheric climate model, providing wind and atmospheric pressure, with a hydrodynamic model to estimate water levels. This model chain delivered the residual of total water level and tidal cycle. To assess a future climate, two IPCC SRES scenarios A2 and B2 were considered (Bindoff et al, 2007). Additionally, two sea-level rise (SLR) scenarios, the moderate 0.5m and the extreme 1m according to IPCC estimates, were linearly added to the storm surge elevations.

For an adequate assessment of losses and their probabilities, a so-called hazard set or a set of events representative for each climate scenario (present day and future) was constructed. The hazard set included real (dynamically modelled) water levels, as well as equally probable but not simulated water levels. For the hourly storm surge time series along the coasts of the North Sea, a total of 200 locations were analysed to select the high water events. Several requirements had to be fulfilled:

- retention of spatial patterns of water levels for each storm;
- independence of the events; and
- exceeding a certain threshold by water levels, at least in some locations, to select only potentially hazardous events.

The time series were divided into equal time intervals (120 hours) and water-level maxima were identified for each location. If the maxima for at least ten locations exceeded the corresponding thresholds (here, mean annual 99.99 percentile), which was equivalent to the high water event along at least 100km of the coastline, the entire time interval was considered as a storm event and the maxima for all locations were included in the event set. The resulting set of water levels was found to follow a normal distribution.

A set of random events based on the fitted normal distribution parameters and correlation of event sets from different locations was generated. In total, the sets of 20,000 events each were constructed to reflect the historical conditions associated with the period of 1958 to 2002 (hindcast), present-day scenario conditions for 1961 to 1990 (control); and two future development scenarios for 2071 to 2100 (A2 and B2). In combination with the SLR scenarios, these built an ensemble of present-day and future possible high water-level events at the coastline and their probabilities.

It is often the case for recent global and regional atmospheric models that extreme conditions were underrated (e.g. wind speeds were underestimated and sea-level pressure overestimated) for control simulations with respect to the reanalysis (e.g. Woth et al, 2006). Together with differences in the hydrodynamic models and model set-ups for the reanalysis and scenarios, this led to an underestimation of storm surge levels for the control simulation and, presumably, for the future scenarios. To give an example, the control and future simulation suffer from the fact that external surges generated beyond the model area (i.e. the North Atlantic) are not included in the simulations. However, as the absolute values (e.g. storm surge heights) were crucial for this study, and not only the differences between future and present-day values, it was important to calibrate the control dataset to the reanalysed dataset and to change the scenarios correspondingly. The method based on the use of cumulative distribution function (CDF) was applied – namely, the CDF of control water levels for each location was linearly fitted to the corresponding CDF from the hindcast. Under the assumption of the stationary relationship, the same transformation coefficients were applied to the scenario CDFs, so the possible deficiencies of the model were treated equally for control and future scenarios.

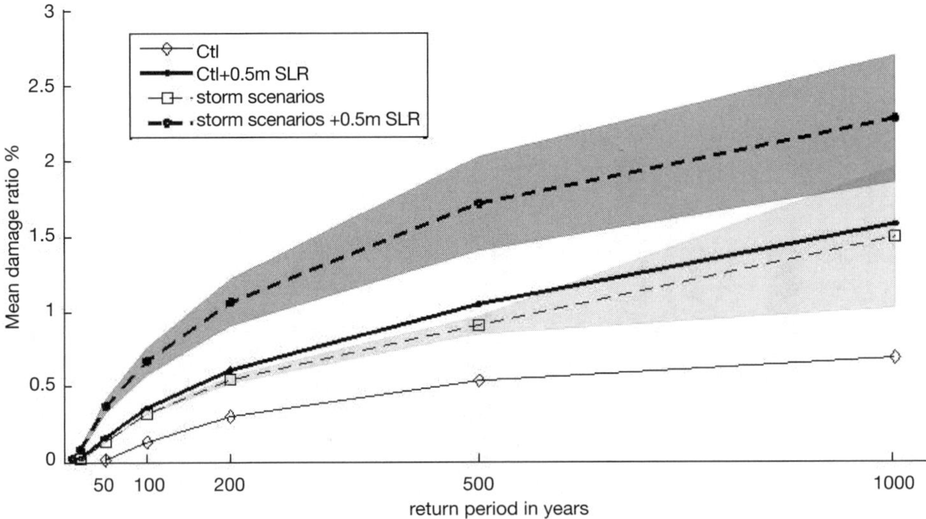

Figure 6.4 *Mean damage ratio with respect to the total values used for modelling for Germany*

Note: Shading indicates the variance due to storm surge climate associated with different future development scenarios (A2 and B2). Light grey = only storm surges; dark grey = storm surges with 0.5m sea-level rise.
Source: Gaslikova et al (2011)

The results of the study indicated the general increase of losses for all future scenarios and all countries, although the rate of changes varied significantly between countries, but also between climate change scenarios. Furthermore, the model uncertainty spanned the range of expected losses. The mean damage ratio (expected loss divided by total value) presented as a function of return period is called a loss frequency curve. Figure 6.4 shows the loss frequency curves for Germany estimated for present-day (control) and future scenarios considering only changes in the storm surge climatology, as well as both with additional 0.5m SLR. All future estimated losses differ significantly from present-day ones. However, the uncertainty due to different future scenarios is rather high (shaded areas). Losses only due to changes in surge frequency are comparable with those associated with 0.5m SLR superimposed upon the present-day climate. A combination of both (changed storm surge and SLR) almost doubles the effect. Figure 6.5 shows the increase in annual expected losses (in percentage) for the A2 scenario and different SLR with respect to the present-day situation for each country and all countries together. The uncertainty associated with the model parameterizations, particularly coastal protection, is shown as error bars. The rate of change of expected losses for the same scenario varies between the countries (e.g. the losses for Denmark are expected to experience a several times larger increase than for The Netherlands). For The Netherlands, the changes associated only with the storm surge climate are

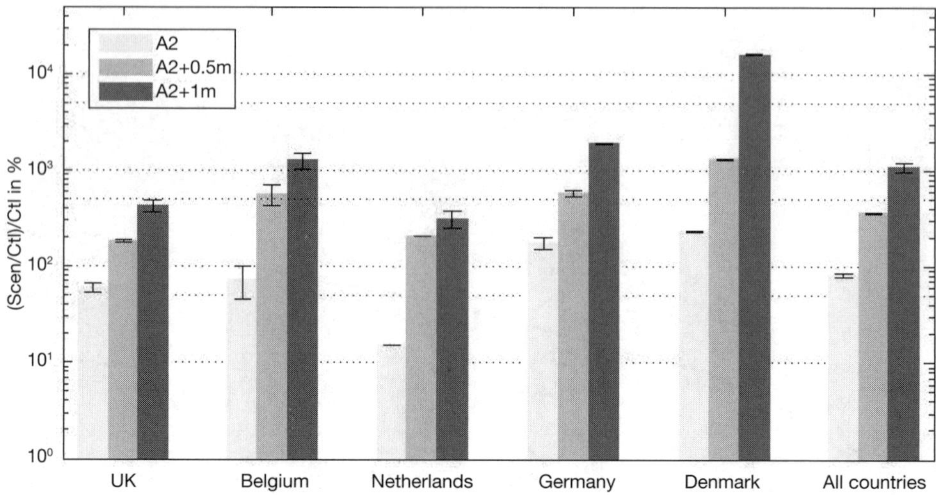

Figure 6.5 *Increase in annual expected losses from the current to the future climate represented by the A2 scenario (in percentage)*

Notes: Error bars represent uncertainty due to variation in the coastal protection parameter. Currently, the annual expected loss burden from surge events in the study area is approximately 0.6 billion Euros.
Source: Gaslikova et al (2011)

very small, whereas the SLR has a more pronounced impact upon the annual expected losses. A possible reason could be that an intensification of storms for these scenarios was mainly found for winds from the south-west direction: these do not cause high surge along the Dutch coast (Sterl et al, 2009). The loss changes are estimated between 15 (no SLR) and 313 per cent (1m SLR) with respect to present-day conditions. For other countries, storm surges alone start to be important. Enhanced by the SLR, they can cause a significant increase of flood losses in the entire region.

In summary, the general demand is that global studies resolve more details and processes (e.g. protection and more complex inland flood incorporation, more detailed property distribution maps, etc.), whereas more accurate but very local studies provide well-validated and well-tested methodology and generalize results for larger areas without losing accuracy. Gaslikova et al (2011) made an attempt to merge regional coverage (e.g. insurance or country-wise economic loss assessment) with local features. Although no socio-economic scenarios were explicitly included within the model, significantly different results were obtained for various countries. These can be attributed to non-uniform changes in storm surge climate for different parts of the North Sea, as well as diverse exposure due to both geographical properties and the characteristics of coastal protection.

6.9 Conclusions

In this chapter we describe important concepts in flood risk assessment and modelling, where flood risk is defined as the probability of flooding multiplied by the potential consequences of a flood. Flood risk assessment is useful since it allows us to estimate current risk, as well as the change in risk that can be expected in the future; to assess the effectiveness of adaptation measures in reducing flood risk; and to create flood risk maps to raise public risk awareness. Flood risk assessments require data on both the flood hazard (e.g. inundation depth and flow velocity) and the consequences of flooding (e.g. economic damage and loss of life). Flood damage includes damage such as physical, economic, psychological or environmental damage. These damages are divided between direct and indirect damages, and between tangible and intangible damages. Flood risk is usually assessed in flood damage models, which combine information on flood hazard characteristics, such as inundation depth, with data on exposure and vulnerability characteristics. Flood damage is calculated for flood events of several return periods, from which a damage-exceedance probability curve can be derived. Flood risk can then be considered as the average total expected flood damage per year, represented by the integral of the area under the damage-probability curve. In order to evaluate the evolution of risk over time, climate change and socio-economic scenarios can be used. Climate change scenarios enable us to obtain information on changes in hazard characteristics, while socio-economic scenarios provide data to determine the changes in exposure and vulnerability characteristics. Different case studies are described in this chapter in order to show how different researchers approach risk analysis in practice. An important facet of risk analysis concerns the identification and quantification of uncertainties. In fact, the methods used to collect and evaluate the data, and the quality and availability of the input data themselves, lead to uncertainties in the results. If flood risk assessments are to be used in flood management decision-support and awareness-raising campaigns, it is important to recognize and analyse the uncertainties related to the processes, and to quantify the reliability of the risk analysis. The chapter ends with the description of an assessment of the impacts of climate change in the North Sea region, which is provided in order to give an example of the assessment of the impacts of flooding and the associated uncertainties using scenarios.

To conclude, even though flood risk and flood damage assessments have received growing attention during recent years, the scientific field of flood risk assessment still lags behind the better-developed fields of hydrology and hydraulics (Büchele et al, 2006; Merz et al, 2010). Given the importance of information on flood damage and risk for developing and supporting sound flood management strategies and adaptation planning, further advances seem necessary. Aspects that are repeatedly mentioned in this respect are the lack of (publicly available) flood damage data, as well as missing information on data quality.

Acknowledgements

We thank the Knowledge for Climate programme (HSINT02; Themes 1 and 4) and the Climate Changes Spatial Planning programme (A20) for supporting this chapter.

References

Aerts, J. and Droogers, P. (eds) (2004) *Climate Change in Contrasting River Basins: Adaptation Strategies for Water, Food, and Environment*, CABI, London

Aerts, J., Sprong, T. and Bannink, B. A. (2008) *Aandacht voor Veiligheid*, 009/2008, 3-196, Leven met Water, Klimaat voor Ruimte, DG Water, The Netherlands

Aerts, J. C. J. H., Major, D. C., Bowman, M. J., Dircke, P., Marfai, M. A., Abidin, H. Z., Ward, P. J., Botzen, W., Bannink, B. A., Nickson, A. and Reeder, T. (2009) *Connecting Delta Cities: Coastal Adaptation, Flood Risk Management and Adaptation to Climate Change*, VU University Press, Amsterdam, The Netherlands

Apel, H., Aronica, G. T., Kreibich, H. and Thieken, A. H. (2009) 'Flood risk analyses: How detailed do we need to be?', *Natural Hazards*, vol 49, pp79–98

Bateman, I. J., Carson, R. T., Day, B., Hanemann, M., Hanleys, N., Hett, T., Jones-Lee, M., Loomes, G., Mourato, S., Ozdemiroglu, E., Pearce, D. W., Sugden, R. and Swanson, J. (2002) *Economic Valuation with Stated Preference Techniques: A Manual*, Edward Elgar, Cheltenham, UK

Bates, P. D., Dawson, R. J., Hall, J. W., Horritt, M. S., Nicholls, R. J., Wicks, J. and Hassan, M. A. A. M. (2005) 'Simplified two-dimensional numerical modelling of coastal flooding and example applications', *Coastal Engineering*, vol 52, no 9, pp793–810

Belz, J., Brahmer, G., Buiteveld, H., Engel, H. and Grabher, R. (2007) *Das Abflussregime des Rheins und seiner Nebenflüsse im 20*, Jahrhundert, Analyse, Veränderungen, Trends, Internationale Kommission für die Hydrologie des Rheingebietes, Bericht no I-22 der KHR, Germany

Bindoff, N., Willebrand, J., Artale, V. A. C., Gregory, J., Gulev, S., Hanawa, K., Quere, C. L., Levitus, S., Nojiri, Y., Shum, C., Talley, L. D. and Unnikrishnan, A. (2007) 'Observations: Oceanic climate change and sea level', in IPCC (ed) *Climate Change 2007: The Physical Science Basis. Contribution of Working Group I to the Fourth Assessment Report of the Intergovernmental Panel on Climate Change*, Cambridge University Press, Cambridge, UK, and New York, NY

Bouwer, L. M., Crompton, R. P., Faust, E., Höppe, P. and Pielke, Jr., R. A. (2007) 'Confronting disaster losses', *Science*, vol 318, p753

Bouwer, L. M., Bubeck, P. and Aerts, J. C. J. H. (2010) 'Changes in future flood risk due to climate and development in a Dutch polder area', *Global Environmental Change*, vol 20, no 3, pp463–471

Brown, C. A. and Graham, W. J. (1988) 'Assessing the threat to life from dam failure', *Water Resources Bulletin*, vol 24, no 6, pp1303–1309

Bryan, B., Harvey, N., Belerio, T. and Bourman, B. (2001) 'Distributed process modeling for regional assessment of coastal vulnerability to sea level rise', *Environmental Modeling & Assessment*, vol 6, pp57–65

Büchele, B., Kreibich, H., Kron, A., Thieken, A., Ihringer, J., Oberle, P., Merz, B. and Nestmann, F. (2006) 'Flood-risk mapping: Contributions towards an enhanced assessment of extreme events and associated risks', *Natural Hazards and Earth System Sciences*, vol 6, no 4, pp485–503

Chowdhury, A. M. R., Bhuyia, A. U., Choudhury, A. Y. and Sen, R. (1993) 'The Bangladesh cyclone of 1991: Why so many people died', *Disasters*, vol 17, no 4, pp291–304

Clausen, L. and Clark, P. B. (1990) 'The development of criteria for predicting dambreak flood damages using modeling of historical dam failures', in W. R. White (ed) *International Conference on River Flood Hydraulics, Hydraulics Research Ltd, 17–20 September*, John Wiley and Sons Ltd, Chichester, UK, pp369–380

Dasgupta, S., Laplante, B., Meisner, C., Wheeler, D. and Yan, J. (2009) 'The impact of sea level rise on developing countries: A comparative analysis', *Climatic Change*, vol 93, pp379–388, doi:10.1007/s10584-008-9499-5

Dawson, R., Sayers, P., Hall, J., Hassan, M. and Bates, P. (2005) 'Efficient broad scale coastal flood risk assessment', in J. McKee Smith (eds) *Coastal Engineering 2004: Proceedings of the 29th International Conference*, Lisbon, Portugal, 19–24 September 2004, World Scientific, New Jersey

de Bruijn, K. M. (2005) *Resilience and Flood Risk Management*, PhD thesis, Technische Universiteit of Delft, The Netherlands

DeKay, M. L. and McClelland, G. H. (1993) 'Predicting loss of life in case of dams failure and flash flood', *Risk Analysis*, vol 13, no 2, pp193–205

de Moel, H. and Aerts, J. C. J. H. (2009) 'Uncertainty and sensitivity assessment of flood risk assessments', *Eos Transactions AGU*, Fall meeting supplement, vol 90, no 52, ppNH52A-05

de Moel, H., van Alphen, J. and Aerts, J. C. J. H. (2009) 'Flood maps in Europe: Methods, availability and use', *Natural Hazards and Earth System Sciences*, vol 9, pp289–301

DINAS-COAST Consortium (2006) *Dynamic Interactive Vulnerability Assessment (DIVA)*, Potsdam Institute for Climate Impact Research, Potsdam, Germany, www.civil.soton.ac.uk, accessed 23 August 2010

FEMA (Federal Emergency Management Agency) (2005) *Effects of Long and Short Duration Flooding on Building Materials*, DFO, FEMA, DR-1539/1545/1551/1565, Orlando, FL

Few, R. (2003) 'Flooding, vulnerability and coping strategies: Local responses to a global threat', *Progress in Development Studies*, vol 3, no 1, pp43–58

FLOODsite (2007) *Evaluating Flood Damages: Guidance and Recommendations on Principles and Methods*, www.floodsite.net, accessed 28 June 2010

Fuchs, S., Spachinger, K., Dorner, W., Rochman, J. and Serrhini, K. (2009) 'Evaluating cartographic design in flood risk mapping', *Environmental Hazards*, vol 8, pp52–70

Funnemark, E., Odgaard, E., Svendsen. V. N. and Amdal, T. (1998) 'Consequence analysis of dam breaks', in L. Berga (ed) *Dam Safety*, Balkema, Rotterdam, pp329–336

Füssel, H. M. (2007) 'Vulnerability: A generally applicable conceptual framework for climate change research' *Global Environmental Change*, vol 17, pp155–167

Gaslikova, L., Schwerzmann, A., Raible, C. C. and Stocker, T. F. (2011) 'Future storm surge impacts on insurable losses for the North Sea region', *Natural Hazards Earth System Sciences*, vol 11, pp1205–1216

Gewalt, M., Klaus, J., Peerbolte, E. B., Pflügner, W., Schmidtke, R. F. and Verhage, L. (1996) *EUROflood. Technical Annex 8: Economic Assessment of Flood Hazards*, Regional Scale Analysis–Decision Support System (RSA-DSS)

Haehnel, R. B. and Daly, S. F. (2002) *Maximum Impact Force of Woody Debris on Floodplain Structures*, US Army Corps of Engineers Engineer Research and Development Center, Technical report ERDC/CRREL TR-02-2

Hall, J. W., Dawson, R. J., Sayers, P. B., Rosu, C., Chatterton, J. B. and Deakin, R. (2003) 'A methodology for national-scale flood risk assessment', *Water and Maritime Engineering*, vol 156, pp235–247

Hall, J. W., Sayers, P. B. and Dawson, R. J. (2005) 'National-scale assessment of current and future flood risk in England and Wales', *Natural Hazards,* vol 36, pp147–164

Horritt, M. S. and Bates, P. D. (2002) 'Evaluation of 1-D and 2-D numerical models for predicting river flood inundation', *Journal of Hydrology*, vol 268, pp87–99

ICPR (International Commission for the Protection of the Rhine) (2001) *Rhine 2020: Program on the Sustainable Development of the Rhine*, Conference of Rhine Ministers, Koblenz, Germany

IPCC (Intergovernmental Panel on Climate Change) (2000) *Special Report on Emission Scenarios*, Cambridge University Press, New York, NY

IPCC (2007a) *Climate Change 2007: The Physical Science Basis. Contribution of Working Group I to the Fourth Assessment Report of the Intergovernmental Panel on Climate Change*, Cambridge University Press, Cambridge, UK

IPCC (2007b) *Climate Change 2007: Impacts, Adaptation and Vulnerability. Contribution of Working Group II to the Fourth Assessment Report of the Intergovernmental Panel on Climate Change*, Cambridge University Press, Cambridge, UK

Jonkman, S. N., Vrijling, J. K. and Vrouwenvelder, A. C. W. M. (2008) 'Methods for the estimation of loss of life due to floods: A literature review and a proposal for a new method', *Natural Hazards*, vol 46, no 3, pp353–389

Jonkman, S. N., Maaskant, B., Boyd, E. and Levitan, M. L. (2009) 'Loss of life caused by the flooding of New Orleans after Hurricane Katrina: Analysis of the relationship between flood characteristics and mortality', *Risk Analysis*, vol 29, no 5, pp676–698

Kahn, J. R. (2005) *The Economic Approach to Environmental and Natural Resources*, third edition, Mason, Thomson South-Western, US

Kron, W. (2005) 'Flood risk = hazard x values x vulnerability', *Water International*, vol 30, no 1, pp58–68

Kuik, O., Buchner, B., Catenacci, M., Goria, A., Karakaya, E. and Tol, R. S. J. (2008) 'Methodological aspects of recent climate change damage cost studies', *The Integrated Assessment Journal*, vol 8, no 1, pp19–40

Kundzewicz, Z. W. and Schellnhuber, H. J. (2004) 'Floods in the IPCC TAR perspective', *Natural Hazards*, vol 31, pp111–128

Lekuthai, A. and Vongvisessomjai, S. (2001) 'Intangible flood damage quantification', *Water Resource Management*, vol 15, no 5, pp343–362

Loonen, W. and Koomen, E. (2009) *Calibration and Validation of the Land Use Scanner Allocation Algorithms: PBL Report*, Netherlands Environmental Agency, Bilthoven, The Netherlands

Maaskant, B., Jonkman, S. N. and Bouwer, L. M. (2009) 'Future risk of flooding: An analysis of change in potential loss of life in South Holland (The Netherlands)', *Environmental Science and Policy*, vol 12, pp157–169

Madsen, H. and Jakobsen, F. (2004) 'Cyclone induced storm surge and flood forecasting in the northern Bay of Bengal', *Coastal Engineering*, vol 51, pp277–296

Marco, J. B. (1994) 'Flood risk mapping', in G. Ross, N. Harmancioglu and V. Yevjevich (eds) *Coping with Floods*, Kluwer Academic Publishers, Dordrecht, The Netherlands

Mastin, M. C. and Olsen, T. D. (2002) *Fifty-Year Storm-Tide Flood-Inundation Maps for Santa de Aqua, Honduras*, USGS, Tacoma, US

McGranahan, G., Balk, D. and Anderson, B. (2007) 'The rising tide: Assessing the risks of climate change and human settlements in low elevation coastal zones', *Environment and urbanization*, vol 19, no 1, pp17–37

Merz, B. (2006) *Hochwasserrisiken. Grenzen und Moeglickeiten der Risikoabschaetzung*, E. Schweizerbart'sche Verlagsbuchhandlung, Stuttgart, Germany

Merz, B. and Thieken, A. H. (2009) 'Flood risk curves and uncertainty bounds', *Natural Hazards*, vol 51, pp437–458

Merz, B., Kreibich, H., Thieken, A. and Schmidtke, R. (2004) 'Estimation uncertainty of direct monetary flood damage to buildings', *Natural Hazards and Earth System Sciences*, vol 4, pp153–163

Merz, B., Thieken, A. H. and Gocht, M. (2007) 'Flood risk mapping at the local scale: Concepts and challenges in flood risk management in Europe – innovation in policy and practice', in S. Begum, M. J. F. Stive and J. W. Hall (eds) *Flood Risk Management in Europe: Innovation in Policy and Practice*, Springer, Dordrecht, The Netherlands

Merz, B., Kreibich, H., Schwarze, R. and Thieken, A. (2010) 'Review article: Assessment of economic damage', *Natural Hazards and Earth System Sciences*, vol 10, pp1697–1724

Messner, F. and Meyer, V. (2005) *Flood Damage, Vulnerability and Risk Perception: Challenges for Flood Damage Research*, UFZ Discussion Paper 13/2005

Meyer, V. and Messner, F. (2005) *National Flood Damage Evaluation Methods: A Review of Applied Methods in England, The Netherlands, The Czech Republic and Germany*, UFZ Discussion Paper, Leipzig, Germany

Middelmann-Fernandes, M. H. (2010), 'Flood damage estimation beyond stage-damage functions: An Australian example', *Journal of Flood Risk Management*, vol 3, pp88–96

Milly, P. C. D., Wetherald, R. T., Dunne, K. A. and Delworth, T. L. (2002) 'Increasing risk of great floods in a changing climate', *Nature*, vol 415, pp514–517

Mitchell, R. C. and Carson, R. T. (1989) *Using Surveys to Value Public Goods: The Contingent Valuation Method*, Resources for the Future, Washington, DC

Nicholls, R. J. (2002) 'Analysis of global impacts of sea-level rise: A case study of flooding', *Physics and Chemistry of the Earth*, vol 27, pp1455–1466, doi:10.1016/S1474-7065(02)00090-6

Nicholls, R. J. (2004) 'Coastal flooding and wetland loss in the 21st century: Changes under the SRES climate and socio-economic scenarios', *Global Environmental Change*, vol 14, pp69–89, doi:10.1016/j.gloenvcha.2003.10.007

Nicholls, R. J., Hanson, S., Herweijer, C., Patmore, N., Hallegatte, S., Corfee-Morlot, J., Château, J. and Muir-Wood, R. (2008) *Ranking Port Cities with High Exposure and Vulnerability to Climate Extremes: Exposure Estimates*, OECD Environment Working Paper no 1, ENV/WKP(2007), OECD, Paris, France, www.oecd-ilibrary.org/environment/ranking-port-cities-with-high-exposure-and-vulnerability-to-climate-extremes_011766488208, accessed 25 December 2009

Oberkampf, W. L., Helton, J. C., Joslyn, C. A., Wojtkiewicz, S. F. and Ferson, S. (2004) 'Challenge problems: Uncertainty in system response given uncertain parameters', *Reliability Engineering and System Safety*, vol 85, no 1–3, pp11–19, doi:10.1016/j.ress.2004.03.002

Parker, D. J., Green, C. H. and Thompson, P. M. (1987) *Urban Flood Protection Benefits: A Project Appraisal Guide (The Red Manual)*, Gower Technical Press, Aldershot, UK

Penning-Rowsell, E. C. and Green, C. H. (2000) 'New insights into the appraisal of flood alleviation benefits (1): Flood damage and flood loss information', *Journal of Water and Environmental Management*, vol 14, pp347–353

Penning-Rowsell, E. C., Fordham, M., Correia, F. N., Gardiner, J., Green, C., Hubert, G., Ketteridge, A.-M., Klaus, J., Parker, D., Peerbolte, B., Pflügner, W., Reitano, B., Rocha, J., Sanchez-Arcilla, A., Saraiva, M. d. G., Schmidtke, R., Torterotot, J.- P., van der Veen, A., Wierstra, E. and Wind, H. (1994) 'Flood hazard assessment, modelling and management: Results from the EUROflood project', in E. C. Penning-Rowsell and M. Fordham (eds) *Floods across Europe: Flood Hazard Assessment, Modelling and Management*, Middlesex University Press, London

Penning-Roswell, E. C., Johnson, C., Tunstall, S., Tapsell, S., Morris, J., Chatterton, J., Coker, A. and Green, C. (2003) *The Benefits of Flood and Coastal Defence: Techniques and Data for 2003*, Flood Hazard Research Centre, Middlesex University, UK

Purvis, M. J., Bates, P. D. and Hayes, C. M. (2008) 'A probabilistic methodology to estimate future coastal flood risk due to sea level rise', *Coastal Engineering*, vol 55, pp1062–1073, doi:10.1016/j.coastaleng.2008.04.008

Rosenzweig, C. and Solecki, W. (2001) *Climate Change and a Global City: The Potential Consequences of Climate Variability and Change – Metro East Coast*, Report for the US Global Change Program, Columbia Earth Institute, New York, NY

Sangrey, D. A., Murphy, P. J. and Nieber, J. L. (1975) *Evaluating the Impact of Structurally Interrupted Flood Plain Flows, Technical Report 98*, Cornell University Water Resources and Marine Sciences Centre, Ithaca, NY

Schotten, K., Goetgeluk, R., Hilferink, M., Rietveld, P. and Scholten, H. (2001) 'Residential construction, land use and the environment: Simulations for the Netherlands using a GIS-based land use model', *Environmental Modeling & Assessment*, vol 6, no 2, pp133–143

Smith, D. I. (1994) 'Flood damage estimation: A review of urban stage-damage curves and loss functions', *Water SA*, vol 20, pp231–238

Smith, K. and Ward, R. (1998) *Floods: Physical Processes and Human Impacts*, Wiley-Blackwell, Chichester, UK

Sterl, A., van den Brink, H., de Vries, H., Haarsma, R. and van Meijgaard, E. (2009) 'An ensemble study of extreme storm surge related water levels in the North Sea in a changing climate', *Ocean Science*, vol 5, pp369–378

Tapsell, S. M., Penning-Roswell, E. C., Tunstall, S. M. and Wilson, T. L. (2002) 'Vulnerability to flooding: Health and social dimensions', *Disaster Medicine and Public Health Preparedness*, vol 360, no 1796, pp1511–1525

Te Linde, A. H., Aerts, J. C. J. H., Bakker, A. M. R. and Kwadijk, J. C. J. (2010) 'Simulating low-probability peak discharges for the Rhine basin using resampled climate modeling data', *Water Resources Research*, vol 46, pW03512, doi:10.1029/2009WR007707

Te Linde, A. H., Bubeck, P., Dekkers, J. E. C., de Moel, H. and Aerts, J. C. J. H. (2011) 'Future flood risk estimates in the Rhine basin', *Natural Hazards and Earth System Sciences*, vol 11, no 2, pp459–473

Thieken, A. H., Müller, M., Kreibich, H. and Merz, B. (2005) 'Flood damage and influencing factors: New insights from the August 2002 flood in Germany', *Water Resources Research*, vol 41, pW12430, doi:10.1029/2005WR004177

Thumerer, T., Jones, A. P. and Brown, D. (2000) 'A GIS based coastal management system for climate change associated flood risk assessment on the east coast of England', *International Journal of Geographical Information Science*, vol 14, pp265–281

UN DHA (United Nations Department of Humanitarian Affairs) (1993) *Internationally Agreed Glossary of Basic Terms Related to Disaster Management*, DNA/93/36, UN DHA, Geneva, Switzerland

USACE (US Army Corps of Engineers) (1996) *Risk Based Analysis for Flood Damage Reduction Studies. Engineering Manual – EM 1110-2-1619*, Engineering and Design, Washington, DC

USBR (US Bureau of Reclamation) (1988) *Downstream Hazard Classification Guidelines*, US Department of the Interior, USBR, ACER Technical Memorandum no 11, Denver, CO

Van Alphen, J., Martini, F., Loat, R., Slomp, R. and Passchier, R. (2009) 'Flood risk mapping in Europe: Experiences and best practices', *Journal of Flood Risk Management*, vol 2, pp285–292

Van den Hurk, B., Klein Tank, A., Lenderink, G., van Ulden, A., van Oldenborgh, G. J., Katsman, C., van den Brink, H., Keller, F., Bessembinder, J., Burgers, G., Komen, G., Hazeleger, W. and Drijfhout, S. (2006) *KNMI Climate Change Scenarios 2006 for the Netherlands*, KNMI Scientific Report WR 2006-01, KNMI, De Bilt, The Netherlands

Veerbeek, W. and Zevenbergen, C. (2009) 'Deconstructing urban flood damages: Increasing the expressiveness of flood damage models combining a high level of detail with a broad attribute set', *Journal of Risk Management*, vol 2, pp45–57

Verburg, P. H. and Overmars, K. P. (2009) 'Combining top-down and bottom-up dynamics in land use modeling: Exploring the future of abandoned farmlands in Europe with the Dyna-CLUE model', *Landscape Ecology*, vol 24, no 9, pp1167–1181

Verburg, P. H., Eickhout, B. and van Meijl, H. (2008) 'A multi-scale, multi-model approach for analysing the future dynamics of European land use', *Annals of Regional Science*, vol 42, pp57–77

Weisse, R., v. Storch, H., Callies, U., Chrastansky, A., Feser, F., Grabemann, I., Guenther, H., Pluess, A., Stoye, T., Tellkamp, J., Winterfeld, J. and Woth, K. (2009) 'Regional meteorological-marine reanalysis and climate change projections: Results for Northern Europe and potentials for coastal and offshore applications', *Bulletin of the American Meteorological Society*, vol 90, pp849–860

Wind, H. G., Nieron, T. M., De Blois, C. J. and De Kok, J. L. (1999) 'Analysis of flood damages from the 1993 and 1995 Meuse floods', *Water Resources Research*, vol 35, pp3459–3465

Woth, K., Weisse, R. and von Storch, H. (2006) 'Climate change and North Sea storm surge extremes: An ensemble study of storm surge extremes expected in a changed climate projected by four different regional climate models', *Ocean Dynamics*, vol 56, pp3–15

Zerger, A. and Wealands, S. (2004) 'Beyond modelling: Linking models with GIS for flood risk management', *Natural Hazards*, vol 33, pp191–208

7

Climate-Resilient Urban Waterfronts

Kristina Hill

7.1 Introduction

Urban waterfronts were once planned and designed as a form of urban competition, when new berths for ships could expand a city's reach into maritime trade and military domination of the seas. In the 16th to 18th centuries, some nations redesigned the waterfronts of their coastal cities as competitive harbours, designed for larger and larger ships as well as better dock facilities. Italian Renaissance humanists believed that there was a set of ideal forms for port cities, and that implementing a rational and beautiful form would both reflect and contribute to making human society more rational and humane (Konvitz, 1978, 1994). As trade expanded rapidly and a new class of wealthy merchants emerged, Venice began to implement some of these humanist urban design ideas as early as the 13th century when a major expansion of global trade occurred. The city became a model for other cities and remains so today when urban leaders claim the title of 'Venice of the North' in Stockholm, or others claim that waterfront improvements will make their city a new Venice.

Throughout history, cities competed for roles in the expanding global trade networks, just as they do today (Hesse, 2006; Laidley, 2007), and waterfront design was seen as integral to competitive success (Minchinton, 1975). All of these changes in waterfronts were part of global industrial and economic trends that have brought an increase in the volume of exchanged goods traded all over the world, and in some cases have resulted in reduced prices of consumer goods. But they have also resulted in urban waterfronts that are divorced from ports and industry as ports continued to develop near the shore – at an increasing distance from city centres. New land-use programmes had to be invented to both alter and take advantage of the unique aspects of a waterside location.

This chapter will describe that history and its contemporary evolution. During recent decades, urban waterfronts have been used as vehicles for attracting new investment into post-industrial urban districts. First these redevelopment projects were targeted at attracting cultural and recreational investments. Then environmental issues were increasingly addressed, such as mitigating harbour pollution and biodiversity losses, and finally as part of a larger urban strategy for mitigating carbon emissions. The role of these investments in adaptation to economic trends is clear. But the role that waterfronts will play in adapting to the increased coastal and riverine flood risks that will come with climate change is less evident.

This chapter focuses on identifying the opportunities for flood-resilient waterfront development within the context of climate change. Contemporary urban waterfronts are discussed from the perspective of economic investments, social change and environmental change. The chapter begins with a brief discussion of the international competition that has been associated with coastal cities for centuries (see section 7.2). Section 7.3 then describes programmes that developed waterfronts as recreational and cultural areas; several key examples of waterfront development and environmental issues are subsequently presented to help clarify the role that these redevelopment projects have begun to play in coastal cities (section 7.4). Finally, section 7.5 addresses the challenge of climate adaptation in waterfront development and how this relates to economic incentives and trends. This is a subject that many cities have been studying, but relatively few have committed to physical actions. Section 7.5 compares urban waterfront adaptation strategies that use different combinations of several elements: a flexible storm-surge barrier, floating structures on calm water, innovative dikes that become districts in themselves, and designs that incorporate flooding within the dynamics of an ambitious, new mixed-use residential and cultural district. The specific configuration of these elements that is applied in a particular city both reflects and produces local conceptions of tolerable risk and goals for resilience.

7.2 International Competition as a Driver for Urban Waterfront Development

The continued increase in the size of container ships as well as military vessels over the 19th and 20th centuries has effectively concentrated ship docks in fewer and fewer cities, most of which have relocated their port facilities farther and farther from their urban core (Konvitz, 1994; Hesse, 2006). London's port is now at the mouth of the Thames, about 30km from its original location at The Pool near London Bridge. Rotterdam, Europe's largest container port, is now located about 35km from its original harbour. On the other side of the world, Los Angeles's port is now also 35km from the city's downtown. New Orleans, which is the US's largest commodity port, has added an offshore facility for oil tankers that allows 'ultra large crude container' ships to unload without even entering a river or harbour, 100km from the other port facilities. And the US military's east coast naval base – now the largest naval base in the world – is located far from a major city in Hampton Roads, Virginia (Washington, DC, is closest, but is more than 225km away).

I note these distances to emphasize that most urban waterfronts are no longer actively involved in either shipping or military use. The rescaling of economic and military ships has isolated these once-crucial locations and led to their decay and disuse. These areas were often transformed by highway construction on what had become inexpensive land, abandoned by heavy industries and warehousing companies. Then, a 40-year-long trend began during the 1970s to repurpose urban waterfronts as destinations for tourists, bringing museums and festivals to the water's edge. New residential use has become part of the mix of uses since the late 1980s (for a Dutch example of how residential uses were introduced, see Korthals-Altes and Tambach, 2008). Higher environmental standards have been applied to the waterfront zone since the 1980s as well, corresponding to the increase in the popularity of using these areas for new luxury housing. Macro-scale economic trends have had important impacts upon this increasing residential use of urban waterfronts (Cooper and McKenna, 2009). Most recently, some cities and nations are reframing waterfronts as flexible zones where adaptations to climate change may be tested and implemented (Ahlberg, 2009).

From a social point of view, waterfronts often represent lost jobs in unionized employment and a transition from utility values to attractiveness values that shifted the audience for waterfront design away from local workers and local or global businesses, and towards tourists who would understand the place through a very different base of knowledge (Bunce and Desfor, 2007; Laidley, 2007). As tourism infrastructure replaced the working waterfront of boats, luxury hotels replaced fishermen's piers and markets in cities such as Boston and Baltimore, San Francisco and Seattle. In other cities, new industries were successfully located at the old docks – as in London's Canary Wharf, a district where companies engaged in international financial transactions replaced physical goods arriving from international ports.

The geographical territory of cities and port authorities has also adjusted to allow cities and countries to compete as ports for international exchange. In some cities where harbours and rivers were too shallow for deep-draught ships, municipal boundaries were extended to allow the city to control a new deep-water port. Rotterdam grew from a roughly circular shape to be extended like an elongated arm, reaching to the North Sea, in order to include a new seaport within its boundaries. Los Angeles added to its shape with a long finger of land pointing south when it annexed its port lands in 1909. Hamburg has not been able to extend its boundary, so it remains an unusual case – bringing large container ships into the city itself up the Elbe River, requiring extraordinary dredging tactics and urban waterfront design measures to accommodate large tidal changes and storm surges. Even there, however, a new port is developing at Wilhelmshaven on the North Sea that may lead to the decline of Hamburg's urban port (Hesse, 2006).

These trends in competitive behaviour and shipping technology have affected different ports in different ways. The vast majority of former working waterfronts, however, no longer have successful commercial fishing fleets, participate in the shipping of goods, provide military berths, or serve passenger cruise lines. In most cities that still have active ports, the actual harbours where ships are served are now located far from their former urban waterfronts. Many industries that once co-located with ports, such as glass manufacturing, metal-working and lumber treatment, have closed in North American and Europe, moving to cheaper labour environments and leaving contaminated soils behind. Warehouse districts became obsolete with the development of containerized shipping, in which containers are simply moved to outdoor storage lots and from there are loaded onto trains, and roll on/roll off shipping, in which wheeled vehicles drive onto ocean-going ships and drive off at their destination (for context, see Upton, 2008; for an example, see Seattle Waterfront Metropolitan Improvement District, 2011). Fisheries industries often moved offshore into canning ships, or declined locally as fish and shellfish stocks were depleted by overharvesting or pollution of local waters. Coal transportation functions were also associated with ports, and as energy sources shifted to nuclear, gas and oil, coal transportation became centralized in a few major ports, rather than a common feature of all ports that served industrial and urban regions. In the developed world, airports have replaced shipping ports as the main hubs for passenger travel. With these global economic changes in labour and energy markets, and a legacy of contamination, overharvesting and changes in production in the fishing industry, waterfronts and ports in the early industrialized countries began to focus more on the potential for water-based tourism.

7.3 Waterfront Tourism and Cultural Centres

From the 1970s to the present, urban waterfronts that had lost their maritime economic functions were redeveloped as tourism destinations. This was not a new idea; the first naval museum originated on Vasilyevsky Island at St Petersburg, Russia, initiated as a room of ship models assembled by Peter the Great in 1709 (www.navalmuseum.ru). During the 1920s and 1930s, many other ship-building and seafaring museums were established around the world – some affiliated with university research programmes.

In Baltimore, Maryland (US), the idea of a waterfront promenade was established in 1960 as a formal goal of the city's planning efforts. That modest concept developed into a large and very successful commercial development that now stretches 11km along the old industrial harbour lands of Baltimore (www.waterfrontpartnership.org/history.aspx). The 'Baltimore model' was followed by many other cities in the US and around the world (Breen and Rigby, 1996). In short, the public sector invested in the public space improvements of a wide brick walkway and a park, but attracted private investments to develop individual sites along that new linear public space. The anchor commercial development is known as Harborplace, and was developed by the Rouse Company and opened in 1980. Harborplace includes retail shops, restaurants and a luxury hotel. The vision of its developer, who was a Baltimore native, was to attract Americans back to their cities to reverse the economic and physical decay that had occurred since the suburbanization of the 1940s. Baltimore is still considered a city with serious social problems and has not returned to its pre-1950 population levels. But its waterfront is still a successful regional attraction, serving local food specialities and bringing suburban capital into the city every day. It is even expanding, bringing more recreational trails to the former industrial waterfront lands farther out from the centre of the city, and engaging in wetland restoration as part of its recreation and biodiversity conservation agenda.

The success of the Baltimore waterfront redevelopment energized urban planners and developers who had been overwhelmed by urban disinvestment in the US and elsewhere. Over the next several decades, Baltimore became a model for other former industrial waterfronts. In the US, success often depended upon finding a single development company that would be willing to join a public–private partnership to reinvest in the private parcels because a single developer with significant capital resources could absorb more political risks and achieve more political leverage (for a well-studied example, see Gillette, 1999, on the Rouse Company's influence on American waterfronts; see also Gordon, 1999, and Zhu, 2001, for discussions and analyses of the extent to which these public subsidies were a response to special interests that shifted development away from the older urban core, rather than a way of increasing total land value). Cities such as Philadelphia either never succeeded in finding a single developer, or followed policies that actually discouraged the formation of one well-funded private development team (McGovern, 2008). Other countries used regional and national financial subsidies to initiate the redevelopment of large waterfront parcels using public money. Barcelona and Stockholm both offer good examples of this type of development in the contemporary practice of waterfront investments (for Barcelona, see Monclus, 2003; for Stockholm, see Rowe and Fudge, 2003, and www.hammarbysjostad.se).

Two important and related phrases have become common in the literature on waterfront redevelopment: 'culture-led' and 'design-led' development. Culture- and design-led development refers to the need for specific qualities of urban space, architecture, landscape architecture and cultural facilities such as museums to be integral to a planned urban redevelopment, rather than simply designating new land uses and allowing the market to supply standards of quality based on individual business and marketing plans (sometimes referred to as 'market-led' development; see Zhu, 2003, for a contemporary Chinese example of market-led urban redevelopment). Baltimore provided the earliest example of a design-led approach, using a high standard of commercial architecture to create an anchor for future investments that would benefit from a geographical association with mid- to high-end retail brands. The private developer who built Baltimore's financially successful Harborplace referred to these developments as festival markets (Hannigan, 1998), and the model became very financially successful in Boston and New York City as well during the 1980s.

In the Barcelona model of design-led waterfront development, the process was much less commercialized. It positioned designers as artists, and public space as a sculptural/ social work of art, which could attract both tourists and businesses by the perceived beauty of the public spaces themselves – not just of the private spaces (Monclus, 2003).

London's riverside development during the 1980s and 1990s followed the 'culture-led' approach, but by investing public money into the location of new cultural facilities along the waterfront. In that case, the South Bank of the Thames had been an informal entertainment district since the Middle Ages. The Festival of Britain during the 1950s incorporated more areas of what had become industrial land, and the recent wave of cultural investments – including museum expansions such as the Tate Modern, located in a repurposed turbine factory building – has continued to treat that bank of the Thames as a place for elite culture as well as circus-type attractions, such as the London Eye (Brown and Reed, 2003).

In a different type of culture-led approach, the city of Providence, Rhode Island (US) commissioned a sculptor to build lines of wood-fuelled bonfires placed on the surface of its urban canals, supported by piles. The lighting of the bonfires corresponds to coordinated festival events in a semi-annual programme known as WaterFire, which attracts performance artists of all types and is linked to concerts and gallery exhibitions as well (www.waterfire.org). This example represents a blend of design-led and culture-led models since it involves the design of a strong visual element of public space by an artist, linked to a periodic arts festival. The intermittent use of a dynamic element, such as a fire sculpture, to generate a spectacle along the waterfront is a more recent approach to waterfront tourism and cultural development that fits well with contemporary planning and design interest in incorporating dynamic systems and processes within designed space (see section 7.5, which addresses climate change adaptations on the waterfront).

7.4 Environmental Changes and Residential Uses

In many urban waterfronts, water quality improvements have coincided with or preceded new investments in mixed-use developments that include housing. Boston, Massachusetts (US), provides a case in point.

7.4.1 The Boston Waterfront, 1972–2006

Boston invested in water quality improvements around the same time that it began to engage in mixed-use waterfront redevelopment. In 1973, a group of citizen advocates formed the Boston Harbor Association, initially made up of members of the League of Women Voters and the Boston Shipping Association (www.tbha.org). This group has advocated for a public walkway along the entire Boston waterfront, with a total planned length of 46.9 linear miles. Around the same time, Baltimore developer James Rouse invested in the redevelopment of an aging group of 18th-century market buildings along the Boston waterfront near Long Wharf, known as Faneuil Hall and Quincy Market (Gillette, 1999). Like Harborplace in Baltimore, this project was extremely successful as an urban entertainment, restaurant and shopping district. It was followed by other private investments, encouraged by city government, including the Charlestown Navy Yard (primarily high-end residential condominiums) and Rowe's Wharf – a luxury hotel, marina, restaurant and water taxi terminal for airport travellers.

The investments in Boston Harbor's water quality, however, were not made by choice (Savage, 1995). Boston Harbor is the largest seaport in New England, and in 1988 it was named 'America's dirtiest harbor' (Penna and Wright, 2009) – primarily because of a long history of inadequate sewage treatment as well as the dumping of industrial wastes (including polychlorinated biphenyls, or PCBs) into the harbour. In 1972, the national government passed a new law requiring secondary sewage treatment. Boston only had modified primary treatment, but instead of immediately proceeding with an upgrade of its sewage plants, it tried to buy time with a study of the sewer system and by applying for a waiver from the new standards. These delays were not resolved until 1982, when the national Environmental Protection Agency (EPA), a neighbouring city and an activist group brought a lawsuit against the city of Boston. The suit extended into a second court case in 1985 and was resolved that year with a court-monitored clean-up process, meaning that the timetable would be set and overseen by a judge. At that point, Boston was the only developed city in the world to remove sludge from its sewage effluent and then dump it into the harbour, along with the effluent, the very next day (Savage, 1995).

Since 1985, the city and a new regional water management agency have cooperated to address the long history of pollution, placing the financial burden on the customers of the water system in that region (MWRA, 2010). In addition to extensive sewage treatment improvements, the clean-up also entailed creating several experimental cells for isolation of contaminated sediments in the harbour by placing a layer of clean dredged material on top of the contaminated sediments – a technology now used in other US and international harbours. Recent assessments have shown that since improvements finally went online during the mid 1990s, the waters of the harbour are improving in quality. Sea life has begun to return, nutrient levels have declined, and the caps on contaminated sediment areas appear to be functioning as designed (Maciolek et al, 2010).

At the same time, a major transportation infrastructure project was undertaken along the waterfront that placed a high-speed expressway into a 3.5 mile (5.6km) underground tunnel (the Central Artery Project, known in the local vernacular as the Big Dig). This effort to reconnect the city's downtown to its waterfront was conceived during the 1970s and finally broke ground with federal and state funding in 1991. It was finished in 2006 at a cost of US$14.6 billion. Together with investments in the sewer system, some local harbour advocates estimate that approximately US$20 billion were spent to provide access and water-quality improvements to Boston's waterfront. The completion of the Big Dig project has led to many more private development projects along the city's waterfront; at the same time, water quality has improved and everything from benthic organisms to eelgrass beds and shellfish are beginning to return to the shoreline. Private investment in the waterfront has totalled an estimated US$2.2 billion between 1987 and 2004, including 1000 hotel rooms, 2700 new residential units, 93,000 square metres of industrial space, 26,000 square metres of retail space, and 180,000 square metres of entertainment, cultural and institutional space (SHSB, 2004). Several different plans for continued waterfront development will soon be compared by local decision-makers and private developers, now that the downtown is once again connected directly to the waterfront zone without the barrier of a partially elevated highway.

7.4.2 Stockholm's urban waterfront

Stockholm offers several significant cases of residential and mixed-use waterfront development that were initiated during and after a series of new investments in improving coastal water quality. With a city built on 14 islands at the head of an archipelago and the mouth of a lake, 'waterfront' is a term that applies to many urban projects. The contemporary example focused on here is the Hammarby-Sjöstad district of Stockholm, a mixed-use development initiated by the city's investments in cleaning up an industrial brownfield site and building 11,000 new housing units and a specially integrated infrastructure system, with a target of 25,000 inhabitants and 10,000 jobs by 2015 (Fränne, 2007).

Stockholm is about 30 per cent water by area, 30 per cent parks, and the rest is close to a waterfront. It was recently ranked as the city with the least pollution in Europe (Cushman and Wakefield, 2010). With almost 80 per cent of its workers in the service sector, Stockholm no longer has much industry to pollute the air (US Bureau of Labor Statistics, 2010). Since the 1970s, upgrades to sewage treatment and the installation of treatment facilities for storm water running off the surfaces of major arterial roads have made inner-city waterways safe for swimming and fishing. Stockholm Water was formed as a public company in 1989, with the city as majority owner. Its stated goal has been to contribute to the sustainable development of Stockholm (Stockholm Vatten AB, 2010). In urban planning circles, the city is known for its water-quality progress (Johansson and Wallstrom, 2001). Once quite polluted, people are now encouraged to fish and swim in central-city areas. Stockholm has used this transformation as the basis for hosting a World Water Week event and offering the Stockholm Water Prize, the 'Nobel prize' of water-related design and engineering. Water Week is part of a larger Swedish effort to lead clean water technology services globally, through the Swedish International Development Cooperation Agency (Sida) as well as industry outreach (Stockholm International Water Institute, 2010).

In addition to the priority accorded to water in decision-making and in the aspirations of the city, one of the extreme differences between Stockholm and Boston is in their processes for land-use planning. Boston, like most American cities, relies on very general, non-binding public plans that are actually driven in large part by the initiative of private investors. As Stockholm's *2009 City Plan* states: 'Swedish municipalities have extensive authority over local land use, sometimes referred to as a "planning monopoly." Land use and building is regulated through legally binding municipal detailed development plans. Building permits are issued in compliance with detailed plans for new building, renovations and additions' (City of Stockholm, 2009, p10). Stockholm has used its land-use authority to not only determine what type of land use will be assigned to a location, but to coordinate infrastructure planning and integration to a very high degree.

This infrastructure coordination has been used in the Hammarby-Sjöstad district redevelopment project to innovate by altering not just urban space, but also urban metabolism. In this conceptualization of urban systems, wastes from one process are seen as fuel for another – with the ultimate goal being a kind of hyper-efficiency in which residents and businesses in a district would reuse almost all wastes and prevent most material and energy losses from the district 'system' so that it can sustain itself on very limited inputs over time. The specific target was to reduce the amount of carbon dioxide (CO_2) emissions and waste produced from the district by 50 per cent compared to other Stockholm districts built during the early 1990s. The integration of energy systems, solid waste systems, transportation systems and sewage waste systems is impressive – for example, waste heat energy from sewage treatment is used to heat water pipes for a district heating system. Biogas is produced from wastewater sludge, and some of the gas is used to power gas stoves in 1000 apartments, while the rest of it is used to power buses in the district. This integrated infrastructure system has come to be known as the 'Hammarby model' and is the primary focus of most international attention on this urban case (Sandelin, 2008).

From a waterfront perspective specifically, the Hammarby development took some key steps that were not part of the concept for other developments that were conceived either before or during the same decade as Hammarby – and, in fact, are still rare components of the newest waterfront districts. First, the designers used public parks to manage some of the storm-water runoff from the development – cleaning it and adding oxygen to it before it enters the waters of the Baltic Sea. A long canal is used to slow the water down to filter it of most metal and phosphorus pollutants, while a sculptural cascade acts as a channel to return the filtered water back to the sea with a higher concentration of dissolved oxygen from the aeration received while the water flows over rough steps. Roadway runoff is also filtered locally, using underground containers, before it is discharged to the sea (Fränne, 2007). The use of local landscapes to both perform these functions and present them to the public was unprecedented in a waterfront project before Hammarby, although the techniques were used in other non-waterfront parks and urban landscapes. The value of doing it at the waterfront derives from the immediacy of the lessons it communicates about how cities can either pollute water or clean it as it passes through.

Hammarby is also special in its attention to biodiversity potential on the waterfront. Two separate vegetated bridges, known as 'ecoducts', cross an automobile expressway next to the district. These provide crucial links for wildlife to move from Stockholm's regional park system to the conserved and restored wetlands at the water's edge at Hammarby. Standing dead trees were retained in the wooded parcels around the district to provide food for birds that eat insects, which tends to be rare in cities where wood-lands are carefully managed for human enjoyment and safety. And at the water's edge, not only are intertidal wetlands an important part of the final design, but they are also placed strategically so that they can provide linear corridors for birds, fish and other animals. The edges of the urban wharves either have wetlands in a continuous line or boat moorings, but not a mix of the two along one edge (Ingebritt Liljeqvist, Landskap AB, pers comm, 2007). This is smart ecological design, and again is unprecedented in areas that are urban enough to include numerous boat moorings.

The model developed at Hammarby, however, is probably more accurately described as mitigation than as adaptation to climate change. According to its own technical evaluations, the development has achieved some impressive performance goals, with 80 per cent of all commutes taken on foot, by bike or by public transport. The model also succeeded in reducing the use of new material in construction by 50 per cent, and runs about 20 buses on local biogas production from wastes. The development uses 50 per cent less water per housing unit and generates 25 to 30 per cent less CO_2 than other similar developments from the early 1990s (Brick, 2008). If there are problems, some note (Khakee, 2007) that they result from the planners' projections of residents' age as primarily an older generation that would be selling single-family homes to move to condominiums and apartments at the waterfront. In fact, the development attracted a group that is predominantly 25 to 45 years old (Future Communities, 2010). The rents are relatively high and the economic resources of the residents are also substantial. Sweden ceased to strictly regulate the housing market during the 1990s, and there have been several changes of government with different philosophies about the use of housing subsidies since that time. Anecdotally, some commentators have noted a lack of local retail in the district that requires more trips to commercial areas; but this seems to be in transition as the population increases to the ultimate target levels. At least one case study reviewer notes that all of the units designated for retail had been rented by 2010 (Future Communities, 2010).

The city planners have noted that residents' behaviour matters more in the achievement of waste reduction goals than they anticipated, and they are trying to incorporate that lesson within their plans for the next big Stockholm waterfront project at the Stockholm Seaport site. The interesting dilemma that they face is that Sweden's conceptualization of the problem is largely technological – they plan to market their expertise in infrastructure and energy adaptations internationally, so technology changes are emphasized as a key strategic component of the urban waterfront projects. But if residents' and workers' behaviour prevents them from achieving their performance goals, then the new Swedish model will have to expand public education, incentives and participation beyond its current scope (Pandis and Brandt, 2009). This represents, on the one hand, a simple extension of existing public education programmes and, on the other, a radical rethinking of the top-down technological approach that was used at Hammarby-Sjöstad and now also at Stockholm Seaport.

Unlike Hammarby-Sjöstad, the Stockholm Seaport project also lists climate adaptation goals as part of its programme, mentioning several measures to reduce the local urban heat island (e.g. green roofs, heat-retaining building walls, increased vegetation) and adapt to more extreme rainfall and flooding events (e.g. more space to detain floodwaters, waterproof basements) (Ahlberg, 2009). The rate of glacial rebound, which lifts the elevation of the land, is highest in northern Scandinavia and reduces the vulnerability of some Baltic coastal areas to an accelerated sea-level rise. The Swedish Meteorological Institute identifies the rate of sea-level rise as increasing in southern Sweden, currently averaging 3mm per year (SMHI, 2010); but Sweden's national study of climate change vulnerability and adaptation notes that very little change in the rate of sea-level rise or the frequency of storm surge events is expected in the Stockholm area specifically (Ministry of the Environment, 2007). The planners for Stockholm Seaport have so far assumed that this dynamic will not be structurally significant and that current sea-wall heights will be adequate. Stockholm's planners have instead focused on the threats of increased temperatures at night, increased flooding from more intense rainfall events, and flooding from increased flows out of Lake Mälaren to the west of the city.

7.5 Waterfront Adaptations to Extreme Flood Events

In cities that have identified specific strategies to adapt to climate changes and increasing flood risk, such as Hamburg, Rotterdam and London, innovations have begun to be implemented. Both London and Rotterdam invested relatively early in moveable storm surge barriers after flooding events made the threat immediate. These two cities are working with assumptions that suggest their current barriers will be adequate for the 21st century, and are now looking at new measures to address flooding driven by upriver and regional extreme rainfall events. Hamburg is a very different case of adaptation since it has decided against building a barrier in favour of designing a new high-end district of the city to be able to accept flooding.

7.5.1 Hamburg: A residential district designed to flood

The HafenCity project in Hamburg takes advantage of the unique political and geographical setting of this city of 4.3 million regional inhabitants to establish a very different model of how to live on a waterfront. Hamburg is Germany's second largest city, and the second largest shipping port in Europe, with Germany's highest gross domestic product (GDP). Its full name, the Free and Hanseatic City of Hamburg, reflects its history as part of a significant port city network that has existed since the medieval Hanseatic League, as well as its contemporary status as an official city-state in the German federal system. The mayor has responsibilities and prerogatives more like a state minister than the usual municipal leader. It was Mayor Henning Voscherau who developed the first concept of the HafenCity in the early 1990s and began to quietly acquire key parcels for public control (HafenCity Hamburg GmbH, 2010a).

The city eventually acquired most of the buildable land in an area extending from an old warehouse district along the bluff of the city's edge to the edge of the Elbe River. Sales of this land funded much of the approximately 2.2 billion Euros of public investment in the project. About 6.6 billion Euros in private funds have been invested since then. The development includes 157ha of land and water, 5800 homes and more than

45,000 jobs, and 10.5km of new waterfront comprising promenades, squares and parks. The development company responsible for managing the project is the HafenCity Hamburg GmbH, which is owned entirely by the city of Hamburg (Jaeger, 2008; HafenCity Hamburg GmbH, 2010b).

The HafenCity development's approach to flooding was apparently conceived early in its master planning, when a Dutch–German design team led by Kees Christianse and ASTOC won the international competition for the master plan and proposed raising buildings up to accommodate storm surge flooding that occurs once or twice each year (HafenCity Hamburg GmbH, 2010a). All new buildings are built on top of mounds of compacted till to establish base elevations at 8m above sea level, which is above annual flood levels. Ample underground parking is provided in this strategy, as well as continuous views of the water – unobstructed by the protective dikes that might otherwise have been needed. In addition, the first floor of the buildings is designed to be wave and debris resistant, and near the historic warehouse district, a few buildings use waterproof steel garage doors on the upland side and masonry batter walls on the water side. No residential units are allowed on the first floors, only parking and service businesses. An emergency walkway and bridge system is located at the second-storey level, above 8m, which is available to replace the ground-level circulation for pedestrians in flood events. Private cars are meant to be stored in the waterproof garages as needed. Even the new parks are designed to handle wave battering, with minimal woody vegetation and lots of hard surfaces. A floating sunbathing space on a wooden deck provides an alternative surface for urban recreation, and rides out floods by being attached to heavy pilings like boat moorings in the working harbour of Hamburg (HafenCity Hamburg GmbH, 2010c).

Unlike most contemporary high-volume ports, the container ships that call on Hamburg's Port Authority come directly into the city's historic harbour. Extensive dredging makes it possible for very deep-draught ships to make the 120km journey from the North Sea (Hamburg Port Authority, 2009). A continuous line of dikes meant to protect farmland all along that riverside from flooding intensifies both the daily tidal maxima and the less common storm surges. Future trends in sea-level rise, storm surge patterns and reductions in the level of mean low tide that have been observed over the last 50 years may necessitate major changes to the dike systems along the Elbe. The Port Authority is already initiating efforts to move the dikes back to accommodate tidal inflow from the North Sea, and potentially reduce tidal energy that today brings enormous amounts of sediment into the Elbe that is not removed by the lower energy of tidal outflow and must be dredged out. Risks from unexploded ordinance and contaminated sediments from local sources, as well as the headwaters of the Elbe in the Czech Republic, and everywhere in between, must be managed as dredging occurs (Hamburg Port Authority, 2009). New biodiversity goals set by the European Union's Natura 2000 policy directive apply to the marine environment, and will also need to be addressed as the Port Authority manages the river channel (European Commission, 2010).

Hamburg's HafenCity itself is designed for today's flooding. It is likely that it will also be ready for higher water as long as relative sea level does not rise more than 1m or so, based on the elevations used for emergency walkways in the district. It is interesting, however, that there is currently no information about climate adaptation on the official website of the HafenCity Hamburg GmbH (HafenCity Hamburg GmbH, 2010e). There is also an active project by authorities in the state of Lower Saxony to develop a major new deep-sea container port at Wilhelmshaven, known as the JadeWeserPort (www.jadeweserport.de), which may turn into a direct threat to Hamburg's role in container-ized cargo shipping, and could alter the economics and thematic focus of waterfront development in the city. Eventually, a shift to a deep-water port on the North Sea for the region could lead Hamburg to abandon its unusual model of absorbing floodwaters instead of building a barrier, after the model of Rotterdam and London.

In order to address its other sustainability goals, the HafenCity includes a district heating system with bio-methane fuel cells, key to plans for low CO_2 emissions, with targets of 175g per kilowatt hour (kWh) for the western section of the development and 89g/kWh for the eastern section. At least 30 per cent of the new buildings are required to meet a 'gold standard' within a Hamburg-specific rating system (HafenCity Hamburg GmbH, 2010d). But the social side of sustainability remains in question (Schaer, 2010). The development is self-described on its website as 'upmarket' and contains a large number of luxury housing units as well as some co-operative housing groups with somewhat lower rents. A recent study by Bunce (2009) documents the extent to which the use of building rating systems and energy efficiency statistics have made the public more accepting of gentrification trends in Toronto's waterfront redevelopment, in which new construction is not sufficiently subsidized to provide equivalent rents for previous residents of redeveloped districts. As Dale and Newman have pointed out (2009), energy-efficient housing may be more about commodification and consumerism than about sustainability and citizens since these new efficient buildings typically are the exception rather than the rule in urban areas, and have little impact upon overall energy efficiency in spite of their designation as model districts.

7.5.2 Other international examples

In Rotterdam, the harbour areas mid-way between the old city centre and the new deep-water port are available as large redevelopment sites. The city has chosen to develop the actual water areas themselves with new housing and mixed-use buildings that would float on stable multistructure platforms provided with infrastructure services. This approach is possible because the Maeslant Barrier (Maeslantkering) and a large number of upstream dams prevent turbulent floods. Water rises in the river, but without waves and debris – allowing structures to float in a stable way in the old harbour areas, outside the city's protective dike system. The Rotterdam case is in some ways a hybrid of the Hamburg and London strategies in that it uses both a barrier and new construction outside the historical protective envelope of the city's dike system. But instead of being designed to flood, Rotterdam's plan is for new districts that float.

The Stadshaven district of Rotterdam is a former shipping harbour. The city plans to build up to 1200 new housing units that would actually float on the water in this district, along with retail and restaurants to serve the population on and around the harbour area. A floating pavilion has been constructed by Dura Vermeer in the harbour as a test of technologies, with the capacity to hold 500 visitors (Rotterdam Climate Initiative, 2010). If this technology or a subsequent version of it succeeds, then many similar areas with very calm water may be accessible as new urban districts, and Dutch construction companies will be well positioned to build them in cities all over the world. The extent to which such districts will require storm surge barriers or permanent dike structures to keep the waters calm is not clear; but it is likely to be an application best suited to coastal zones with either very good protections or very few storms.

The Rotterdam floating pavilion also contains systems that treat the wastewater from the three hemispherical structures, and draws its energy from solar panels and heat exchangers that use the harbour water as a reservoir. These systems make it relatively self-sufficient, and link to trends in the urban infrastructure industry to explore decentralized waste and energy systems that operate at a district or even a parcel scale. Trends towards decentralization, and the technologies they produce, could have an enormous impact upon the level of experimentation that is possible at the district scale – particularly in coastal cities, which may need to consider compartmentalization as part of their adaptation strategy as sea levels, rainfall intensities and groundwater levels all rise over the next decades.

London's approach has been to focus on the tributaries of the Thames, particularly in east and south-east London. Their hydrologic modelling work suggests that extreme rainfall events will increasingly cause flooding along the tributaries, in districts that have historically been some of the poorest parts of the city and region (Entec UK Limited, 2005). Plans are proposed to create storm-water parks along these tributaries, including at the Olympic site being prepared for 2012 (Mayor of London, 2006; Design for London, 2011). Recent residential development has occurred along old industrial canals in northern and eastern London, and very large housing developments are being proposed along the River Thames, known collectively as the Thames Gateway Project (the delivery agency is the London Thames Gateway Development Corporation). More than 1 million housing units are planned north and south of the River Thames east of London by 2016. Private investments in these residential developments will be required to provide financial support for new regional park networks that are intended to detain floodwaters along the tributaries, as well as provide a regional recreational and ecological network (ODPM–Defra, 2004). Significant public investments in open space in that region have already occurred (e.g. see Greening the Gateway: Kent and Medway, 2011).

New fluctuating water conditions will be created in these parks, some of which are likely to have urban edges, while others will be surrounded by parkland. Fluctuating water conditions are difficult to incorporate within traditional canal or lake designs since they create large mudflats unless the designers find ways of either absorbing changes in water elevation on vertical surfaces such as walls, or include wetlands that can be exposed at low water levels. All of these efforts will be contentious to implement since many of the areas needed for water storage are already occupied by existing sports parks or other public landscapes that have local constituencies of users. While building a 'green infrastructure' of parks with functional purposes beyond recreation is an old idea, doing it to adapt to climate change is not a common goal. That added layer of functionality, planned for the next century of climate dynamics, is what makes the Thames waterfront and river valley developments interesting as an adaptation.

In the US, cities have been much slower to consider and implement adaptation concepts. Most of the US east coast and Gulf coast is subsiding, and in the Louisiana Delta the rates of subsidence alone are predicted to lower the ground by as much as 1m over the next 100 years. Most of the west coast is rising in elevation, thanks to geological uplift, but is affected by Pacific storms generated by El Niño events that come with storm surges of 30cm or more. In spite of – or perhaps because of – its relatively small storms, the San Francisco Bay area is best known for its efforts to identify strategies to adapt to climate change, specifically. In a recent design competition, the Bay Area Conservation and Development Commission and a group of other agency partners received proposals to combine new wetlands with green corridors and waterfront public zones as part of stepping development back from the water's edge (Rising Tides Competition, 2009). Questions about whether wetlands can accumulate sediment and rise in elevation fast enough to keep pace with an accelerated sea-level rise make it possible that these measures cannot simply rely on natural processes and will need to be supplemented by human actions in some significant way.

The US city that is spending the most to adapt is New Orleans on the Gulf coast. New Orleans is continuing to adapt to address vulnerabilities created by three major factors: its growth into very low-lying areas, subsidence accelerated by a lowered water table, and its extreme climate context in a hurricane region. The city expanded its residential territory dramatically in the 20th century without building a sufficiently high level of flood protection from hurricanes that can bring fast-moving 9m or 10m storm surges and sustained winds of 400 knots. These challenges are larger in magnitude than the predicted acceleration in sea-level rise, although clearly that will make living with hurricanes even more difficult over the coming century. A new surge barrier is being constructed that was designed by Dutch companies as part of new measures to achieve a predicted 1/100-year standard of protection (for a review of New Orleans' recovery efforts since 2005, see Glavovic, 2008). In order to prevent loss of life in extraordinary storms, the levee system is combined with mandatory evacuation and – increasingly – elevated houses (Hill, 2009).

Unfortunately, the treatment of waterfronts in New Orleans is primarily 'engineering led' because of the high risks of flooding in most of the expanded city. Concrete flood walls prevent residents from even seeing the water along most canals. The lakeshore is a potential recreational resource, but it suffers from damaging wave action in major storms and is not developed as a significant public space. Pollution from the city's storm water has created an anoxic 'dead zone' off of the city's lakeshore at the best location for public use, and may create hazardous conditions for swimming as well. While this city can be considered the leading edge of adaptation in the US, there are many questions about whether it is really ready for climate change. Urban planning goals have been focused on bringing the protection system up to its stated standard, and have not projected into the future to consider changes in hurricane intensity or accelerated sea-level rise, in spite of the fact that non-urban port facilities along the Gulf of Mexico are known for their attention to climate adaptation. A new effort is under way to create a water plan for the city that would renew the public value of all of its waterfronts, including an international competition to bring the best design ideas from other cities.

In Asian countries, waterfronts are being redeveloped with a capital-intensive but apparently long-lasting strategy of building a very wide dike, so wide that it is essentially a ridge – and so wide that it appears to be impossible for it to fail as a flood control embankment, although it could theoretically be overtopped by an extreme flood event. During the 1990s, Japan used infrastructure construction as part of an internal economic stimulus plan. After a dike failed in 1986 in Ibaraki Prefecture, a super-dike was built along the Yodo River in Osaka and the Edo and Arikawa rivers in Tokyo, among others, and now there are plans to extend this strategy to many of Japan's urban waterfronts.

The great advantage of a super-dike, in which the width is established as 30 times the height of the dike is that it is so structurally stable that new housing, parks and roadways can be built on top. The Japanese moved landownership rights up vertically with the land surface, and gained income for the project by selling the former embankment zone to new private landowners (Arakawa-Joryu River Management Office, 2011). The big advantage for residents – in addition to flood prevention – is that they would have a water view after construction, instead of the embankment view that they had before. This is a fundamental change in the structure of waterfronts for cities below sea level in which most buildings near the water do not have water views at ground level. More than 872km of super-dikes are planned now for Japanese cities. The major drawback seems to be that it can take four years for the construction of the super-dike to be ready for buildings to be built on top, which means that residents have to relocate during that period. There are also on-going discussions about whether the super-dikes may be unnecessarily large and expensive, and should not be extended as planned (*Japan Times Blog*, 2010).

In Taiwan, about 900m of Taipei's central Danshui riverfront has been proposed for alteration with a park, replacing an old 8m high concrete flood wall with a new zigzagging concrete wall that is covered with large enough berms of earth that the entire structure approximates the width of a Japanese super-dike, although it only alters a segment of the flood wall at one location (Gerfen, 2009).

In general, riverfronts seem to have been places of more frequent experimentation with new approaches than marine waterfronts. An excellent review by Spits et al (2010) compared development projects on rivers in three north-west European countries to identify links between waterfront development approaches and flood management policies. They note that during the 1990s, the Dutch national policy of making 'room for the rivers' was adopted. Initially, the concept was to prevent building in the floodplains. Within ten years, these authors note, the policy was altered to allow experiments in flood-adapted buildings to be implemented – as long as the resulting construction actually increased the amount of area available for river water at the expense of the developers.

7.6 Conclusions

The three waterfront cases of Hamburg, Boston and Stockholm are illustrations of how waterfronts can change while addressing sustainability issues. These projects have thus far led primarily to a change from lower-income residents and industries to higher-income residents. Existing buildings were demolished to build new ones, and in the two European cases the new buildings and infrastructure incorporated ambitious technological goals for reduced CO_2 intensity and use of new building materials. The Swedish case added reduced water use and waste generation. In the Stockholm case, human behaviour is known to have limited the success of the technological approach, thanks to careful monitoring of actual performance. No such performance study exists yet in the scientific literature for the Hamburg case, and the Boston case includes monitoring of environmental impact reductions upon the harbour itself, but includes no articulated goals for building or social performance.

Evans et al (2009) pointed out that these large urban waterfront redevelopment projects typically emphasize speed and efficiency of governance rather than a nested complex system of agency checks and balances that might allow multiple scales of investment and redevelopment to occur in the same complex urban space. Rather than dramatically simplifying governance with a public corporation or single government entity, the developers' equivalent of microfinance might be used to reinvest in existing structures' energy performance and durability, providing housing types that are more likely to support a mix of residents with different income levels in a regenerated urban district.

On the ecological side, very few waterfront developments have taken the approach of trying to establish new positive impacts upon specific species of cultural and commercial value. Aquatic ecologists recognize that the loss of intertidal wetlands and sub-tidal seagrass meadows produces critical stresses in estuary systems that should be acting as nurseries for fish and shellfish populations. Yet, very few examples exist of cities that have tried to bring intertidal or sub-tidal habitat into their waterfront designs, in part because the promenade emphasizes a more manicured aesthetic that includes water and the reflections of buildings and boats, but not grassy mudflats that might smell of sulphur at low tide. Stockholm's Hammarby project included some well-placed linear wetlands; but while they are critical to ecological integrity, they are not extensive. Hamburg includes no local habitat areas in its HafenCity parks or promenades along the water's edge. Boston measures impacts upon seagrass in the harbour, but does not actively make space for either seagrass beds or saltmarshes along the shoreline.

From an aesthetic point of view, many designers now seek to introduce dynamic processes into the way in which they propose to structure space. Tidal fluctuations are typically hidden on the surface of vertical sea walls, even in developments that are considered models of sustainability. And yet, as a result, many waterfront visitors and residents who are not sailors do not understand or appreciate the dramatic changes of inundation that can be associated with daily tidal changes, or the seasonality of tidal maxima and minima. They are also unable to appreciate the diverse species supported by intertidal and sub-tidal vegetation, from crabs to birds and juvenile fish. These types of shoreline environments may be better suited to kayak recreation than to motorboats; but that, too, is likely to be a trend. Adults and especially children have so much more to learn from walking through a saltmarsh and finding a crab shell than from walking the paved surface of a promenade, 2m or more above the surface of the water. Now that the water quality of these urban shorelines is substantially improved, our next frontier is to understand the changes in the biological environment that will come with climate change. How will we situate ourselves among other forms of life if we can't even see them living near us?

The cities that are currently doing the most to implement adaptations to future flooding are cities that have already experienced severe flooding over the last 50 years or so. Dutch cities and the German city of Hamburg have taken the most aggressive approach to actually combining buildings and other elements of urban structure with controlled water flows, typically on rivers rather than on marine waterfronts. Cities in coastal areas with high risks of hurricanes are focused primarily on providing flood defences for those potentially extreme events to protect new low-lying districts, and less on climate adaptation.

Many more cities, states and regions are studying scientific models of flood dynamics, calling for initial conceptual proposals, and beginning to make plans, as in the San Francisco Bay case, or New York City or Shanghai, but have not yet made commitments (see Sanchez-Arcilla et al, 2008; Cruce, 2009; Gong and Yu, 2009; Hamid, 2009). Most European coastal cities are also monitoring sea-level rise but have not yet implemented adaptive responses (Tol et al, 2008), with the exception of the cities mentioned here. Many cities and regions are only beginning to discover the transboundary management issues that will be raised by climate change in coastal areas (Michel and Pandya, 2009). It will be very important as these cities and states begin to make adaptive commitments to observe whether they are choosing to pursue a rigid policy-led adaptive response, or one that encourages experimentation and innovation in urban waterfront design, as in the Dutch examples, in particular. There is clearly much more to learn and no substitute for experience.

Plate I *Number of coastal and offshore ports per country*

Source: Lloyds List (2009) *Ports of the World 2009*, Informa and Maritime, London, UK

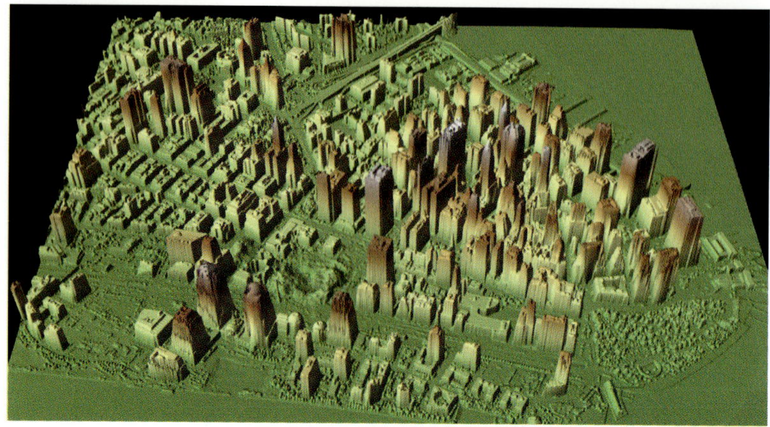

Plate 2 *LIDAR remote-sensing image of the Manhattan Business District, 27 September 2001*

Note: The imager is flying over the Hudson River, looking to the east. The image emphasizes the vulnerability of lower Manhattan to inundation.
Source: NOAA/US Army JPSD

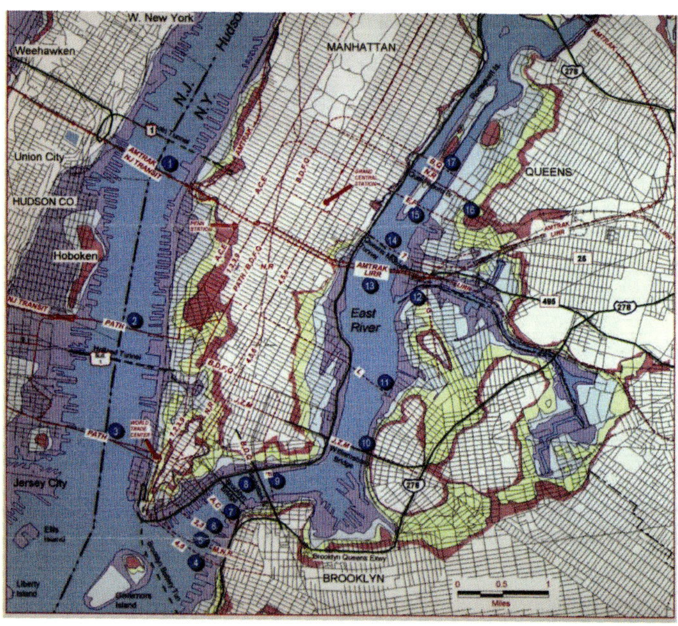

Plate 3 *Predicted worst-case scenario inundation areas in lower Manhattan and environs resulting from various categories of hurricane*

Key:

Colour code	Winds	Maximum speed (mph)	Surge at The Battery (feet)
Purple	Flood area of a category 1 hurricane	74–95	10.5
Light blue	Additional flood area of a category 2 hurricane	96–110	16.6
Yellow	Additional flood area of a category 3 hurricane	111–130	23.9
Red	Additional flood area of a category 4 hurricane	131–155	28.7

Source: US Army Corps of Engineers (1995) *Metro New York Hurricane Transportation Study, Interim Technical Data Report 74*, New York, NY

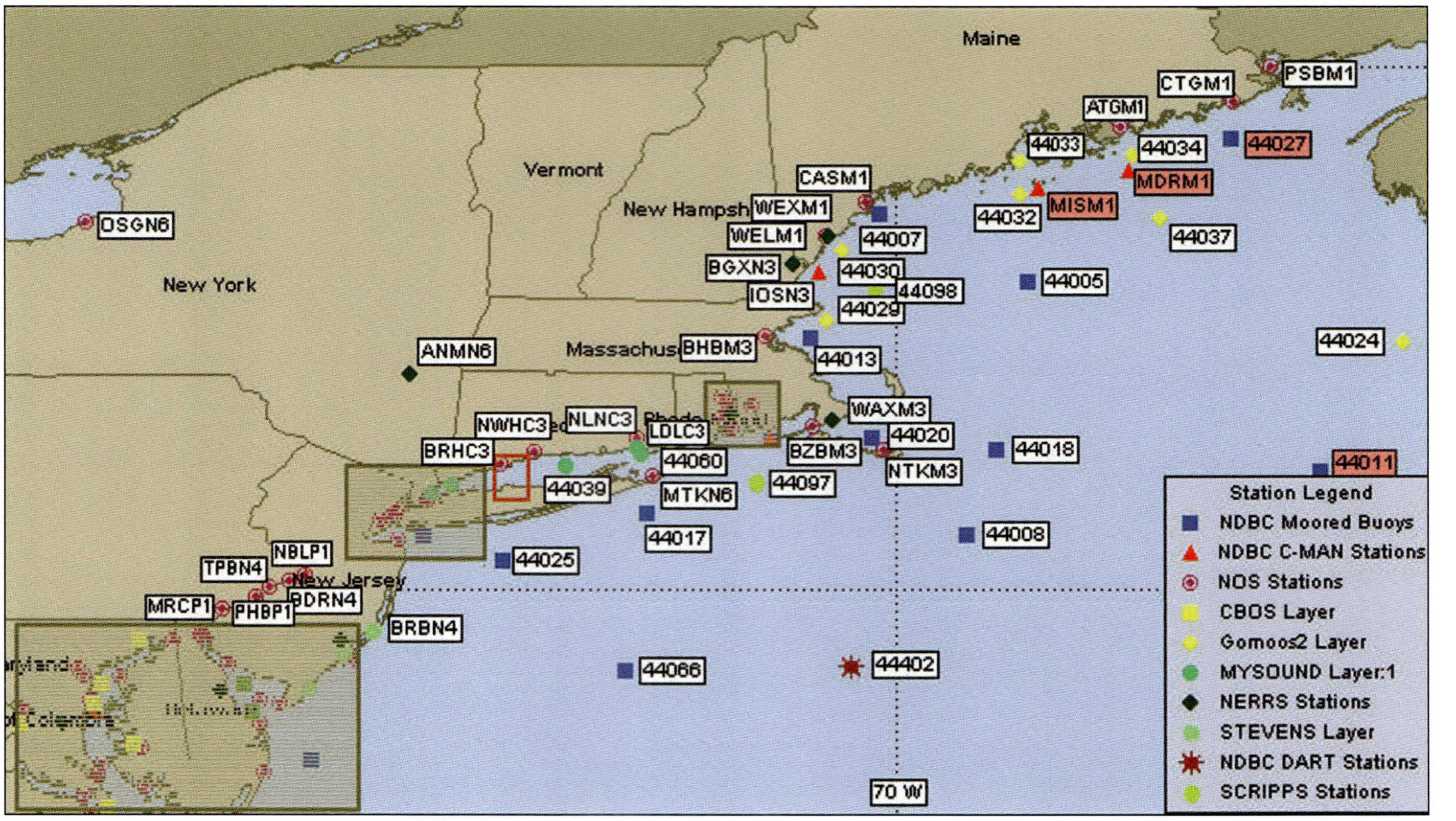

Plate 4 *Locator map of the National Data Buoy Center (NDBC) array of offshore weather buoys on the north-eastern seaboard of the US*

Source: www.ndbc.noaa.gov

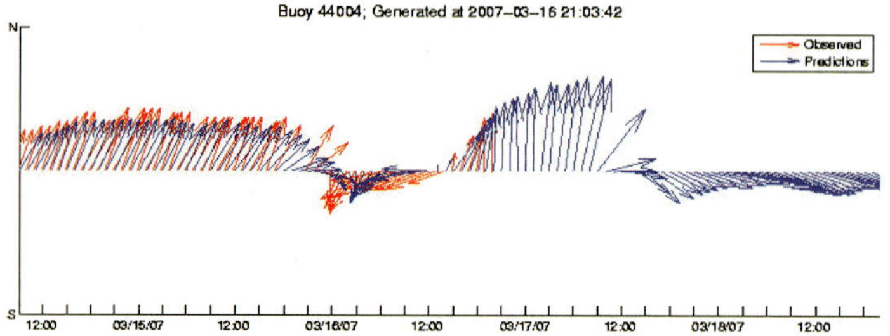

Plate 5 *Observed (red arrows) and predicted (blue arrows) wind speeds and direction for buoy 44004 during the 16 March 2007 storm*

Note: Location was 200 nautical miles east of Cape May, New Jersey.
Source: http://www.ndbc.noaa.gov/station_page.php?station=44004

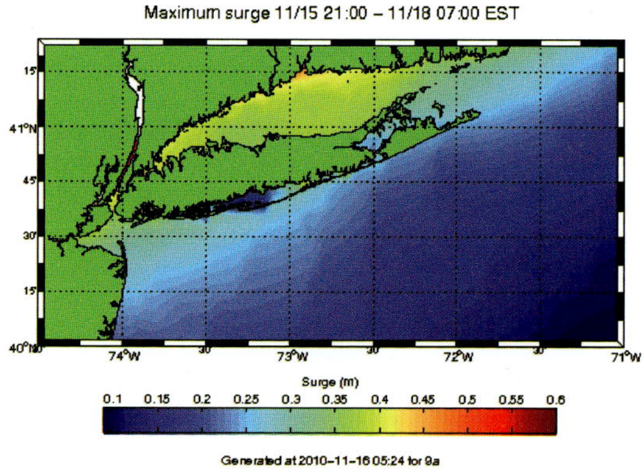

Plate 6 *An example of the maximum predicted surge for the Stony Brook Storm Surge Model (SBSS) run for the period of 15–18 November 2010*

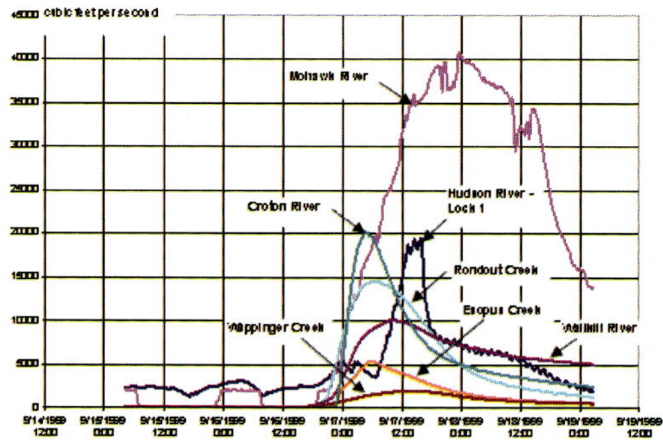

Plate 7 *Gauged river flows over the lower Hudson watershed during extra-tropical storm Floyd, September 1999*

Note: The major flows arrived at Green Island, opposite the city of Troy, New York, from the Mohawk River and peaked near midnight, 17 September; the Upper Hudson River at Lock 1 peaked earlier that day. The other major flow is from the Croton River, 58km (36 miles) north of The Battery. According to the US Geological Survey (USGS), these sources constitute 76 per cent of the drainage areas.
Source: R. V. Allen, pers comm, 2002, cited in Bowman, M., Colle, B., Flood, R., Hill, D., Wilson, R. E., Buonaiuto, F., Cheng, P. and Zheng, Y. (2005) *Hydrologic Feasibility of Storm Surge Barriers to Protect the Metropolitan New York–New Jersey Region*, Final Report, Marine Sciences Research Center Technical Report, Stony Brook University, New York, NY, p106

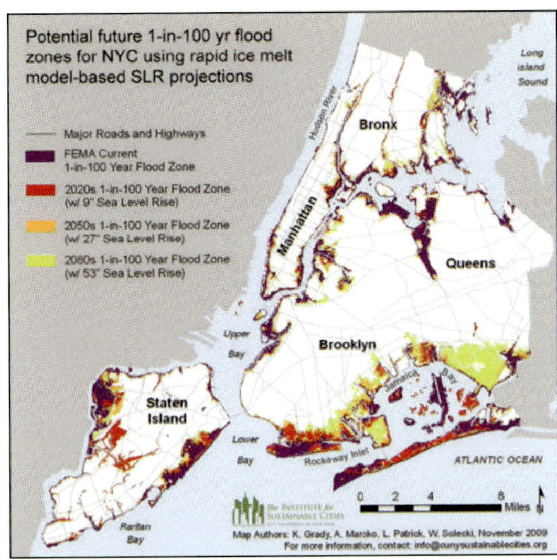

Plate 8 *Current 1/100-year flood zone for New York City and potential future 1/100-year flood zones under various scenarios of climate change*

Source: NPCC (2010) *Climate Change Adaptation in New York City: Building a Risk Management Response*, New York Academy of Sciences, Wiley-Blackwell, New York, NY

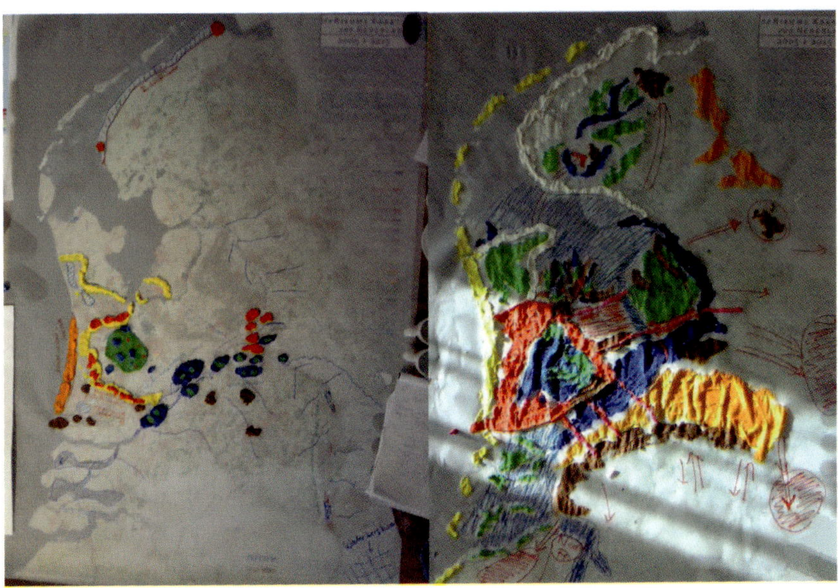

Plate 9 *The maps/visions produced during The Netherlands exercise*

Note: Vision NL1 (left): in this vision (based on moderate climate assumptions) there are no additional settlements below sea level and several retention areas are constructed along (future) river bottlenecks. Several water barrages, so-called terps (hillocks), floating infrastructures and other innovative technical constructions are foreseen.

Vision NL2 (right): in this vision (based on extreme climate assumptions) parts of The Netherlands are abandoned and given back to 'water/nature'. New roads and other infrastructure are built 5m above sea level and the coastline has been strengthened by means of sand supplements. In addition, a super-dike has been constructed around the cities of Rotterdam and Amsterdam, and the 'Green Heart' is transformed from an agricultural area into a big lake.

Source: Photo provided by Tjeerd Stam

Plate 10 *The maps/visions produced during the Groningen exercise*

Note: Vision G1b 'Generous Groningen' (left): in this vision, the coastal defence line moved northwards to the current location of the Frisian Islands. Large-scale freshwater reservoirs are created to supply new and existing agricultural lands. The current function of the province as a large-scale agricultural producer is maintained. In addition, the eastern part of the province will be transformed for biomass, wind and solar energy production. The west part of Groningen is an open-air museum for recreation and tourism.
Vision G2 'Natural elevation' (right): the basis of this vision is a compartmentalized inundation of the lower-lying part of the province. The inundation of different compartments with a thin layer of calmly flowing water causes sedimentation processes to expand the land area along with the sea level, eventually making it again suitable for agriculture. A super-dike prevents dike breaches. A new high-speed railway line from Amsterdam to Hamburg demarcates the inundation areas in the south, as it is built on a dike.
Source: Photo provided by Susan van 't Klooster

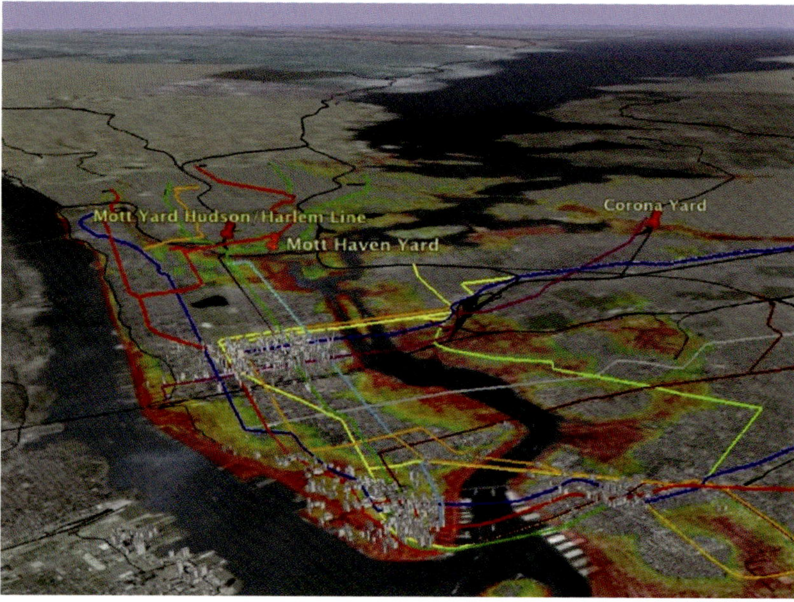

Plate 11 *Current estimated flood areas for part of New York City for hurricanes of different magnitudes*

Note: Coloured areas are worst-track storm-surge flood zones for Saffir–Simpson category 1 (SS1) in red, SS2 in yellow, SS3 in brown and SS4 in green. Coloured lines are subways; black lines are rail systems. The three labelled areas are the locations of case study sites in Rosenzweig et al (2007b).
Source: Lamont-Doherty Earth Observatory, Google Earth and NYSEMO (for coloured flood zones and New York City subway lines); Rosenzweig, C., Horton, R., Major, D. C., Gornitz, V. and Jacob, K. (2007b) 'Climate component', in Metropolitan Transportation Authority, *August 8, 2007 Storm Report*, 20 September (for map)

Plate 12 *Soft adaptation approaches to controlling flooding in New York Harbor*

Note: The light green areas represent potential wetlands that can act as buffer areas.
Source: Nordenson, G., Seavitt, C. and Yarinsky, A., with Cassell, S., Hodges, L., Koch, M., Smith, J., Tantala, M. and Veit, R. (2010) *On the Water: Palisade Bay*, Metropolitan Museum of Art, New York, NY

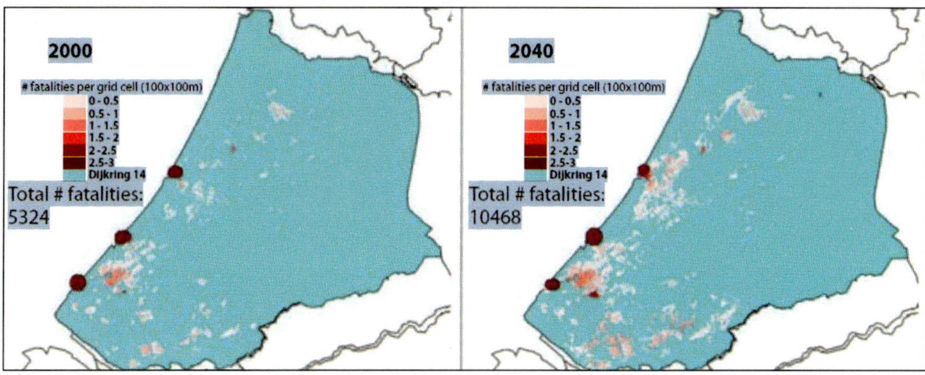

Plate 13 *The projected effects of growth in urban development in low-lying polders north of Rotterdam, by 2040, on the potential number of casualties in the province of South Holland in case of dike breaches (three locations)*

Note: Dijkring = ring dike.

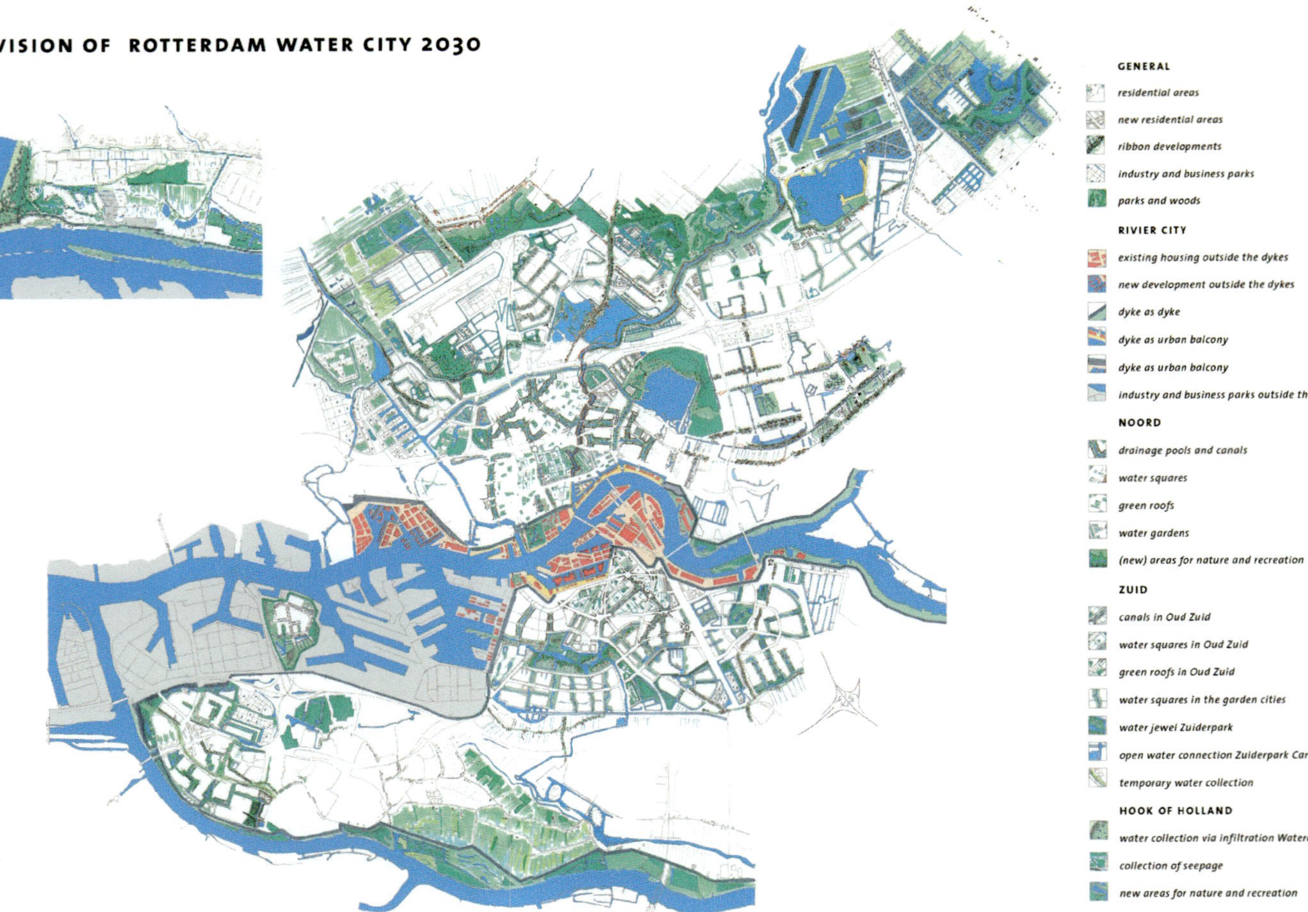

Plate 14 *A map of Rotterdam and locations where additional investments in the water system are needed to become climate proof in 2030*

References

Ahlberg, I. (2009) 'Integrating sustainable solutions from city level to urban city districts', City of Stockholm, Development Administration, www.stockholm.se/norradjurgardsstaden, accessed 21 December 2010

Arakawa-Joryu River Management Office (2011) 'Super levee', Tokyo, Japan, www.ktr.mlit. go.jp/arajo/english/how/flood_con/business/super.html, accessed 2 January 2011

Breen, A. and Rigby, D. (1996) *The New Waterfront: A Worldwide Urban Success Story*, McGraw-Hill, New York, NY

Brick, K. (2008) 'Follow-up of environmental impact in Hammarby-Sjostad', Grontmij AB, Stockholm, Sweden, www.hammarbysjostad.se/inenglish/.../Grontmij%20Report%20eng. pdf, accessed 21 December 2010

Brown, N. and Reed, G. (2003) *London's Waterfront: The Thames from Battersea to the Barrier*, Burke's Peerage and Gentry, London

Bunce, S. (2009) 'Developing sustainability: Sustainability policy and gentrification on Toronto's waterfront', *Local Environment*, vol 14, no 7, pp651–667

Bunce, S. and Desfor, G. (2007) 'Introduction to "Political ecologies of urban waterfront transformations"', *Cities*, vol 24, no 4, pp251–258

City of Stockholm (2009) *City Plan*, p10, http://international.stockholm.se/PageFiles/165973/ Cityplan_short.pdf, accessed 21 December 2010

Cooper, J. and McKenna, J. (2009) 'Boom and bust: The influence of macroscale economics on the world's coasts', *Journal of Coastal Research*, vol 25, no 3, pp533–538

Cruce, T. (2009) *Adaptation Planning: What US States and Localities are Doing*, Pew Center on Global Climate Change, Arlington, VA, www.pewclimate.org/publications/workingpaper/ adaptation-planning-what-us-states-and-localities-are-doing, accessed 2 January 2011

Cushman and Wakefield (2010) *2010 European Cities Monitor*, www.europeancitiesmonitor.eu/ wp-content/uploads/2010/10/ECM-2010-Full-Version.pdf, accessed 16 November 2010

Dale, A. and Newman, L. (2009) 'Sustainable development for some: Green urban development and affordability', *Local Environment*, vol 14, no 7, pp669–681

Design for London (2011) *East London Green Grid, Lea Valley, Area 1*, London, www. designforlondon.gov.uk/uploads/media/ELGGarea1.pdf, accessed 2 January 2011

Entec UK Limited (2005) *Thames Gateway London Partnership*, Entec UK Limited, London

European Commission (2010) *Natura 2000 in the Marine Environment*, http://ec.europa.eu/ environment/nature/natura2000/marine/index_en.htm, accessed 27 December 2010

Evans, J., Jones, P. and Krueger, R. (2009) 'Organic regeneration and sustainability, or, can the credit crunch save our cities?', *Local Environment*, vol 14, no 7, pp683–698

Fränne, L. (2007) *Hammarby-Sjöstad Miljö Bok, English Version*, GlasshusEtt, Stockholm, Sweden, www.hammarbysjostad.se/ineglish/pdf/HS_komb_eng_dec_2008, accessed 2 January 2011

Future Communities (2010) *Hammarby Sjostad, Stockholm, Sweden, 1995–2015: Building a 'Green' City Extension*, www.futurecommunities.net/case-studies/hammarby-sjostad-stockholm-sweden-1995-2015, accessed 21 December 2010

Gerfen, K. (2009) '2009 P/A Awards: Taipei Waterfront, Taipei, Taiwan/Stan Allen Architect', *Architect Magazine*, 17 January

Gillette, H. (1999) 'Assessing James Rouse's role in American city planning', *American Planning Association Journal*, spring, pp151–166

Glavovic, B. (2008) 'Sustainable coastal communities in the age of coastal storms: Reconceptualising coastal planning as "new" naval architecture', *Journal of Coastal Conservation*, vol 12, pp125–134

Gong, Y. and Yu, F. (2009) 'Shanghai leads China's fight of climate change and low-carbon development', *China View*, 15 December, http://news.xinhuanet.com/english/2009-12/15/content_12651242.htm, accessed 2 January 2011

Gordon, I. (1999) 'Internationalisation and urban competition', *Urban Studies*, vol 36, nos 5–6, pp1001–1016

Greening the Gateway: Kent and Medway (2011) *About Parklands*, www.gtgkm.org.uk/gateway-parklands/about-parklands/, accessed 2 January 2011

HafenCity Hamburg GmbH (2010a) *HafenCity*, www.hafencity.com, accessed 27 December 2010

HafenCity Hamburg GmbH (2010b) *HafenCity: Facts and Figures*, www.hafencity.com/en/overview/hafencity-facts-and-figures.html, accessed 21 December 2010

HafenCity Hamburg GmbH (2010c) *HafenCity Sustainability: Flood Secure Bases instead of Dikes*, www.hafencity.com/en/concepts/flood-secure-bases-instead-of-dikes-safe-from-high-water-in-hafencity.html, accessed 21 December 2010

HafenCity Hamburg GmbH (2010d) *HafenCity Sustainability: Clean Thermal Energy for a New Part of Town*, www.hafencity.com/en/concepts/clean-thermal-energy-for-a-new-part-of-town.html, accessed 21 December 2010

HafenCity Hamburg GmbH (2010e) *HafenCity: The Genesis of an Idea*, www.hafencity.com/en/overview/hafencity-the-genesis-of-an-idea.html, accessed 21 December 2010

Hamburg Port Authority (2009) *Annual Report 2009*, www.hamburg-port-authority.de/en, accessed 27 December 2010

Hamid, M. (2009) 'Climate change in the Arab world: Threats and responses', in *Troubled Waters: Climate Change, Hydropolitics, and Transboundary Resources*, Report prepared for The Stimson Center, pp45–62, www.stimson.org/images/uploads/research-pdfs/Troubled_Waters-Complete.pdf, accessed 2 January 2011

Hannigan, J. (1998) *Fantasy City: Pleasure and Profit in the Postmodern Metropolis*, Routledge, London and New York

Hesse, M. (2006) 'Global chain, local pain: Regional implications of global distribution networks in the German North Range', *Growth and Change*, vol 37, no 4, pp570–596

Hill, K. (2009) 'New Orleans since Katrina', *Topos*, vol 68, pp29–33

Jaeger, F. (2008) *Waterfront Living and Working: Hamburg's HafenCity*, City of the Future, Goethe Institute, www.goethe.de/kue/arc/dos/dos/sls/sdz/en3356905.htm, accessed 23 December 2010

Japan Times Blog (2010) 'Yen for living: Flood control destroying neighborhoods to save them', http://blog.japantimes.co.jp/yen-for-living/flood-control-destroying-neighborhoods-to-save-them, accessed 20 December 2010

Johansson, L. and Wallstrom, K. (2001) 'Urban impact in the history of water quality in the Stockholm archipelago,' *Ambio*, vol 30, nos 4–5, pp277–281

Khakee, A. (2007) 'From Olympic village to middle-class waterfront housing project: Ethics in Stockholm's development planning', *Planning, Practice & Research*, vol 22, no 2, pp235–251

Konvitz, J. (1978) *Cities and the Sea: Port City Planning in Early Modern Europe*, Johns Hopkins Press, Baltimore, MD

Konvitz, J. (1994) 'The crises of Atlantic port cities, 1880 to 1920', *Comparative Studies in Society and History*, vol 36, no 2, pp293–318

Korthals-Altes, W. and Tambach, M. (2008) 'Municipal strategies for introducing housing on industrial estates as part of compact-city policies in the Netherlands', *Cities*, vol 25, pp218–229

Laidley, J. (2007) 'The ecosystem approach and the global imperative on Toronto's central waterfront', *Cities*, vol 24, no 4, pp259–272

Maciolek, N. J., Dahlen, D. T. and Diaz, R. J. (2010) *2009 Boston Harbor Benthic Monitoring Report*, Massachusetts Water Resources Authority, Report 2010-18, Boston, MA

Mayor of London (2006) *East London Green Grid Primer and Draft Supplementary Planning Guidance to the London Plan*, November, London

McGovern, S. (2008) 'Evolving visions of waterfront development in postindustrial Philadelphia: The formative role of elite ideologies', *Journal of Planning History*, vol 7, no 4, pp295–326

Michel, D. and Pandya, A. (eds) (2009) *Troubled Waters: Climate Change, Hydropolitics, and Transboundary Resources*, Report prepared for the Stimson Center, www.stimson.org/images/uploads/research-pdfs/Troubled_Waters-Complete.pdf, accessed 2 January 2011

Minchinton, W. (1975) 'Pattern and structre of demand, 1500–1750,' in C. Cipolla (ed) *The Fontana Economic History of Europe*, Fontana, London, pp83–177

Ministry of the Environment (2007) 'Sweden facing climate change – Threats and opportunities', Commission on Climate and Vulnerability, Ministry of Environment, Sweden, www.sweden.gov.se/sb/d/574/a/96002, accessed 23 December 2010

Monclus, F.-J. (2003) 'The Barcelona Model: An original formula? From reconstruction to strategic urban projects', *Planning Perspectives*, vol 18, pp399–421

MWRA (2010) www.mwra.state.ma.us/02org/html/whatis.htm, accessed 10 December 2010

ODPM–Defra (Office of the Deputy Prime Minister–UK Department for Environment, Food and Rural Affairs) (2004) *Creating Sustainable Communities: Greening the Gateway Implementation Plan*, London, www.communities.gov.uk/documents/regeneration/pdf/146685.pdf, accessed 2 January 2011

Pandis, S. and Brandt, N. (2009) *Utvärdering av Hammarby Sjöstads miljöprofilering – vilka erfarenheter ska tas med till nya stadsutvecklingsprojekt i Stockholm?*, KTH Avdelningen för Industriell Ekologi, Kungliga Tekniska Hogskolan, Stockholm, Sweden, TRITA-IM 2009:03, http://djurgardsstaden.info/web/page.aspx?pageid=75497, accessed 21 December 2010

Penna, A. and Wright, C. (2009) *Remaking Boston: An Environmental History of the City and Its Surroundings*, University of Pittsburgh Press, Pittsburgh, PA

Rising Tides Competition (2009) 'About the competition', San Francisco Bay Area Conservation and Development Authority, San Francisco, CA, www.risingtidescompetition.com/risingtides/Home.html, accessed 2 January 2011

Rotterdam Climate Initiative (2010) *Floating Pavilion in the Center of Rotterdam*, www.rotterdamclimateinitiative.nl/en/100_climate_proof/projects/floating_pavilion?portfolio_id=19, accessed 28 December 2010

Rowe, J. and Fudge, C. (2003) 'Linking national sustainable development strategy and local implementation: A case study in Sweden', *Local Environment*, vol 8, no 2, pp125–143

Sanchez-Arcilla, A., Jimenez, J., Valdemoro, H. and Gracia, V. (2008) 'Implications of climatic change on Spanish Mediterranean low-lying coasts: The Ebro Delta case', *Journal of Coastal Research*, vol 24, no 2, pp306–316

Sandelin, A. (2008) 'Hammarby Sjöstad – living green in central Stockholm', www.sweden.se/eng/Home/Society/Sustainability/Reading/Hammarby-Sjostad---living-green-in-central-Stockholm

Savage, A. (1995) 'Boston Harbor: The anatomy of a court-run cleanup', *Boston College Environmental Affairs Law Review*, vol 22, issue 2, pp365–411

Schaer, C. (2010) 'Hamburg's new quarter: The challenge of making HafenCity feel neighborly', *Der Speigel*, www.spiegel.de/international/germany/0,1518,714008-2,00.html, accessed 21 December 2010

Seattle Waterfront Metropolitan Improvement District (2011) 'Seattle waterfront history', www.seattlewaterfront.org/index.cfm?section=history, accessed 3 January 2011

SHSB (Save the Harbor, Save the Bay) (2004) *The Leading Edge: Boston Harbor's New Role in the City's Economy*, SHSB, Boston, MA

SMHI (Swedish Meteorological and Hydrological Institute) (2010) www.smhi.se/en/theme/sea-level-1.11009, accessed 21 December 2010

Spits, J., Needham, B., Smits, T. and Brinkhof, T. (2010) 'Reframing floods: Consequences for urban riverfront developments in Northwest Europe', *Nature and Culture*, vol 5, no 1, pp49–64

Stockholm International Water Institute (2010) www.siwi.org/history, accessed 21 December 2010

Stockholm Vatten AB (2010) 'Mission', www.stockholmvatten.se/en/About-Stockholm-Vatten, accessed 20 December 2010

Tol, R., Klein, R. and Nicholls, R. (2008) 'Towards successful adaptation to sea level rise along Europe's coasts', *Journal of Coastal Research*, vol 24, no 2, pp432–442

Upton, D. (2008) *Another City: Urban Life and Urban Spaces in the New American Republic*, Yale University Press, New Haven, CT

US Bureau of Labor Statistics (2010) 'Percent of employment in services in Sweden', in *International Comparisons of Annual Labor Force Statistics*, US Department of Labor, www.bls.gov/fls/flscomparelf.htm, accessed 20 December 2010

Zhu, J. (2001) 'Commercial real estate capital in the restructuring of downtown Baltimore', *RURDS*, vol 13, no 1, pp73–81

Zhu, J. (2003) 'Urban physical development in transition to market: The case of China as a transitional economy', *Urban Affairs Review*, vol 36, no 2, November, pp178–196

8

Innovative Flood Defences in Highly Urbanized Water Cities

Bianca Stalenberg

8.1 Impacts of Climate Change and Economic Development upon Flood Risk

Since the 20th century, the flood risk in river cities worldwide has increased significantly due to economic development and population growth. For instance, Tokyo (the capital of Japan) has grown from about 3.7 million inhabitants in 1920 to more than 12.8 million people in 2008 (Statistics Bureau of Japan, 2010), and the population in the Dutch city of Rotterdam has grown from 0.3 million people in 1900 to 0.6 million people in 2010 (COS, 2010). These developments have increased the potential consequences of a flood event because more people and property are at risk of flooding. As a result, economic damage and social disruption caused especially by storm surge and fluvial floods are higher. Moreover, the transformation of rural land into urban land has caused an increase in the probability of pluvial floods. Since the end of the 19th century, in many cities canals have been transformed into roads and train tracks and the construction of residential districts has caused the transformation of green areas into paved areas that contain dwellings and apartment blocks. These land-use changes have reduced the ability of the subsoil to absorb rainwater.

In addition to economic development and population growth, flood risk in urbanized coastal areas is influenced by climate change. It is expected that the probability of flooding will increase due to the effects of climate change. For the global average, warming in the last century has occurred in two phases. Between 1910 and 1940, the increase in temperature has been 0.35°C, and this increase has been 0.55°C from the 1970s to the present. A decrease in temperature of about 0.5°C occurred during the period of 1940 to 1970 (Kroonenberg, 2008). This implies that temperature has increased by 0.4°C during the last century. The Intergovernmental Panel on Climate Change's 2007 report predicts that global average temperature will have increased between 1.1°C and 6.4°C in 2100 (IPCC, 2007). The predicted increases in temperature will most likely increase the amount of precipitation and evaporation. More and heavy downpours increase the discharge of rainwater and most likely increase river discharges as well. In addition, climate change is projected to cause sea-level rise, which will increase flood risk.

Climate change and, especially, economic development during the previous decennia have triggered urgency in improving the current flood defence system in water cities around the globe. Additionally, the disastrous flooding caused by Hurricane Katrina has shocked the world. Cities such as Rotterdam (The Netherlands), Tokyo (Japan), New York (US) and Venice (Italy) are facing a real challenge. Moreover, the uncertainties induced by climate change and economic development in the future will probably require improved flood defences in water cities. At the same time, the appearance and functionality of urban waterfronts change over time due to the changing preferences of local residents and policy-makers. These urban redevelopment projects need to be granted sufficient freedom to be successfully executed. Unfortunately, the improvement of flood defences and the redevelopment of urban riverfronts are extremely challenging and often result in conflicts between the stakeholders involved. Flood protection and urbanization seem to have conflicting objectives. The objective of this chapter is, therefore, to find a solution for this conflict so that flood protection and urbanization can be synergistically combined in the shared realm of an urban waterfront.

8.2 Conflict between Flood Control and Urban Planning

An urban waterfront contains a number of structures which make specific urban uses possible, such as residential, recreational and industrial uses. In order to function properly, all urban structures require a certain space and protection against fluvial floods. Flood-retaining structures can protect these urban structures from fluvial floods by retaining floodwater. The area behind these flood-retaining structures is protected against floods. However, to function properly, all flood-retaining structures need a certain space. Both types of structures have to share the available space in an urban waterfront, which could result in conflicting demands for scarce space.

Focusing on the challenges faced by urban flood control, the main challenge is to avoid visual hindrance of the urban uses that are present. A quay wall is a flood-retaining structure that is very often seen in urban areas. If the top of the quay wall is equal to the ground surface of the urban waterfront, then the quay wall will not cause any hindrance to existing urban uses. However, if heightening of the quay wall is needed, then this would probably evoke resistance from local residents. Strolling along the boulevard will become less attractive when the river view is blocked by a quay wall and the value of properties will decrease if the river is no longer visible from, for instance, the living room (see Figure 8.1). The physical accessibility of the urban waterfront is also negatively affected by the improvement of the quay wall. Vertical expansion in many cases leads to a decrease in the value of the area as access to the shore and riverbanks is hindered. The impact of a dike improvement is generally larger than adjustments that are made on a quay wall. Vertical expansion not only leads to a less attractive view of the river; heightening of a dike is only possible if the body of

Figure 8.1 *Flood wall in Tokyo adjacent to the Sumida River, Japan*

Source: Bianca Stalenberg

the dike is widened as well. In other words, the dike needs more space. However, this space is often not present in an urban waterfront because of the high occupation rate by urban structures. Hence, horizontal expansion of the flood defence frequently leads to the removal of urban structures, such as taking down buildings or diversion of roads. As a result, improvement of flood defences is difficult in high-density urban areas.

Focusing on the challenges faced by urban planning, the main challenge is to preserve sufficient space for inspection, maintenance and operation of the flood-retaining structures. Ideally, some space between the flood-retaining structure and urban structures is preserved to enable inspection and maintenance. Flood controllers also have specific demands about what can and cannot be integrated with the flood-retaining structure in a way that the strength of the structure is guaranteed. Urban planners of the municipalities need to take these demands and wishes of the flood controllers into account in their designs, which often leads to a compromise. The challenge in urban waterfronts, therefore, is to seek a solution that can reduce or overcome the difficulties that occur when flood defences are improved or when urban waterfronts are redeveloped in a shared realm. The question is whether it is possible to create and maintain synergy between flood protection and urban uses in an urban waterfront, while taking economic development and climate change into account?

8.3 Examples of Realized Multifunctional Flood Defences

Today, the challenge is to develop new flood control strategies and urban plans which are appealing to the public, and which can cope with the effects of climate change. The application of multifunctional flood defences can contribute to tackling this challenge. Although the concept of combining flood protection with urban activities is not new, examples of such structures are rather scarce. Of course, a dike combined with a road is a well-known example; but other examples are more difficult to find. Nevertheless, most flood defences are still singular structures due to the demands of flood controllers for easy access for operation or maintenance work. However, the following examples of different international cities demonstrate that multifunctional flood defences are feasible.

8.3.1 Dordtse Wand in the city of Dordrecht, The Netherlands

The Noordendijk in the Dutch city of Dordrecht, The Netherlands, is a dike along a branch of the River Rhine on which a road is situated. According to the municipality, the houses along the dike were outdated and the area needed to be revitalized. These renewal activities gave flood controllers the opportunity to simultaneously improve the flood defence (Water Board de Groote Waard, undated). This project was a joint effort of the municipality and the water board. Technically, part of the existing dike has been replaced by a concrete L-shaped retaining wall which is incorporated within the foundation of the new buildings, with integrated parking space (see Figures 8.2 and

Figure 8.2 *Photograph of multifunctional flood defence in Dordrecht, The Netherlands*

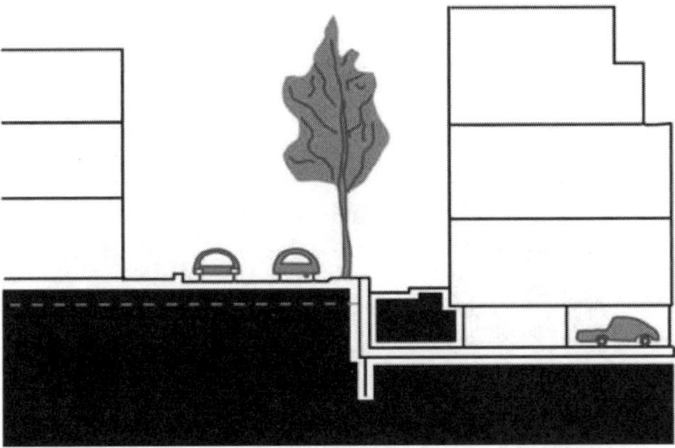

Figure 8.3 *Cross-section of multifunctional flood defence in Dordrecht, The Netherlands*

Source: Water Board de Groote Waard (undated)

8.3). The urban functions in this area remain the same. The flood retaining structure is designed for a functional lifespan of at least 100 years, which is much longer than the usually applied functional lifespan of 20 to 30 years of urban structures. However, the design did not take uncertainties into account caused by climate change or economic development, which could change the future flood safety level.

8.3.2 Super-levee in the city of Tokyo, Japan

A super-levee is especially constructed for extreme events in highly urbanized areas, such as in Tokyo, Japan. In Tokyo, natural hazards such as typhoons and earthquakes are part of everyday life, and historical flood events show that the consequences of a breach of a conventional dike can be severe. A super-levee is a very wide levee (> 200m) that is practically unbreakable and is designed especially for the protection of highly urbanized areas against extreme events. Moreover, due to its width, it can withstand flood events with a long duration much better than a conventional dike. A super-levee has a mild inner slope of 1/30, so a super-levee with a height of 10m will have a width of about 300m (Arakawa-Karyu River Office and MLIT, 2006). A super-levee is resistant to overflow, seepage and earthquakes, which is one of the main reasons for its construction. Additionally, a super-levee integrates both the requirements of flood control and the interests of urban planners (Takahasi and Uitto, 2004). Super-levee projects are mostly implemented in conjunction with urban redevelopment, land rezoning projects or other urban planning activities. The gentle inner slope, in combination with open recreational spaces at crest level, provides an open view towards the river. The restored accessibility of the river leads to an increased urban value of the region. In Tokyo, super-levees are often combined with parks and high-rise buildings (see Figures 8.4 and 8.5).

Figure 8.4 *Photograph of multifunctional flood defence in Tokyo, Japan*

Source: William Verbeek

Figure 8.5 *Cross-section of multifunctional flood defence in Tokyo, Japan*

Source: Bianca Stalenberg

8.3.3 HafenCity in the city of Hamburg, Germany

The German city of Hamburg is situated approximately 110km from the mouth of the River Elbe. At this moment, the former harbour areas are being transformed through the HafenCity redevelopment project. The goal of this project is to develop large numbers of residential blocks combined with offices and tourist attractions (Pols et al, 2007).

Figure 8.6 *Photograph of multifunctional flood defence in Hamburg, Germany*

Source: Jeroen Aerts

This redevelopment is allowed on the condition that flood defences are improved. Here, houses are elevated, quays are heightened and singular protection measures are applied to enable sufficient flood protection (see Figures 8.6 and 8.7). Every ground floor which is located below the level of 9m above the German water-level standard needs to be dry-flood proofed. Windows need to withstand high water pressure, and manually operated steel shutters have to prevent damage to the glass windows that could be

Figure 8.7 *Cross-section of multifunctional flood defence in Hamburg, Germany*

Source: Pols et al (2007)

caused by floating debris. Ground floors are used for car parks, restaurants and offices. Elevated evacuation routes are also important in the HafenCity, so an escape route is always guaranteed. This project shows that the integration of urban functions and flood protection is possible if additional regulations are formulated that allow stakeholders (including urban planners and flood managers) to seek their mutual interest in the project, co-develop the design and jointly execute the implementation.

8.4 Innovative Multifunctional and Adaptable Flood Defences

Currently, 'invisible' flood-retaining structures such as movable flood defence gates or temporary stop logs are used to limit the conflict between the desire for a certain urban quality and the desire for sufficient flood protection. Gates are often made of steel in combination with a steel or concrete case. During a flood event the gates are rotated by, for instance, hydraulic equipment. Support structures with rubber profiles secure a watertight connection between the gates and the adjoining urban structures. Stop logs retain water that would otherwise flow through gaps in a flood defence during flood conditions. Support structures are constructed on each side of the gap in which the stop logs are manually placed during flooding. Stop logs used to be made of wood, but now mainly consist of aluminium. Support structures with rubber profiles secure a water-tight connection. These types of flood-retaining structures do not hinder urban activities during normal conditions, which can be seen as a benefit. However, the presence of support structures does not achieve a higher urban quality. In this sense, they do not contribute to existing urban activities in the area. In addition, they are subject to human error, which is an unpredictable failure mechanism. The concept of adaptable flood defences (the AFD concept) does have the potential to contribute to the urban quality of urban waterfronts and simultaneously improve flood-retaining structures. The AFD concept contains two key features: multifunctionality and adaptability.

8.4.1 Feature 1: Multifunctionality

The AFD concept has the potential to create physical synergy by developing innovative structures, which combine urban functions and flood protection into one multifunctional structure. Structures such as car parks, buildings, dwellings and roads can be designed or transformed with the additional ability of protecting the hinterland against floods. The public does not feel impeded in their activities and plans by the flood defence due to its hidden and discrete appearance. On the contrary, flood defence structures in the urban waterfront are no longer seen in a negative light, but as a contribution to urban quality: an urban waterfront is protected against fluvial floods up to the desired design flood event, with the consequent increase in urban quality of life.

The 'adaptable building' (Stalenberg, 2005) was designed as an alternative solution for the Dutch city of Deventer. The city was built on a sand dune which still protects

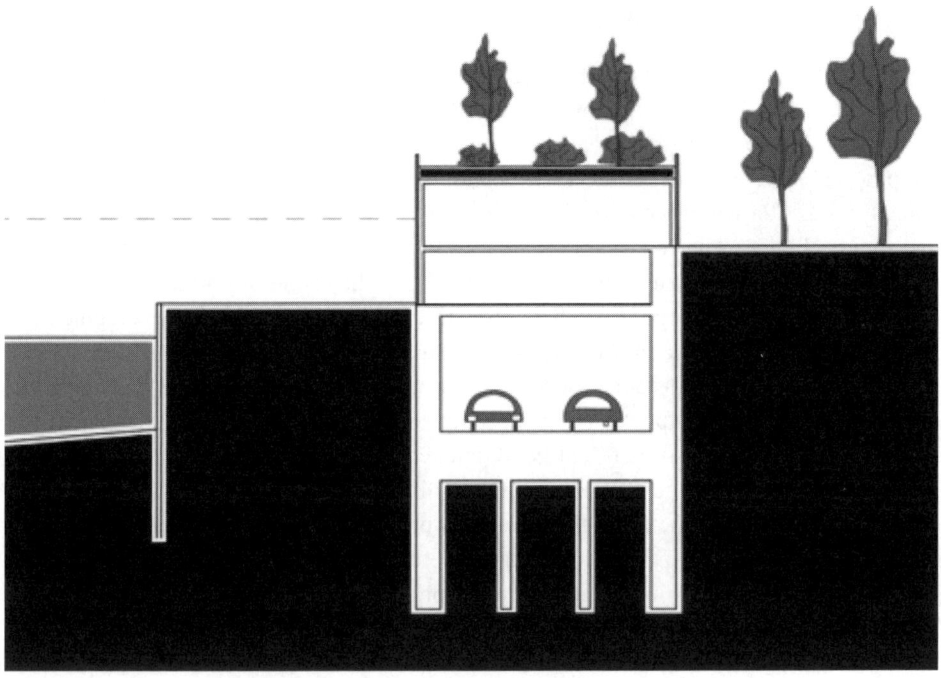

Figure 8.8 *Example of multifunctionality*

Source: Bianca Stalenberg

large parts of the city against fluvial floods. The design illustrates the principles of multifunctionality within the AFD concept. It combines flood protection with several urban functions, such as transportation and recreation (see Figure 8.8). Flood protection is established by a waterproof tunnel wall in combination with an adapted building front. The working principle is similar to a quay wall in combination with a flood wall. The walls are made of concrete and are combined with an optional cut-off to ensure a waterproof structure. The building front has sufficient height to retain flood levels. Doors and casement windows along the riverside below the flood-retaining height are not encouraged due to the probability of human error when applying flood-retaining structures such as stop logs to close gaps. Toppling is a frequent failure mechanism of quay walls, but is of less concern here. The flood-retaining walls are part of a tunnel and building, which provide additional stability against toppling. The tunnel is also used for transportation and is part of the foundation in combination with a deep foundation for the upper building. As a result, the tunnel is a hybrid structure. The same applies to the building, which is used as a flood defence and restaurant. Additionally, the roof is used as a public park.

8.4.2 Feature 2: Adaptability

Creating physical synergy in an urban waterfront should not be a snapshot action, but a sustainable affair. It is, therefore, important to take time into account. The AFD concept applies to multifunctional flood-retaining structures that are adaptable so that the physical synergy between flood protection and urban functions can be maintained and adjusted. Adaptability can be defined as the capacity of a structure to absorb minor and major change. The 'adaptability' feature of the AFD concept works both ways. First, flood controllers who apply the AFD concept have the opportunity to take external influences into account, such as climate change and economic development. It also enables them to extend the technical lifespan of the flood-retaining elements of a multifunctional flood defence. Second, urban planners who apply the AFD concept take alterations of urban functions into account, and extend the technical lifespan of the urban elements of a multifunctional flood defence.

Functional lifespan and technical lifespan differ for flood-retaining structures and urban structures. The functional lifespan of an urban structure is about 20 to 30 years, while the technical lifespan of an urban structure is about 50 years (Gijsbers et al, 2009). For a flood-retaining structure, functional lifespan and technical lifespan are both about 100 years. Note that there is a distinction between the design lifespan and the actual lifespan; here, only the design lifespan is used. In order to reach the technical lifespan of 100 years of an adaptable flood defence, adaptability of the urban elements is essential.

Possible types of adaptability are extension, overcapacity and refurbishment, which are discussed below. All types of the adaptability of the AFD concept could lead to more sustainable structures, which anticipate uncertainties of, for instance, climate change or change in urban functions.

8.4.2.1 First type of adaptability: Extension

This type implies an easy expansion of a multifunctional flood defence in the future by constructing a larger foundation than initially needed. The structural redundant capacity will enable an increase in the volume of the multifunctional and adaptable structure when, for instance, the river discharge increases or the economic value of the hinterland increases (see Figure 8.9). The advantage is that the structure does not have to be replaced by another one; a combination of flood-retaining elements and urban elements can be added to the initial structure. The flood-retaining elements ensure the desired flood safety level, whereas the urban elements ensure the desired urban quality. This type of adaptability preserves the multifunctional character of the structure as well as the synergy between both functions. When applying extension, one should ensure that a distinction is made between the load-bearing structure and other parts, such as the façade, during the design phases of the initial structure. Currently, this type of adaptability is mainly applied to office buildings, but is perfectly applicable to multifunctional flood defences.

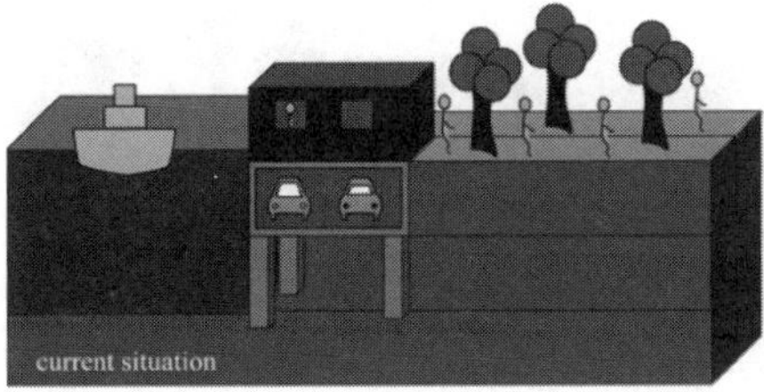

Figure 8.9 *Type of adaptability: Extension*

Source: Bianca Stalenberg

8.4.2.2 Second type of adaptability: Overcapacity

This type implies oversizing in such a way that expansion or reconstruction of a multi-functional flood defence due to increasing flood risk becomes unnecessary in the short term. The entire structure can retain higher flood levels than is initially demanded by the hydraulic boundary conditions. This is achieved by constructing urban elements with flood-retaining abilities when the dimensions of these urban elements are larger than the initially needed dimensions for flood protection purposes (see Figure 8.10). Oversizing does not adversely affect the urban quality of a waterfront because the adaptable and multifunctional structure contains both the technical function of flood protection and urban functions. In other words, the structure is also used for urban activities.

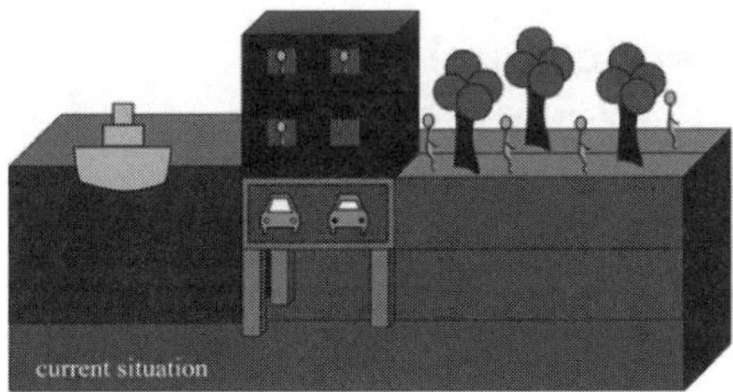

Figure 8.10 *Type of adaptability: Overcapacity*

Source: Bianca Stalenberg

8.4.2.3 Third type of adaptability: Refurbishment

This type implies refurbishment of a multifunctional structure so that the technical lifespan can be extended without alteration of functions, which applies to both urban elements and flood-retaining elements. Elements that tend to exceed their technical lifespan are replaced by taking the technological possibilities and building regulations

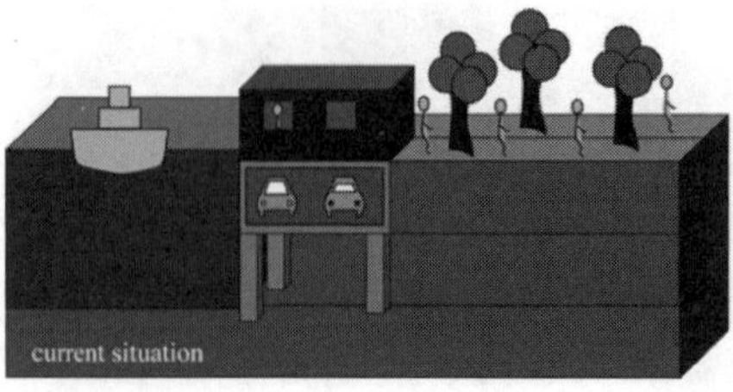
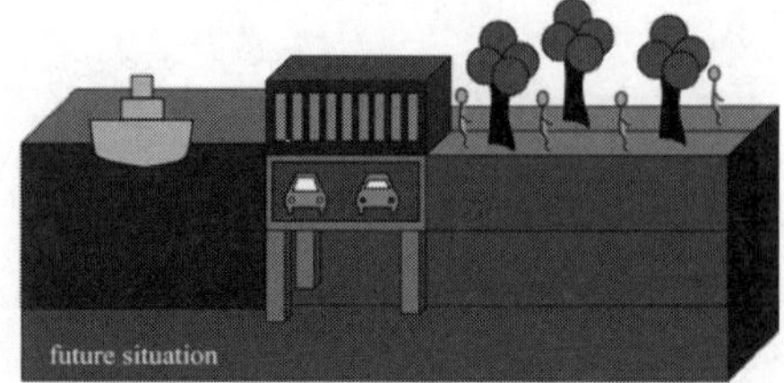

Figure 8.11   *Type of adaptability: Refurbishment*

Source: Bianca Stalenberg

of the future into account. Refurbishment is purely driven by the technical perform-ance of the structure and not by external factors, such as climate change or economic development (see Figure 8.11). Refurbishment ensures that existing multifunctional structures do not have to be demolished and rebuilt. This type of adaptability is espe-cially promising when urban elements tend to exceed their technical lifespan (50 years) in comparison with the technical lifespan of flood-retaining elements (100 years). When applying refurbishment, one should ensure that a distinction is made between the load-bearing structure and other parts, such as the façade, during the design phases of the initial structure in order to ensure that it remains possible to adjust the multifunctional structure in the future. Hence, the load-bearing structure should not be connected to other parts of the structure so that it remains possible to adjust elements without adversely affecting the base.

8.5 Assessment of Multifunctional and Adaptable Flood Defences

The previous examples of multifunctional flood defences demonstrate that these in-novative flood defences are also feasible in practice. In this section, an assessment of technical, financial, and legal and process aspects is carried out in order to investigate the feasibility of the AFD concept in a more generic way, and is partly based on the work presented in Stalenberg (2010).

8.5.1 Technical feasibility

No flood-retaining structure can provide 100 per cent safety against floods. These struc-tures are designed according to a specific flood event for which the government has decided on an upper boundary regarding flood safety measures. Additional safety can be realized by taking damage mitigation measures. The reliability of a flood-retaining structure depends upon movability. A flood defence could contain only immovable elements or only movable or temporary elements. In theory, each of these types of flood-retaining structures should be equally reliable; in practice, they differ. Human error can cause failure of the flood-retaining structure before the desired safety level is reached. For instance, stop logs could be misplaced or forgotten. Therefore, flood defences that contain only immovable elements are more reliable than flood defences that also or only contain movable elements which are manually operated, or temporary elements that are manually placed. This also applies to multifunctional and adaptable flood defences.

An interesting question is whether the urban elements of the multifunctional flood defence can contribute to the strength and stability of the flood retaining elements? In other words: can they increase the technical performance of the flood-retaining elements? Quay walls are often applied in urban areas for flood protection. An important failure mechanism of a quay wall is toppling. This occurs when the embedded length of the structure is not deep enough to create a counter-moment to resist the differences in horizontal forces. Additional urban elements can reduce toppling because they provide additional stability. The horizontal dimensions of the entire structure are larger and the structure weighs more (see Figure 8.12). However, dikes are a different story. This type of flood defence is often applied in urban areas with lesser population densities. An important failure mechanism of a dike is piping underneath the soil body. Piping is the result of seepage flow through the subsoil, which causes erosion behind the flood-retaining structure. Additional urban elements could decrease the probability of piping when the saturation length is lengthened by, for instance, impermeable structures on the inner slope of the dike. On the other hand, these urban elements could also increase the probability of other failure mechanisms, such as instability of the dike. Dikes are gravity/weight structures which derive their stability from the friction between the dike body and the subsoil. Non-soil elements in the dike body could decrease its stability. For instance, trees can jeopardize the dike's stability. The same applies to buildings with a deep foundation. This is a point of concern and should be taken into account during the design stage.

A lack of awareness by the public of the multifunctionality of the adaptable flood defence could jeopardize its technical performance. Today, people might be informed

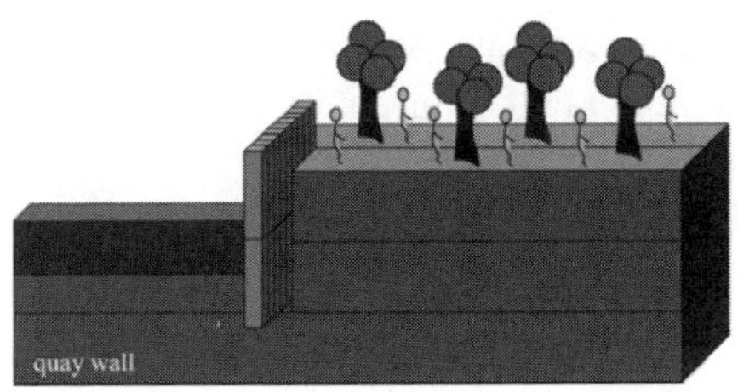

Figure 8.12 *Contribution of urban elements to stability of flood-retaining elements*

Source: Bianca Stalenberg

about the special status of the structure; but whom and what keeps this awareness alive? Over time the true meaning of the structure might be lost, which could result in failure of the flood-retaining elements of the multifunctional and adaptable flood defence, causing a flooded hinterland. The example of the Dutch city of Dordrecht shows how this awareness could be kept alive. Local residents had used stop logs, which were supposed to block floodwater in the doorways of buildings, for timber wood or for firewood. Luckily, this malfunctioning of the flood defence turned up during a routine Water Board check; this is why the Water Board organizes a simulation of a flood event every year in order to keep the awareness of flooding alive amongst the residents.

8.5.2 Technical sustainability

In theory, the AFD concept creates (technical and functional) sustainable designs with a long lifespan, achieved through the 'adaptability' of urban and flood-retaining elements (extension, overcapacity and refurbishment).

Looking at the urban elements, a question arises as to whether the technical lifespan of the flood-retaining elements need to be the same as the technical lifespan of the urban elements. As stated earlier, the technical lifespan of, for instance, a building and a quay wall differ. Urban functions tend to have a lower technical lifespan (50 years) than flood-retaining structures (100 years); the functional lifespan is often even lower (30 years). The municipality today might have a different policy concerning the waterfront of their city than in, for instance, 20 to 30 years. Flexible urban elements enable urban planners to alter the waterfront's functionality before its technical lifespan is reached. This adaptability enables the extension of the functional lifespan of the urban elements of a multifunctional and adaptable flood defence to match the technical lifespan of these urban elements. The replacement of flexible urban elements can also extend the technical lifespan of these urban elements to match the technical lifespan of the flood-retaining elements. As a result, the lifespan of the urban elements and flood-retaining elements might differ. However, alteration of urban elements is a challenge: they have to be altered without negatively affecting the technical performance of the flood-retaining elements. In order to overcome the difficulties of adapting urban elements without hindering flood-retaining elements, the load-bearing structure should be partitionable and should not be interwoven with elements such as façade, roof covers and installations. This type of design is not yet common in the current building technology and is an important issue when applying the AFD concept.

Looking at the flood-retaining elements, the question is whether every type of adaptability is fit for every occasion? The performance of flood-retaining elements is influenced by external influences such as climate change and economic development, and by the technical lifespan. The expected increase in flood levels can be tackled with the application of adaptability type 1 (extension) and adaptability type 2 (overcapacity). The structural redundant capacity of adaptability type 1 (extension) could cope with every expected increase in flood level by enabling an increase in the volume of the multifunctional and adaptable flood defence. It is clear that a small increase in the flood level of, for example, 0.5m needs a smaller structural redundant capacity than a large increase in the flood level of, for instance, 3m. The challenge is that an increase in the flood level has an effect on the additional functionality of the added flood-retaining elements. Weak spots such as doors and windows need special attention. In addition, the flood-retaining ability of urban elements of adaptability type 2 (overcapacity) can cope with every expected increase in flood level. The benefit is that the multifunctional and adaptable flood defence does not have to be replaced by another structure to cope with this increase. However, weak spots such as doors and windows also need special attention during the design stage or need alteration after increasing flood levels.

8.5.3 Financial feasibility

The AFD concept faces financial challenges posed by its multifunctionality and adaptability. The construction of a multifunctional flood defence is generally more expensive than the construction of a conventional flood defence because it contains both flood-retaining elements and urban elements. Governments tend to finance only the flood-retaining elements of a multifunctional structure, and not the entire structure. Therefore, the construction of a multifunctional flood defence could only be financially successful if it were realized through the financial support of different stakeholders. For instance, adding dwellings to a multifunctional flood defence could enable the financial feasibility of the multifunctional structure. As in any other new housing development projects, these dwellings are likely to be built by property developers who will sell them to private households.

If the additional urban facilities can be cost effective, then the construction of the adaptable flood defence will not be more expensive than the construction of a monofunctional conventional flood defence. An example of a successful joint project can be found in the Dutch city of Zwijndrecht in which a quay wall is combined with parking and apartment blocks (see Figures 8.13 and 8.14).

Project developers may be interested in a joint project because the presence of water could create an added value in residential areas. Research shows that the presence of water in residential environments may be good value for money (Kauko et al, 2003). They state that the attraction (or lack of it) of a location can either increase (or reduce) the house price, compared to the price of a similar dwelling situated at an 'average' location. A distinction is made between positive effects, such as proximity, location with a view and size of the water area, and negative effects, such as floods, droughts and water pollution. According to the reviewed studies in Kauko et al (2003) and Goetgeluk et al (2005), water can have a conservative added value of 10 to 15 per cent for the seashore, 5 to 10 per cent for river locations and 5 per cent for lakes. Flood events can have a negative impact of 0 to 22 per cent (Goetgeluk et al, 2005).

Furthermore, when focusing on the adaptability feature of the AFD concept, the aim of the concept is to provide a positive financial end result by investing more money

Figure 8.13 *The Westkeetshaven joint project in the city of Zwijndrecht,*
The Netherlands

Source: Bianca Stalenberg

in the flood-retaining elements at the initial phase so that money can be saved in the future. The question is whether it is worthwhile investing in structural redundancy, like an oversized foundation of the structure, or to construct urban elements with the ability to retain floods so that the need for future adaptations that are triggered by external influences, such as climate change and economic changes, is minimal or not necessary at all. Another option is to invest the least possible in a 'fit-for-purpose' design of a mono-functional conventional flood defence, such as a quay wall, and to reinvest if the structure needs improvement or needs to be replaced by an entirely new one. The

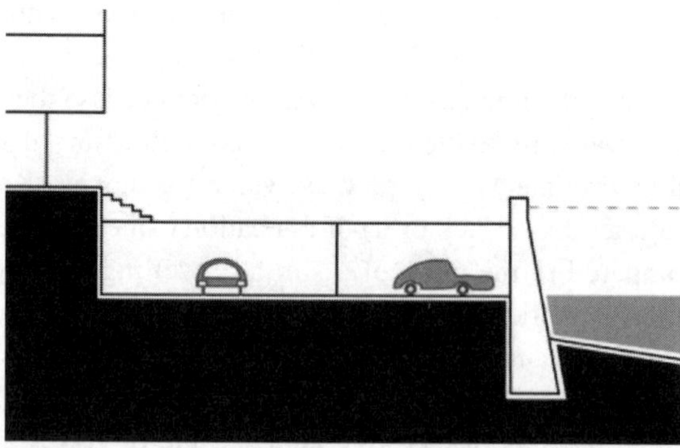

Figure 8.14 *The Westkeetshaven joint project in the city of Zwijndrecht,*
The Netherlands: Cross-section

Source: Municipality of Zwijndrecht (1998)

financial success of an adaptable flood defence depends on the costs of adaption due to increasing flood levels and increasing population rate or economic development, compared to the costs of a mono-functional and conventional flood defence without adaptation capacities.

In order to make the AFD concept more financially appealing than a conventional mono-functional flood defence, both flood-retaining elements and urban elements should have the ability to adapt. Therefore, flood management organizations (mostly coordinated by the government) need to be willing to pay for the extra costs of the flood-retaining elements or of the urban elements that have the ability to retain flood-water and to ensure the adaptability of those elements. A type of adaptability 'extension' is financially feasible if the flood risk increases at a rate such that the conventional (mono-functional) flood defence has to be rebuilt, whereas the multifunctional and adaptable flood defence only needs improvement. However, the financial support of stakeholders involved during improvement of a multifunctional and adaptable flood defence is important when the costs of adapting the flood-retaining elements are approximately similar to the costs of improving a mono-functional conventional flood defence, and additional costs have to be made to alter the urban elements. The same applies to multifunctional flood defences with 'overcapacity'. The extra costs during construction are bound to be so high that this type of flood defence is only financially feasible when the mono-functional conventional flood defence needs to be rebuilt. Improving a conventional type is often cheaper than constructing a multifunctional flood defence with overcapacity.

It is crucial not only to look at costs, but also at the benefits. The added value of multifunctional and adaptable flood defences to the urban quality of a waterfront could be an important criterion in a cost–benefit analysis. Additionally, urban structures could generate revenue after construction of a multifunctional structure. It could, therefore, well be that a multifunctional and adaptable flood defence turns out to be the best solution. Other additional benefits can be found in the environmental value of waterfronts, such as buffers zones and retention areas.

8.5.4 Legal and process aspects: An example from The Netherlands

The existing legislation on flood control can be a challenge when implementing innovative solutions. For instance, existing legislation in The Netherlands is unfit to cope with multifunctional and adaptable flood defences. Governmental reports only provide instructions on how to test mono-functional flood defences. As a result, it is currently not possible to officially test if multifunctional and adaptable flood defences fulfil the required safety level. Many Dutch flood controllers see this as an obstacle for successful implementation. They do not want to construct a multifunctional structure if it is uncertain whether the structure will be accepted by the Dutch national government as a reliable flood defence during the test phase in the future. This often leads to a reserved mindset of flood controllers about the application of the AFD concept. Adding multifunctional flood-retaining structures to existing Dutch governmental reports is, therefore, essential for more common use of the AFD concept. Currently, this is work in progress. The expectation is that these innovative structures will be applied on a larger scale when multifunctional flood defences are included in the governmental reports.

Processes for urban planning differ from processes for flood control. In The Netherlands, urban planning can be seen as a continuous process, whereas improvement of flood defences is mainly induced by a compulsory check of the entire flood defence system, which is conducted by various water boards every six years. Maintenance of flood defences does have a more continuous character, but has significantly less impact upon the surrounding environment than improvement. Also, the existing processes and legislation hardly encourage municipalities and water boards to work together. Urban planners mainly use the Act on Spatial Planning, whereas flood controllers primarily use the Act on Flood Control. Urban planners are obliged to add a section about water in every development plan to guarantee, for instance, sufficient storage of storm water; but this is seen more as a nuisance than as an opportunity to create both quality and safety in a city. In return, flood controllers are obliged to carry out an environmental impact assessment in the case of development or improvement of a flood defence; but this does not directly involve urban functions in, or the urban quality of, the concerning area. Additionally, there is hardly any overlap in the experience of both stakeholders. Each stakeholder has its own vocabulary, which makes it difficult to understand each other. Therefore, in order to make the implementation of the AFD concept a success, it is necessary for urban planners and flood controllers to invest in exploring each other's world, and to invite each other to join in on separately initiated projects. This could be accomplished in different ways. For instance, students from civil engineering and students from architecture could execute joint projects to understand each other's wishes and demands by taking these into account during the design process. This could reduce miscommunication between water boards and municipalities in the future. In addition, (partial) integration of the policies concerning flood control and urban planning could help to create the synergy desired in highly urbanized river cities. One way to accomplish this is to merge governmental ministries, something that recently occurred in The Netherlands.

Ownership of a multifunctional and adaptable flood defence creates a challenge. Flood-retaining elements are most likely owned by the Water Board and the urban elements by the municipality, property developers or private households. Who decides what happens to which part of the structure? How do you avoid deterioration of flood-retaining elements by public or private usage of urban elements? How do you ensure the desired urban quality of the urban elements when flood controllers need to inspect, maintain or operate the flood-retaining elements? For each multifunctional and adaptable flood defence, it should be clear who owns and who is responsible for which part. In the case of flood-retaining dwellings that are owned by private people, a solution could be to incorporate this in contracts.

8.6 Conclusions and Recommendations

It is possible to create and maintain synergy between urban functions and flood protection in urban waterfronts, taking climate change and economic development into account by applying the concept of multifunctional and adaptable flood defences (the AFD concept). This concept has the potential to create a physical synergy between flood protection and urban functions due to its multifunctionality. This physical synergy can be maintained by the adaptability of the concept, which takes possible changes in flood risk into account. The main recommendation is, therefore, to consider the adoption of the AFD concept in those cases in which it is difficult to improve urban flood defences, or in which redevelopment of an urban waterfront is difficult due to conflicts between different stakeholders.

The AFD concept can be applied under certain conditions. For instance, flood-retaining structures which contain only immovable elements are preferred over those which contain movable or temporary elements. This is due to the possibility of human error. Moreover, one should keep in mind that the financial feasibility of the AFD concept is at risk in the event of a lack of social support by inhabitants or financial support by stakeholders when future improvements of the flood-retaining elements are necessary. However, the success of the AFD concept does not solely depend upon the costs, but also upon the benefits; so the surplus value that is given to the urban quality is an important criterion when considering adopting a multifunctional and adaptable flood defence. It is, therefore, recommended to conduct research into the costs and benefits of the AFD concept in order to gain more insight into financial feasibility. In order to make the implementation of the AFD concept a success, it is also necessary for urban planners and flood controllers to invest in exploring each other's world and to invite each other to join in on separately initiated projects. In The Netherlands, the current processes for urban planning differ from processes for flood control, leaving hardly any overlap in the experience of both stakeholders. Additionally, the legislation concerning flood defences should incorporate multifunctional flood defences in order to facilitate the common usage of the AFD concept.

More information on the adaptable flood defences concept is available on the website www.urbanriverfronts.com. This website also contains a downloadable version of the dissertation *Design of Flood-Proof Urban Riverfronts* (Stalenberg, 2010), upon which this chapter is based.

References

Arakawa-Karyu River Office and MLIT (2006) *The Arakawa: River of the Metropolis: A Comprehensive Guide to the Lower Arakawa*, Arakawa-Karyu River Office Ministry of Land, Infrastructure and Transport, Japan

COS (Centrum voor Onderzoek en Statistiek) (2010) *Publiekszaken Rotterdam*, COS, www.cos. rotterdam.nl, accessed 9 November 2010

Gijsbers, R., Lichtenberg, J. and Erkelens, P. (2009) 'A new approach to flexibility-in-use: Adaptability of structural elements', SASBE2009, TU Delft, Faculty of Architecture, The Netherlands

Goetgeluk, R., Kauko, T. and Priemus, H. (2005) 'Can red pay for blue? Methods to estimate the added value of water in residential environments', *Journal of Environmental Planning and Management*, vol 48, no 1, p17

IPCC (Intergovernmental Panel on Climate Change) (2007) *Climate Change 2007: The Physical Science Basis*, IPCC, Cambridge University Press, Cambridge, UK

Kauko, T., Goetgeluk, R., Straub, A. and Priemus, H. (2003) 'Presence of water in residential environments: Value for money?', in *The 10th European Real Estate Society (ERES) Conference*, ERES, Finland

Kroonenberg, S. (2008) *The Human Measure: The Earth in Ten Thousand Years*, Atlas, The Netherlands

Municipality of Zwijndrecht (1998) *Zoningplan Westkeetshaven*, The Netherlands

Pols, L., Kronberger, P., Pieterse, N. and Tennekes, J. (2007) *Flood Risk as Spatial Task*, NAi and Ruimtelijke Planbureau, The Netherlands

Stalenberg, B. (2005) 'Optimal design of multifunctional flood defences in urban areas: Case study Deventer (NL)', in *International Symposium on Stochastic Hydraulics 2005*, IAHR, The Netherlands

Stalenberg, B. (2010) *Design of Flood-Proof Urban Riverfronts*, PhD thesis, Delft University of Technology, The Netherlands

Statistics Bureau of Japan (2010) *Japan Statistical Yearbook 2010*, Japan

Takahasi, Y. and Uitto, J. I. (2004) 'Evolution of river management in Japan: From focus on economic benefits to a comprehensive view', *Global Environmental Change*, vol 14, no 1, pp63–70

Waterboard de Groote Waard (undated) *Dordtse Wand: A Unique Project in Flood Management*, The Netherlands

9

The National Flood Insurance Program (NFIP) and Climate-Resilient Waterfront Development in New York City

Wouter Botzen and Jeroen Aerts

9.1 Introduction

The recently published report of the New York City Panel on Climate Change (NPCC) indicates that climate change poses a challenge for waterfront development in New York City (NYC), given the uncertain risks of sea-level rise and more flooding (NPCC, 2010). The NPCC states that NYC is vulnerable to coastal storm surges, which are associated with either late summer–autumn hurricanes or extra-tropical cyclones in the winter period ('nor'easters'). Climate change and sea-level rise is projected to increase the frequency and intensity of flood events, and although these events are relatively rare, it is relevant to address flood risks in current and planned urban development activities, such as waterfronts, because they can cause considerable losses (Aerts et al, 2009).

In order to address climate change in waterfront development, NYC has embarked on a climate adaptation programme (PlaNYC) outlining the policies needed to anticipate the impacts of climate change. As part of this policy, the Department of City Planning is currently preparing *Vision 2020: New York City Comprehensive Waterfront Plan* for the over 500 miles (805km) of NYC waterfront, defined as New York Harbor and its tributaries, creeks and bays. *Vision 2020* will set long-term goals to guide waterfront development. An integral part of the vision is to improve resilience to climate change and sea-level rise.

This chapter will provide input for the NYC's *Vision 2020* by assessing how flood insurance can contribute to waterfront development that is more resilient to climate change. A well-designed flood insurance programme with risk-based premiums can, in theory, be beneficial for societal welfare by giving a price signal of risk that prevents the uneconomical use of floodplains, and by incentivizing the undertaking of cost-effective risk-reducing or 'mitigation' measures, while providing compensation for flood damage to inhabitants of flood-prone areas (Burby, 2001; Botzen and van den Bergh, 2008, 2009). In this chapter, mitigation measures are defined as measures that limit (flood) damage which is consistent with the terminology in actuarial sciences, but differs from the climate sciences where it refers to reductions in greenhouse gas emissions. In the US, the federally operated National Flood Insurance Program (NFIP) is an important programme for achieving risk reduction because it imposes the minimum requirements for local governments' flood-zoning and flood-building codes, and it provides incentives to homeowners to invest in risk reduction beyond these minimum standards. This chapter assesses how the NFIP can be geared to cope with future challenges that are expected to considerably change the risk landscape, such as climate change and increased developments in flood-prone areas.

The remainder of this chapter is structured as follows. Section 9.2 describes the NFIP and federal flood-damage mitigation policies. Section 9.3 examines the strengths and weaknesses of the NFIP in the face of climate change. Section 9.4 gives recommendations for improving the NFIP. Concluding comments are provided in section 9.5.

9.2 Description of the Current NFIP and Federal Mitigation Policies

The federal government provides flood insurance through the National Flood Insurance Program (NFIP) in the US, which insures a value of about US$31.6 billion in New York State (NYS) and US$8 billion in NYC. The programme was initiated by the US Congress in 1968, and several amendments took place in 1969, 1973, 1994 and 2004 in order to improve its performance. The programme was established during the 1960s to ensure that local government planning and land-use management decisions give adequate recognition to flood hazards and meet insurance needs in flood-prone areas (US Congress, 1966a, 1966b, cited in Burby, 2001). The NFIP enables homeowners to purchase insurance coverage against flood damage under certain conditions that most commercial insurance companies refuse to cover in standard insurance policies.

The standard flood insurance policy of the NFIP covers direct material damage caused by floods, flood-related erosion as a result of waves or currents of water, and mudslides. The general coverage limits are US$250,000 for residential buildings and US$500,000 for non-residential buildings, such as shops and businesses. Risks to infrastructure, life and agriculture are outside the scope of the NFIP. In addition, wind damage (e.g. caused by hurricanes) is not covered through the NFIP; but this coverage can be purchased from private insurance companies, although flood losses caused by hurricanes fall under the NFIP coverage. In 2007, the total value of this private insurance coverage (thus excluding the NFIP, but including wind damage) in NYC and Long Island was more than US$2.3 trillion, which illustrates the large size of the value at risk (LeBlanc and Linkin, 2010).

Table 9.1 *The stakeholders of the National Flood Insurance Program (NFIP) and their main responsibilities*

Federal Emergency Management Agency (FEMA)
- Sets flood insurance premiums.
- Identifies flood-hazard areas.
- Makes flood risk maps.
- Designs and reviews construction criteria and floodplain management by communities.
- Funds mitigation projects.
- Provides flood insurance.

State governments
- Authorize and assist building regulation of local governments.

Local governments
- Regulate building in floodplains.

Private insurers
- Market insurance policies.
- Handle the process of paying out claims.

The NFIP operates like a public–private partnership. Table 9.1 lists the main stakeholders of the NFIP and their responsibilities. The Federal Emergency Management Agency (FEMA) administers the programme and sets flood insurance premiums, identifies flood-hazard areas, makes flood risk maps and provides design standards for constructions in floodplains. The state governments authorize and assist building regulations in flood-prone areas of local governments. In principle, local governments of communities can decide, on a voluntary basis, whether they want to join the NFIP. Insurers are the agents who sell the policies and underwrite the risk for which they are compensated; but the NFIP has the final financial responsibility and bears the risks. The NFIP operates like a national type of risk pool, and has never been reinsured by private insurance. The government operates like a reinsurer of last resort and Congress can lend money to the NFIP if it experiences deficits; in principle, this borrowed money should be repaid in the future. Reserves are built up by premiums. Most of the time premiums match losses, but not during times of catastrophes, such as Hurricane Katrina. Reinsurers, such as Swiss Re, have no stake in the NFIP. In addition to the NFIP, private flood insurance is available that is especially attractive for industry and large commercial entities, for which the coverage limit of US$0.5 million of the NFIP is too low. Apart from the NFIP, the federal government may provide federal disaster assistance to areas that have experienced a severe natural disaster, such as a hurricane or earthquake; but this *ad hoc* relief is not directly tied to the flood insurance programme.

9.2.1 Amendments made to the NFIP over time

Table 9.5 in this chapter's concluding appendix lists the main amendments made to the NFIP since 1968 and the problems they aimed to overcome. In 1969, an amendment was needed because it became clear that the NFIP did not have the capacity to conduct all the detailed flood studies and flood risk mapping in a reasonable timespan. These studies were required in order to allow communities to enter the programme. To solve this problem, it was decided that communities could enter the NFIP in what is called 'an emergency phase', in which preliminary flood risk studies are sufficient to enter the programme. In addition, the local government needs to agree to make best effort to control building in order to reduce flood risk.

The second amendment occurred with the passing of the Flood Disaster Protection Act in 1973 (US Congress, 1973). This act came to pass because few communities entered the NFIP and the market penetration of flood insurance was low until that time. The incentives of the insurance programme to encourage the adoption of re-quired floodplain building regulations were also insufficient. The amendment aimed to increase communities' participation by declining grants in aid for construction in flood-hazard areas to local governments which were not in the NFIP, and refusing them federal disaster assistance. Moreover, the amendment required homeowners in the 1/100-year floodplain to have flood insurance if they had federally backed mortgages in order to stimulate market penetration. These amendments to the NFIP proved to be effective in engaging the participation of communities in the programme, and increased the market penetration rate in the 1/100-year flood zones (Burby, 2001).

In 1994, the National Flood Insurance Reform Act aimed to further increase the market penetration of flood insurance and limit the experience of high losses by pro-moting risk reduction (US Congress, 1994). Market penetration was stimulated by strengthening the monitoring and enforcement of the requirements to purchase flood insurance, limiting disaster assistance for households without flood insurance, and by requiring that households in the 1/100-year flood zones who received disaster assist-ance should purchase and maintain flood insurance. Indeed, it has been empirically verified that the purchase requirements did increase flood insurance demand (Kriesel and Landry, 2004). Moreover, the Community Rating System (CRS) was established, which is a voluntary programme that rewards communities who invest in risk reduction with premium discounts for the policyholders in their community. The majority of NFIP policyholders participate in the CRS (Burby, 2001), and it has been found that CRS par-ticipation reduces the flood claims of communities (Michel-Kerjan and Kousky, 2010).

The 1994 act, moreover, established the Flood Mitigation Assistance Program (see Table 9.4 for more details) that provides mitigation grants to states and local governments. In addition, the 'increased costs of compliance' coverage aims to help homeowners pay the costs of making their property comply with the NFIP floodplain standards and regulations. This grant, given in the form of additional coverage, will pay the homeowner who has NFIP flood insurance a maximum of US$30,000 to comply with state or local floodplain management laws and ordinances (which include NFIP regulations) that affect the repair or reconstruction of a structure that has been damaged by flooding. Mitigation activities that are financed by this fund include elevation, flood-proofing, relocation or demolition of the building. The fund has only been applied to a limited extent, with total expenditures of US$59 million for 3209 'increased costs of compliance' claims until 2006 (Wetmore et al, 2006). This is mainly caused by strict eligibility requirements so that only specific expensive mitigation measures can be funded.

A recurrent problem for the NFIP is that certain existing buildings in a floodplain suffer flood damage repeatedly over time because they were built in very high-risk areas. About 1 per cent of the properties insured by the NFIP are defined as repetitive loss properties, and these have suffered about 38 per cent of all flood claims since 1978 (Bingham et al, 2006). The 2004 reform started a pilot programme to mitigate the flood damage of properties that have suffered repetitive losses. For this purpose, the Repetitive Flood Claims Grant Program and the Severe Repetitive Loss Program were established (see Table 9.4 for more details).

9.2.2 Requirements for purchasing NFIP insurance

Individuals and businesses in the 1/100-year floodplain in NFIP communities are required to have NFIP flood insurance as a prerequisite for receiving any type of federal financial assistance (including federal disaster assistance) and federally backed mortgages (including those provided by the main mortgage providers Fannie Mae and Freddie Mac). The objective of this requirement is to stimulate the market penetration of flood insurance. The amount of insurance coverage that needs to be purchased equals the amount of the outstanding mortgage loan, but obviously cannot exceed the maximum NFIP coverage. There are no obligations for contents coverage unless contents serve as collateral for a loan. The regulation is enforced by eight agencies that regulate the banking and mortgage lending industry. These include the Comptroller of the Currency, the Federal Reserve Board, the Federal Deposit Insurance Corporation, the Office of Thrift Supervision, the Farm Credit Administration and the National Credit Union Administration. Property owners are obliged to retain flood insurance for the entire duration of their mortgage loans. The federal lending regulators can impose monetary penalties upon lenders that have a 'pattern of practice' of violating the regulations. Lenders should assess whether a loan is secured by real estate subject to the mandatory insurance purchase requirement and, if so, notify the borrower that NFIP flood insurance is required, and ensure that flood insurance is maintained throughout the loan duration. If people 'forget' to renew their policy, then the bank notices this and obtains private flood insurance for the property ('force-placed flood insurance') for a much higher rate than the NFIP flood insurance (Tobin and Calfee, 2006). The premium of the force-placed flood insurance is charged to the homeowner again. In this way the homeowner gets an incentive to apply for flood insurance through the NFIP.

9.2.3 Flood-hazard maps and insurance premiums

An important task of FEMA is to produce flood-hazard studies and maps. This task consists of delineating flood-hazard areas and mapping floodways, flood elevations and flood velocity. FEMA defines the floodway as 'the channel of a river or other watercourse and the adjacent land areas that must be reserved in order to discharge the base flood without cumulatively increasing the water surface elevation to more than a designated height'. Floodplains or flood zones are defined as areas that can be inundated by a specific flood event, and are often expressed as having a certain flood probability, such as the 1/100-year floodplain. These flood-hazard studies form the basis of the creation of Flood Insurance Rate Maps (FIRMs). FEMA determines the insurance premiums based on historical losses averaged for the entire US. The insurance rates are divided into a few groups and depend upon a flood-zone classification system. It should be noted that similar standard premiums are charged per flood zone across the US, regardless of whether a community has suffered repeatedly from very high flood losses, or at the other extreme, has never experienced major flooding. The NFIP classifies flood zones according to their flood return period. An important flood zone is the area that is flooded 1 in 100 years, on average, which is defined as the Special Flood Hazard Area (SFHA). The following differentiation has been made to reflect different levels of flood risk – namely, moderate to low risk (zone B, C, X), which is outside the 1/100-year floodplain; high risk (zone A), which is in the 1/100-year floodplain; and the high-risk 1/100-year floodplain in coastal areas (zone V) (see Table 9.2).

Table 9.2 shows the standard premium levels in the different zones for residential insurance policies that cover damage to buildings and contents, only buildings or only contents. This table gives an indication of the degree of premium differentiation across the distinctive flood zones. Actual rates depend upon, amongst other factors, the type of building, number of floors, the level of coverage and the deductible (excess). Policyholders can choose amongst six deductible levels: US$500, $1000, $2000, $3000, $4000 and $5000. The lower the deductible, the higher is the insurance premium charged by the NFIP. Table 9.2 only gives the rates for the maximum coverage, which are US$250,000 for residential buildings and US$100,000 for contents. Two types of policies are available in moderate- to low-risk zones: the preferred risk policy and the standard policy. The premiums of a preferred risk policy are considerably lower and are only applicable to households that meet specific eligibility criteria that depend upon the buildings' entire flood-loss history. Only standard policies are available in the A and coastal V zones that are at higher risk, and have consecutively higher premiums of up to US$5700 in the latter zone. Average flood insurance premiums in NYC are US$776 per year for an average coverage per policy of US$218,563, which is relatively high compared with average coverage in other states, mainly because of the higher average values of the properties in the former (Michel-Kerjan and Kousky, 2010).

Existing buildings in floodplains that were erected before the creation of the NFIP flood risk maps pay lower premiums, which are between 35 and 40 per cent of the true

Table 9.2 *Standard annual flood insurance premiums for residential properties in 2010*

Risk level of the area	Zone	Policy	Building and contents	Only building	Only contents
Moderate to low risk	B, C, X	Preferred risk	US$395*	n/a	$228**
Moderate to low risk	B, C, X	Standard	$1489	$911	$618
High risk	A	Standard	$2633	$1620	$1053
Coastal high risk	V	Standard	$5700	$3487	$2253

Notes: *This premium is for buildings with a basement or enclosure, and is higher than those without.
** This premium is lower if only the contents above the ground floor are insured.
n/a = not available.
Source: FEMA (2010)

risk premium (Bingham et al, 2006). This subsidy is granted because it was considered unfair for these homeowners to pay high rates since they may have lacked the knowledge that their building was in a floodplain. Moreover, these subsidies were meant to encourage homeowners to purchase flood insurance, and stimulate communities to join the NFIP (Burby, 2001). The premiums in Table 9.2 are standard premiums, and individual policyholders may obtain reductions in their premium if the lowest floor of their house is located above the baseline flood elevation (BFE) level of the 1/100-year flood. These reductions in premiums can be obtained for both new and existing buildings. New buildings have to be elevated at least to the BFE level according to the NFIP regulations, and only buildings that are further elevated can obtain a premium discount.

Policyholders can, moreover, obtain premium discounts through the Community Rating System (CRS) that rewards communities who invest in risk reduction with premium discounts of up to 45 per cent of the full rates defined by FEMA. These credits benefit all the NFIP policyholders in a community because they receive a percentage reduction in premiums. The CRS assigns a ranking of 10 classes, where a 1 indicates that the community has taken the most significant collective flood mitigation measures, while a 10 is given to communities who have done nothing in this respect. Table 9.3 shows the premium discount that can be obtained per class. Communities can improve their ranking by implementing a range of mitigation strategies, improving flood

Table 9.3 *Community Rating System (CRS) premium discounts per class*

Class	Discount	Class	Discount
1	45%	6	20%
2	40%	7	15%
3	35%	8	10%
4	30%	9	5%
5	25%	10	0%

Source: FEMA (2006)

awareness or facilitating accurate insurance ratings. The CRS discount is not applicable to preferred risk policies.

9.2.4 NFIP requirements and mitigation of flood damage

In principle, communities decide themselves whether they want to join the NFIP, although a few state governments require that their communities participate in the programme (Burby, 2001). The NFIP has established the eligibility requirements related to flood risk reduction that communities have to meet in order to be covered by the flood insurance. For example:

- Local governments have to restrict development in floodways, but they are not required to apply zoning regulations in the remainder of the floodplain.
- The ground floor of new constructions in the 1/100-year flood zone needs to be elevated to the estimated baseline flood elevation (BFE).
- Existing structures should meet this elevation criterion if improvements are made that exceed 50 per cent or more of a structure's market value.
- Mobile homes need to be anchored and are no longer allowed to be built in the floodway.
- Subdivisional proposals, which are areas of real estate that are composed of subdivided lots, are to be reviewed to limit flood risk. For example, the works should be consistent with minimizing flood damage in the flood-prone area; public utilities and facilities should be located and constructed while simultaneously minimizing or eliminating flood damage; and an adequate drainage system should be provided.

The elevation requirements in the 1/100-year floodplain and prohibitions to build in floodways remain in force after the community has entered the programme. The NFIP enforces local compliance with the flood insurance legislation and can conduct site visits to local governments, and state governments can report violations to the NFIP. Moreover, enforcement is checked through the 'rate applications' that are made in a specific community. If a homeowner applies for flood insurance, then a special assessment of the building is required to estimate premiums if the building does not meet the general standards. For example, such a special assessment is needed if the elevation is further below the BFE than can be indicated on the standard application forms for flood insurance. These rate applications are submitted to FEMA, and together form a database which FEMA can use to evaluate the level of compliance of structures with the NFIP standards.

9.2.5 Mitigation grant programmes

Table 9.4 provides the main characteristics of the five most important currently available mitigation grant programmes of the NFIP. The Hazard Mitigation Grant Program can provide mitigation funds to state and local governments and to some private non-governmental organizations (NGOs) after a disaster has occurred, as indicated by a presidential disaster declaration (FEMA, 2009). The main aim of this fund is to promote more hazard-resilient rebuilding after a disaster. The fund is not limited to financing flood mitigation, but also provides funds to mitigate other hazards, such as earthquakes and wildfires. The Flood Mitigation Assistance Program is specifically geared towards mitigating flood damage and provides grants to state and local governments for developing FEMA applications, assessing risk and developing mitigation plans, and implementing projects that 'flood-proof' buildings. The Pre-Disaster Mitigation Program was authorized in 2000. The main objective was to allow the funding of mitigation measures under the Hazard Mitigation Grant Program before disasters occur, which was regarded as being more effective than only providing mitigation funds to areas that have actually suffered from a disaster.

The Repetitive Flood Claims Grant Program funds 100 per cent of those mitigation measures for structures that have one or more NFIP loss claims (see Table 9.4), and is paid out of NFIP premium revenues. Funding will be prioritized for those applications that can result in the largest cost-savings for the NFIP. In addition, the Severe Repetitive Loss Program provides financial resources to state and local governments for mitigation activities, such as elevation, relocation, demolition, rebuilding, flood-proofing and purchasing the property (see Table 9.4). These state and local governments have to match 25 per cent of the funding provided, which may be reduced to 10 per cent if the state has an approved mitigation plan, and if FEMA has determined that it has taken action to reduce the number of severe repetitive loss properties. The FEMA grant is paid out of NFIP premium revenues. If a homeowner with a severe repetitive loss refuses to agree with a reasonable mitigation offer, then the flood insurance premium will be increased to 150 per cent of the chargeable rate of the property at the time of the flood loss. At the maximum, this chargeable rate is the estimated risk premium of the area. The mitigation fund consisted of US$40 million in 2006 and 2007, US$80 million in 2008 and 2009, and US$70 million in 2010. In NYS and the state of New Jersey, respectively, 206 and 509 buildings have been validated as a severe repetitive loss property, and about US$4 million and US$11 million in grants have been provided out of the mitigation funds to 'flood-proof' these buildings in NYS and the state of New Jersey, respectively (US Department of Homeland Security, 2009).

9.3 Strengths and Weaknesses of the NFIP in the Face of Climate Change

Several studies have critically reviewed the performance of the NFIP and recommended major revisions to improve the programme, which mostly focus on more effective incentives and policies for flood risk mitigation (Burby, 2001, 2006; Wetmore et al, 2006; Kunreuther et al, 2009). These studies do not provide an explicit or in-depth assessment

Table 9.4 *Federal Emergency Management Agency (FEMA) mitigation grant programmes*

Mitigation grant programme	Requirements	Eligible recipients	Main measures funded	Yearly financial capacity
Hazard Mitigation Grant Program	Presidential Disaster Declaration	State and local governments; some private NGOs	Elevation of homes and businesses Demolition or relocation of homes Retrofitting buildings Flood control projects for critical facilities Construction of safe rooms for tornado protection	Sliding scale funding of three portions: 1 15% of first US$2 billion of disaster assistance; 2 10% of disaster assistance between US$2 and $10 billion; 3 7.5% of disaster assistance between US$10 and $35.333 billion
Flood Mitigation Assistance Program	Flood Mitigation Plan for project grants	States and communities	Planning grants for assessing risk and developing flood mitigation plans Project grants for elevating, acquiring, demolishing or relocating NFIP insured buildings Technical assistance grants for developing FEMA applications and implementing projects	US$32.3 million in 2010
Pre-Disaster Mitigation Program	Flood Mitigation Plan	States and communities	Elevation and relocation of existing public or private structures Flood control projects for critical facilities Protective measures for utilities Storm-water management projects	US$90 million in 2009

			Vegetation management for natural dune restoration, wildfire or snow avalanche Structural and non-structural retrofitting Construction of safe rooms for public and private structures Voluntary acquisition of real property	
Repetitive Flood Claims Grant Program	One or more claim payments from NFIP ineligible for Flood Mitigation Assistance Program; State Hazard Mitigation Plan	States and communities	Acquisition, structure demolition or structure relocation, with the property deed restricted for open space uses in perpetuity	US$10 million per year
Severe Repetitive Loss Program	Only for residential properties; flood losses that resulted in either within a 10-year period (1) four or more flood insurance claims payments each > $5,000 with at least two of the payments made or (2) two or more flood insurance claims payments that cumulatively exceeded the property value	States and communities	Elevation, relocation or demolition of existing residential properties Flood-proofing measures for historic properties Minor physical localized flood control projects Demolition and rebuilding of properties to at least the base flood elevation (BFE) or greater if required by any local ordinance	US$70 million in 2010

of how the NFIP needs to be geared towards dealing with a potential increase in future flood risk as a result of climate change, as is the focal point of this chapter. Nevertheless, the recommendations of existing studies about how the NFIP can be geared towards achieving risk reduction are especially relevant in the context of the future increase in flood risk due to climate change that is projected for some regions in the US. This chapter provides a novel contribution by focusing on how the NFIP can accommodate, and contribute to, flood-resilient waterfront development in NYC, although many of the issues addressed are more broadly applicable.

9.3.1 Requirements for purchasing insurance through the NFIP

Although the NFIP has been successful in providing many homeowners with flood insurance that would otherwise not be available, the market penetration of flood insurance is rather low. The exact market penetration of the NFIP is difficult to assess; but a study has been made by Dixon et al (2006). According to these authors, only about 49 per cent of single-family homes in the 1/100-year flood zone carry NFIP flood insurance, which is similar to estimates by Kriesel and Landry (2004) for coastal areas in the US. Market penetration is very low (about 1 per cent) in the flood zones without the mandatory insurance purchase requirement. Large regional differences exist in market penetration rates. For example, market penetration is estimated to be only 28 per cent in the north-east of the US, while it is the highest at 60 per cent in the south. This sample excludes NYC and, to our knowledge, there is no specific study about market penetration of flood insurance in NYC. Total NFIP policies in force in NYC were about 37,000 in 2010, which suggests that market penetration is also low in NYC. The market penetration of commercial flood insurance is even more uncertain than that of the NFIP because it is not monitored and reported. Kriesel and Landry (2004) estimate that the market penetration of commercial flood insurance is only about 4 per cent in coastal areas (across all flood zones). Large industrial risks are likely to carry flood insurance coverage.

There are several explanations for the low market penetration of flood insurance. The mandatory purchase requirement of flood insurance only applies to federally backed mortgages of homeowners in the 1/100-year flood zone, while there are no obligations for tenants to purchase flood insurance. This severely limits the scope of the obligation, and only an estimated 50 to 60 per cent of the single-family homes in the 1/100-year zones are subject to the mandatory insurance purchase requirement (Dixon et al, 2006). There are no obligations to purchase flood insurance outside the 1/100-year flood zone. Many homeowners drop the coverage once their loans have been paid off because they are only required to purchase insurance coverage for the portion of the mortgage that is outstanding (Tobin and Calfee, 2006). Compliance with the requirement is estimated at approximately 75 to 80 per cent (Dixon et al, 2006). Enforcement may be complicated because FEMA has no authority over the eight financial institutions that are responsible for checking compliance with the requirements (Wetmore et al, 2006). Several behavioural explanations have been put forward to explain the low voluntary purchase of flood insurance, such as low risk perceptions and awareness by individuals (Browne and Hoyt, 2000).

The low market penetration is an impediment for stimulating flood risk reduction through insurance. The empirical studies of NFIP market penetration mentioned earlier indicate that many households are not insured, which implies that insurance does not impose market discipline by promoting risk reduction or prevent uneconomic use of the floodplain. The low market penetration has, moreover, the adverse effect that the ability to spread risk is impaired, which generally results in higher premiums for the remaining pool of insured. The NFIP insurance may suffer from adverse selection that causes the observed shortfall of premium revenue compared with claims because flood insurance does not generate enough revenue if homeowners with lower risk do not purchase it. Adverse selection in insurance markets occurs if mostly individuals with a high risk choose to purchase insurance, while insurers are insufficiently capable of distinguishing low- from high-risk policyholders and charging the latter a higher premium. Especially in the face of a projected rise in flood risk due to climate change, it is desirable to stimulate the take-up of flood insurance and increase the market penetration in order to improve risk-sharing and make better use of the tools that insurance offers to encourage households to invest in risk reduction. If climate change increases flood risks, then this would imply that more uninsured households would suffer flood damage, which would either increase the need for federal disaster assistance or leave many households uncompensated and in financial distress. A higher market penetration of flood insurance could, therefore, be a way to alleviate the impacts of, and promote adaptation to, climate change.

9.3.2 Flood Insurance Rate Maps (FIRMs) and insurance premiums

An important strength of the NFIP is that it provides a platform for studying flood risk and creating flood-hazard maps for communities in the US, which is a tremendous task given the large geographical area of the country. Indeed, the NFIP delineates flood-hazard areas for most of the US and updates their Flood Insurance Rate Maps (FIRMs), which are publicly available on the internet. An accurate and detailed assessment of the regional flood risk is essential for a well-functioning insurance programme because it is the input needed to set actuarially based premiums that reflect risks. Moreover, such maps provide important information for local floodplain management policies.

According to Burby (2001), the quality of the flood risk studies falls short of achieving an accurate risk assessment for four reasons, which have nationwide implications and do not only apply to NYC. First, there is a lack of funding to keep flood insurance maps up to date by revising them frequently enough so that they reflect changes in local conditions. Second, no maps are created for areas subject to localized storm-water drainage flooding and areas that are vulnerable to the failure of dams and flood control infrastructure. If areas are protected by levees that are high enough to withstand the 1/100-year flood level, then these areas are mapped out of the 1/100-year zone. However, it has been asserted that maintenance of levees is often inadequate, and a variety of failure mechanisms exist that cause the flood probability to exceed the 1/100 level, which would provide a rationale for taking a more prudent approach and mapping these areas more carefully (Bingham et al, 2006). Third, neither have maps been created nor are regulations in place for probable future hazards and changes in risk that could, for example, arise because of watershed development, land subsidence, erosion and sea-level rise caused by climate change. Fourth, the maps are often not detailed enough for incorporating flood hazards in land-use planning and management. For example, our discussions with local planners in NYC revealed that, in practice, a single property can lie partly within and partly outside the 1/100-year flood zone demarcation line. In that case, inspectors need to determine whether the property needs to comply with the flood building codes.

The consequences of the resulting inaccuracy of FIRMs are threefold. First, although premiums partly depend upon the flood zone classifications, premiums do not completely reflect actual risk if the maps are inaccurate. This impairs incentives for homeowners to limit risk (Kunreuther, 2008). Second, the decision not to map certain risks, such as the failure of flood protection, results in an underestimation of flood risk and too low premiums, which contribute to the operating losses incurred by the NFIP and distort incentives for mitigation (Burby, 2001). In the past, the NFIP has experienced considerable shortfalls of premium revenues compared with payouts of claims: in particular, operating losses occurred in the 18 years between 1972 and 2005 (Pasterick, 1998; Burby, 2006). Third, Burby (2001) recognizes that, in general, the spatial (in)accuracy of flood risk maps, the lack of detailed geo-references and the lack of property lines indicating the location of structures are problematic for local government building code and flood-zoning policies.

The inaccuracy of the current flood-hazard maps is especially problematic in the face of the projected rise in flood risk as a result of climate change. If current hazard information is insufficient to steer effective risk reduction by individuals and governments, then the resulting suboptimal flood protection and preparedness may turn out to be very costly when flood risk increases in the future. Moreover, a continuous process of updating maps may be especially important if risk changes over time not only because of socio-economic developments in the floodplains, as has been the case during the past, but also because of changes in the frequency and intensity of the flood hazard due to changes in precipitation, storms and sea-level rise. Accurate and up-to-date FEMA flood-hazard maps are important in steering appropriate waterfront development in NYC. For example, the delineation of the 1/100-year flood zone determines the minimum standards to which new constructions are subject, and whether it is mandatory for new construction financed with federally backed mortgages to carry flood insurance. Moreover, the flood-hazard maps are an input for the insurance rates, which can provide an important price signal that may steer new development and provide homeowners and contractors with incentives to implement mitigation measures.

In general, the premiums of the NFIP do not completely reflect risks (LeBlanc and Linkin, 2010). There are four main reasons for this. First, the premiums of buildings in the 1/100-year floodplain that were constructed before the NFIP came into force are subsidized. Although, the percentage of subsidized policies declined from 83 per cent of total policies in 1985 to approximately 25 per cent in 2004, the number of such policies remains high (Bingham et al, 2006). In principle, premiums are actuarially based for new constructions. Second, the existing premium levels are high enough to ensure that the NFIP can pay its claims in an 'average historical loss year', but not a catastrophic year. This implies that, in general, premiums are too low. The average premium shortfall before Katrina was about US$800 million per year, which needs to be borrowed from the federal government (Bingham et al, 2006). During some years, shortfalls of premium revenue of the NFIP were considerable. For instance, FEMA paid US$19.28 billion in claims during the destructive hurricane season in 2005 compared with an annual US$2.2 billion in premium revenues, which required a large expansion of the borrowing capacity of the NFIP (Michel-Kerjan and Kousky, 2010). Third, the premiums and deductibles remain the same, even after repetitive losses have been suffered (unlike the case with car insurance after many accidents). It is noted here that most repetitive losses occur in the Gulf states. Fourth, the current premium differentiation is based on relatively few risk zones that are the same for the whole nation, with no distinctions made for different states or counties (GAO, 2008). For example, the areas of 1/100 and 1/500 years are very large in certain cases, and there is no differentiation of premiums within the zones.

Incentives for homeowners to implement risk reduction measures are distorted if premiums do not accurately reflect risk, which can be especially troublesome if more investments in mitigation are needed in the future due to risk increases. A useful characteristic of the programme is that homeowners have an incentive to invest in an elevation above the BFE level because this can lower their premiums. Nevertheless, it has been argued that the NFIP should re-evaluate flood insurance premium discounts for buildings in A zones because A zone discounts effectively cease at 1 to 2 feet above the BFE, even though a higher elevation would be desirable (Jones et al, 2006). A future challenge is that the BFE is projected to increase in many places, as the estimates made for NYC show (NPCC, 2010). Another problem with increasing flood risk due to climate change is that updates of FIRMs over time will imply a shift in the zone classification for many properties. The current regulation for such properties is that property owners who are remapped to a more costly zone classification are charged the lower insurance premium of the former flood zone through grandfathering (Bingham et al, 2006). If this policy continues, then it may be expected that the number of properties that pay such subsidized rates would increase considerably in the future.

The Credit Rating System gives communities incentives to implement mitigation measures in return for premium discounts. Although, in principle, this may be a good incentive for communities to increase mitigation investments, concerns have been raised that the CRS contributes to the too-low levels of premiums in some communities (Burby, 2001). Our discussion with government experts in the New York area revealed that the CRS is not attractive for NYC. It is unclear for policy-makers how the credits relate to specific measures and policies, and what the resulting premium reductions are. This has been examined for NYC; but turned out to be administratively very complex and not worthwhile. It could be useful to change this credit system so that its eligibility requirements are better suited to the needs of a densely populated city such as NYC.

9.3.3 NFIP requirements and mitigation of flood damage

The NFIP aims to link flood insurance to improved risk prevention and damage mitigation at both the community and individual level. At least the NFIP forced the local governments to pay attention to managing urban development in a way that reduced flood risk in order to be eligible for flood insurance. It has been estimated that the NFIP standards and mitigation programmes have saved, in total, over US$1 billion of flood losses per year (Sarmiento and Miller, 2006). However, it has been recognized that more can be done in this respect (Burby, 2001). The programme has been rather effective in limiting the vulnerability of new constructions to flood hazards. Pasterick (1998) showed that buildings that were constructed before the enforcement of the NFIP standards suffered approximately six times more flood damage than buildings that were constructed in compliance with the NFIP standards.

Even though the NFIP requirements are quite successful in flood-proofing new buildings, the programme is generally evaluated as being ineffective in limiting new developments in high-risk areas. For instance, during the first 30 years of the NFIP, the number of buildings in such areas has increased by 53 per cent (Burby, 2001). These trends are likely to continue in the future, as is apparent from Figure 9.1, which shows the past and projected increase in the number of buildings in Special Flood Hazard Areas, which is the 1/100 flood zone. Although the number of old buildings constructed before the NFIP (pre-FIRM) is expected to decrease, the total number of buildings in SFHA is increasing due to rapid urbanization in flood-prone areas. It should be recognized that large regional differences exist in the success of steering development away from high flood-risk zones. Fast-growing coastal communities appear to have been least successful in shifting development to non-vulnerable areas (Burby and French, 1985; Burby, 2001), which is probably because of the high opportunity costs of building further inland.

Figure 9.2 shows the cumulative value at risk of all properties in NYC over time between the years 1880 and 2009 (in 2009 US dollars) in the different flood zones. Three interesting observations can be made. Note, however, that there is uncertainty related to these calculations as newer buildings could have replaced older buildings. First, during the 1930s and the 1990s there was a very large increase in the value at risk, which is probably due to an accelerated increase in urban development, in combination with the

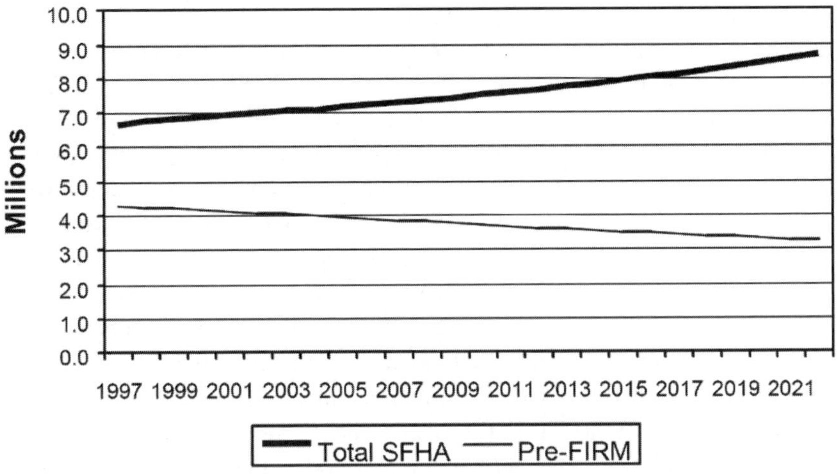

Figure 9.1 *Total historical and projected future number of buildings in Special Flood Hazard Areas (SFHAs) in the US, and pre-Flood Insurance Rate Map (FIRM) buildings*

Source: Wetmore et al (2006)

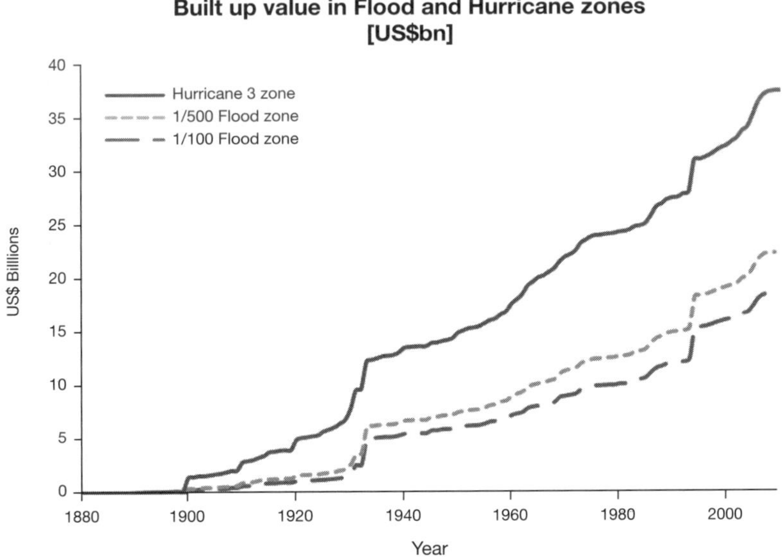

Figure 9.2 *Cumulative values of the properties at risk in the 1/100, 1/500 and Hurricane 3 flood zones in New York City over time (1880–2010)*

Note: The figure shows total built-up values (all in net present 2009 values, in US$ billion).

development of some expensive properties, such as airfields and naval bases. Second, during the 1990s, the figure shows that the increase in the value at risk in the 1/100 flood zone was similar to that in the 1/500 flood zone. This indicates that flood risk management, such as that conducted by the NFIP, was not able to restrain development in the 1/100 floodplain, which has a higher risk than the 1/500-year floodplain. Three, Figure 9.2 indicates that development in the flood zones continues steadily over time. The overall trend in the 1/100-year flood zone is the same as the 1/500-year flood zone, which indicates that the NFIP has been unable to steer urban growth to lower-risk areas in NYC, as is also the case for the US in general.

In contrast to the success of flood-proofing new constructions, the programme did very little to reduce the vulnerability of existing buildings to flood hazards. Very few homeowners in the US have taken measures to limit flood risk in existing buildings (Kunreuther and Roth, 1998). The NFIP can purchase insured properties that have flood damage representing more than 50 per cent of their value, or have been repeatedly damaged. These houses can then be relocated or demolished. Approximately 1400 properties were purchased at a total cost of US$51.9 million between 1968 and 1994 (Pasterick, 1998). Although this seems like a good strategy to deal with old buildings located in very high-risk areas, the programme continues to suffer from the problem of repetitive flood losses affecting existing buildings. For instance, properties suffering from repetitive loss were estimated in 2004 to cost the NFIP about US$200 million annually. NYS ranks in the sixth place of states with the most insured repetitive losses in the US. It has 7141 repetitive loss properties and suffered, in total, US$100.4 million in insured repetitive losses between 1978 and 2002 (King, 2005). A factor that contributes to the problem of severe repetitive losses is that the premiums charged for these properties are not actuarially fair and often too low, which impairs incentives to flood-proof these buildings (US Department of Homeland Security, 2009).

Another problem is that violations of the NFIP mitigation standards by local governments and homeowners are not uncommon and have resulted in considerably more losses during the past than would have occurred if regulations were adhered to (Burby, 2001). Moreover, many buildings that are exposed to flooding are currently unregulated by the NFIP because they are located outside the 1/100-year flood zone. The decision to regulate only the 1/100-year flood zone has been criticized and seems to have been based on political criteria instead of efficiency grounds guided by cost–benefit analyses. This has been acknowledged by FEMA (2006) in its evaluation of the NFIP. Most flood losses in the US are caused by less-frequent flood events (Burby, 2001). The programme could increase its effectiveness by regulating flood zones affected by lower probability floods, such as the 1/500-year floodplain (Association of State Floodplain Managers, 2000).

9.3.4 Mitigation grant programmes

FEMA has been actively involved with funding flood prevention by state and local governments and their efforts to mitigate potential flood damage to structures. The availability of the grant programmes can be a motivation for communities to join the NFIP, while it reduces the costs of flood hazards to the NFIP and society in general. However, there could be scope to expand the grant programmes for mitigation. The budgets of, respectively, the Flood Mitigation Assistance Program and the Pre-Disaster Mitigation Program are, in total, US$32 million and US$90 million, which is not very large when it is considered that this applies to the entire US, while the latter programme also funds non-flood-related mitigation measures. Indeed, more investments in mitigation seem to have a high benefit compared with costs. The National Institute of Building Sciences (2005) has estimated that effective flood-hazard mitigation has an overall benefit–cost ratio of 5 to 1.

The Repetitive Flood Claims Grant Program and the Severe Repetitive Loss Program seem promising initiatives to tackle the problems that the NFIP has experienced with repetitive losses. However, the budgets of these programmes are, respectively, US$10 million and US$70 million per year, which appears to be too small for solving the problem with repetitive losses in a short timespan. The costs to acquire the current properties defined as severe repetitive losses are estimated to be US$1.8 billion (US Department of Homeland Security, 2009). This is considerably higher than the current funding available for the mitigation programme for severe repetitive losses, which had, in total, US$160 million available in 2009. Although this may be enough to gradually reduce severe repetitive loss claims, it should be noted that the number of severe repetitive loss properties tends to increase over time, which may be accelerated by the projected increase in flood risk due to climate change. Another limitation is that many of the grant programmes, such as the Severe Repetitive Loss Program, have a cost-sharing condition for local governments of about 25 per cent before FEMA provides the flood damage mitigation grant. Local governments often lack these financial resources, which results in a limited applicability of the grants (US Department of Homeland Security, 2009).

It seems unlikely, therefore, that the current grant programmes will free up sufficient financial resources to finance the required adaptation policies and measures that will be needed to accommodate rising flood risk due to climate change. Examples of adaptation measures that could prevent the projected rise in flood frequency are beach nourishment, and building dikes, sea walls and storm surge barriers. Moreover, sea-level rise may increase the areas that will be inundated during floods, such as the 1/100-year flood zone, which requires additional mitigation measures for buildings, such as elevating new constructions. There is much uncertainty about the costs of adaptation, which is obviously dependent upon the uncertain impacts of climate change upon flood risk. As an illustration of the possible size of these costs, Nicholls (2003) states in a study for the Organisation for Economic Co-operation and Development (OECD) that costs to adapt to a 1m sea-level rise could be in the order of US$257 billion (2010 values) for the US, considering only direct protection and assuming no changes in storm conditions. This estimate may be considered an underestimation because many cost categories are excluded, such as the costs for flood-proofing buildings. Adaptation costs would be considerably more expensive in some places than in others, partly because current protection levels are already suboptimal, which has been defined as an 'adaptation deficit'. For example, Nicholls (2009) mentions NYC as an example of a city with a large 'adaptation deficit' because parts of the city are currently insufficiently protected against storm surge and flooding with hard infrastructure, such as coastal protection by dikes, high sea walls or barriers. The current grant programmes are, moreover, too limited in scope to provide a meaningful financial contribution towards building new waterfront development in NYC in a flood-proof and climate-change resilient manner because the grants mainly focus on protecting existing buildings from flood damage.

9.4 Suggestions for Improving the NFIP

Section 9.3 identified several weaknesses of the NFIP with respect to dealing with the impacts of climate change and on-going socio-economic developments in flood-prone areas. This section provides several suggestions for improvements, particularly:

- enhancing the NFIP's market penetration;
- improving flood-hazard maps;
- charging insurance premiums that reflect risk;
- extending the duration of insurance contracts;
- addressing climate change in the NFIP requirements;
- broadening mitigation grant programmes.

9.4.1 Enhancing the NFIP's market penetration

The currently low market penetration of the NFIP flood insurance could be increased so that more households would have their flood losses covered by insurance after a flood event, and more buildings would be subject to the NFIP's mitigation requirements. There seems to be some scope to improve the enforcement of the mandatory purchase requirement for homeowners with federally backed mortgages who live in the 1/100-year flood zone. Nevertheless, there would still remain a very low market penetration outside the 1/100-year zone. Raising risk awareness in areas with lower probability floods could bring about a higher demand for flood insurance. Given that the 1/100-year flood zone is expected to increase in the future, consideration should be given to making insurance compulsory in the expected future 1/100-year zone or in a current flood zone with a larger flood-return interval, such as the 1/500-year zone. Expanding the mandatory purchase requirements brings the benefit that more constructions which are expected to be at future risk of flooding would be subject to the NFIP mitigation requirements. Alternatively, coverage against flooding could be made compulsory on all existing building and home contents insurance policies that, for example, cover fire risks, which could result in a very large risk pool and high market penetration. This is the practice in France where flood insurance is covered via a natural disaster insurance coverage called CatNat, which is compulsorily included with property insurance. This requirement has resulted in an almost universal market penetration among both tenants and homeowners (Poussin et al, 2010). If such a requirement were to be introduced in the US, then it would be important to set premiums according to risk in order to provide adequate incentives for mitigation. Furthermore, the French property insurance provides coverage against a variety of natural hazards via the CatNat system, and it could be useful to explore the application of this system in the US. Broad natural hazards coverage could prevent the problems that homeowners in the US have experienced about whether hurricane damage should be covered by the NFIP flood insurance or private windstorm damage insurance, which resulted in long and expensive lawsuits after Hurricane Katrina (Kunreuther and Michel-Kerjan, 2009).

9.4.2 Improving flood-hazard maps

The discussion in section 9.3 highlighted that there is scope to improve the accuracy of existing flood-hazard maps produced by FEMA. More detailed maps would be useful to guide local land-use planning. Moreover, it could be beneficial to map residual risk more extensively, such as areas protected by levees, in order to improve the information about potential flood zones and to ensure that adequate building code regulations are in place in such areas. In addition, a mechanism could be put in place that allows for regular updates of flood-hazard maps, so that changing local conditions such as developments in flood-prone areas are adequately accounted for in estimating the flood zones. At this moment, FEMA is updating its flood maps through its Flood Hazard Map Modernization Initiative, which has been work in progress since 1997 (Morrissey, 2006). The plan is to update all flood-hazard maps and create digital maps that can be used by computer applications. Even though the creation of digital flood maps could lower the costs of updates, concerns have been raised about whether FEMA has the capacity to regularly update maps that reflect the continuously changing conditions in risk in the future (Morrissey, 2006).

In order to implement measures to limit the expected future increase in flood risk due to climate change, it would be useful to make maps that indicate the expected future change in the floodplain or change in risk. For example, it could be useful to know where the future 1/100-year flood zone will be (e.g. in the year 2050). Knowing the future 1/100 flood zone offers the possibility to impose the same requirements that apply to the current 1/100-year zone, such as elevating new buildings, to the future 1/100-year zone. Such a policy could ensure that adequate mitigation measures are undertaken in what is expected to be the future floodplain. Given the long lifetime of new constructions, such regulation would anticipate the current problems with flood-proofing existing properties.

It can be questioned whether FEMA currently has the financial and human resource capacity to produce future-looking flood-hazard maps. A more active role for the insurance industry may be needed to create flood-hazard maps, as is, for example, the case in the UK and Germany (de Moel et al, 2010). Alternatively, local governments with a large exposure to flood risk and a strong adaptation policy agenda, such as NYC, may take the lead in this and produce their own future flood hazard and risk maps. The NPCC (2010) has already made an initial estimate of the future 1/100-year flood zone for NYC, which is shown in Plate 8. This map does not show the exact boundaries of the 1/100-year flood, but is a rough indication of how the current 1/100-year zone may change in the future. There is a high degree of uncertainty about the extent of future flood zones, and more research is needed if future flood maps are to be used as an input in insurance, flood zoning or building regulations. In order to effectively guide adaptation, a more accurate and detailed analysis is in order, while an estimation of the geographical distribution of flood risk is needed to indicate those locations that are especially vulnerable to flood damage. Current and future flood risk in NYC can be estimated more precisely with catastrophe models and the expertise of insurance companies and academics (Leblanc and Linkin, 2010). The knowledge to create such maps has already been developed in other countries: for example, the modelling of current and future flood risk is already common practice in The Netherlands (e.g. Bouwer et al, 2010).

9.4.3 Charging insurance premiums that reflect risk

Section 9.3 indicated that the NFIP insurance premiums do not accurately reflect the actual risk faced by policyholders. It has been well recognized that risk-based premiums can facilitate risk reduction by providing a price signal of risk of settlement in risk-prone areas, and by providing incentives to policyholders for risk reduction (Kunreuther, 1996). Kunreuther et al (2009) propose that premiums that reflect risks are a fundamental guiding principle to develop efficient insurance protection against natural disaster risks. The application of this principle would provide a signal of relative flood damage to those currently residing in flood-prone areas and those who are considering moving to those locations, such as constructions near the waterfront. Indeed, such homeowners are charged higher rates if they move into, for example, a coastal flood zone (see zone V in Table 9.2); but the current crude relation between premiums and actual risk provides an inaccurate or 'noisy' signal of risk, at best. Adequate incentives to policyholders, such as premium discounts, to invest in cost-effective flood risk reduction cannot be provided through insurance, unless premiums reflect the risk of exposure to floods.

The NFIP could set actuarially fair premiums that better reflect risk in order to make more effective use of the insurance mechanism to steer risk reduction by individuals.

This could be implemented in the current programme by charging premiums that are high enough to not only cover losses in an average loss year, but are also, on average, capable of covering losses of catastrophic floods, and by better differentiating premiums according to the actual flood risk faced by policyholders. Moreover, the current subsidies given to premiums of pre-FIRM buildings could be phased out over time (Bingham et al, 2006), and the grandfathering of lower premiums for structures that are rezoned to a higher risk area after a map update could be abolished. Alternatively, a greater role may be given to private insurance companies in the pricing and underwriting of risk, which could result in more actuarially sound rates. This is, for example, current practice in the UK where flood insurance is provided by the private sector without government support (Crichton, 2008). In the UK, premiums are differentiated according to risks, and insurers obviously have an incentive to set premium levels high enough to recoup flood losses. With a higher spatial accuracy of mapping risk, NFIP could improve pricing the risk into actuarially fair premiums.

Changing the premium structure of the NFIP by charging homeowners premiums that fully reflect risk brings about important equity and affordability issues. While the premiums of policyholders with a very low risk would decrease, premiums would increase for policyholders in flood-prone areas with a very high flood risk. This rate of increase for households currently residing in hazard-prone areas may be regarded as inequitable, and low-income households may not be able to afford higher risk-based premiums. Therefore, some form of income support or compensation could be given to these individuals to overcome this issue of equity. However, any treatment should come from public funding and not from subsidies on insurance premiums in order to provide incentives for risk reduction. For instance, tax reductions could be provided to partly compensate for the drop in households' disposable income due to higher insurance premiums. Along the lines of the current food stamp programme in the US, Kunreuther et al (2009) propose the introduction of 'insurance vouchers' provided by the state or federal government that would enable low-income residents to purchase insurance. Nevertheless, it should be noted that such special treatments are only needed in the short term since they should not apply to new residents who locate in flood-prone areas, as otherwise there would be an incentive to settle in hazard-prone areas in order to be eligible for these benefits.

9.4.4 Extending the duration of insurance contracts

The short-term nature of the current flood insurance policies of only one year may restrain active collaboration between insurers and policyholders in reducing exposure to flooding. Several scientific experts have suggested introducing long-term flood insurance contracts that are tied to properties instead of to individuals with a duration of, for example, 5, 10 or 20 years (Kunreuther, 2008; Kunreuther et al, 2009). Such long-term contracts would establish a long-term relation between insurers and policyholders, which gives both incentives to implement cost-effective risk-reducing measures. For example, insurers have an interest in rewarding the undertaking of such measures with premium discounts that the policyholders can receive for a long time horizon – namely, for the duration of the insurance policy. A problem with short-term contracts is that policyholders are not sure that premium reductions which would be provided as a reward for mitigation would continue long into the future, while insurers may be reluctant to stimulate risk reduction of policyholders who are able to cancel their contract each year after it expires. A long-term policy would prevent homeowners from cancelling their flood insurance after a few years if they had not experienced any flood damage, which often occurs in practice (Michel-Kerjan and Kousky, 2010). A long-term insurance contract could overcome such problems and increase the effectiveness of the insurance instrument in encouraging risk reduction. However, a challenge with long-term policies is how to price the future risk if the risk landscape changes due to climate change. The question is whether climate change projections are currently sufficiently reliable to price future flood risk, and further research in this area is required.

9.4.5 Addressing climate change in the NFIP requirements

It could be useful to adjust mitigation requirements in the current 1/100-year zone so that they reflect the future change in water levels that are expected due to sea-level rise. For example, elevation requirements could be based on the likely future BFE level instead of the current level. Moreover, the geographical extent of the 1/100-year zone is likely to increase. The NFIP regulations could be made applicable to the future 1/100-year flood area to promote adaptation in that area. Failing to do this now means that many existing buildings without mitigation measures would exist in the future 1/100 flood zones. Delineating the future flood zones and the potential water level is a complicated and probably expensive task. If this task is not considered to be feasible, then an alternative policy is to regulate not only the current 1/100 flood zone, but also other flood zones, such as the current 1/500-year flood zone. This would be an effective adaptation policy if it is expected that the current 1/500 flood zone is approximately the future 1/100 zone. Moreover, regulating the current 1/500-year flood zone may be justified by the high amount of flood losses that already occur in that zone, and the arbitrary nature of the decision of the NFIP to regulate only the 1/100-year flood zone.

9.4.6 Broadening mitigation grant programmes

The current grant programmes are unlikely to provide sufficient financial resources to keep mitigation efforts in line with the expected change in risk in the future. Sea-level rise requires additional beach nourishment; in some places with considerable capital at risk, additional protection against storm surge may be cost effective, such as sea walls and storm surge barriers. The limited financial capacity of state and local governments may result in a suboptimal level of investments in prevention, while *ad hoc* funding through Congress may be difficult to obtain. An expansion of the current NFIP grant programmes (e.g. by establishing a specific Flood Risk Adaptation Grant Program) could be effective in ensuring a continuous funding source for investments in flood risk adaptation. This fund could finance adaptation measures aimed at flood prevention, as well as damage mitigation. In addition, mitigation grant programmes should be better integrated with on-going and new coastal zone management programmes that are funded by both the federal government and the state.

Detailed cost–benefit studies of such adaptation measures are needed in order to prioritize the most effective measures and projects that need funding. Moreover, new and existing grant programmes could be better geared towards financing low- and high-cost mitigation measures that can be implemented by homeowners or contractors of new waterfront development projects. Further research could examine how much money is needed for investments in flood risk adaptation in the US and assess the financial capacity required for grant programmes that stimulate these investments.

9.5 Conclusions

The NFIP has been quite successful in providing flood insurance to many households in the US that would otherwise not be available. The NFIP is an important programme for achieving risk reduction because it imposes the minimum requirements for local governments' flood-zoning and building codes, and it provides incentives to homeowners to invest in risk reduction beyond these minimum standards. Past experience has shown that the NFIP has been rather effective in limiting the vulnerability of new constructions to flood hazards through flood-proofing measures. However, the programme is generally evaluated as being ineffective in limiting new developments in high-risk areas and in reducing the vulnerability of existing buildings to flood hazards. In addition, many buildings that are exposed to flooding but are located outside the 1/100-year flood zone are not covered by the NFIP. This indicates that there is a need for an assessment of how zoning policies can be better geared towards reducing flood risk. Climate change or other future developments, such as urban development, are not addressed in the NFIP. The general recommendation for the NFIP is the need for a thorough assessment of how the NFIP can be geared towards accommodating, and ameliorating the impacts of, increased flood risk. Moreover, there seems to be considerable scope to improve the current programme even if flood risk would not increase as a result of climate change. In this respect, several issues that could be addressed by the NFIP have been identified in this chapter, such as enhancing the NFIP's market penetration; improving flood-hazard maps; charging insurance premiums that reflect risk; extending the duration of insurance contracts; addressing climate change in the NFIP requirements; and broadening mitigation grant programmes.

An important recommendation for improved cooperation between the NFIP, the NYC Department of Buildings and the NYC Department of City Planning is to take into account future flood risk maps (e.g. the future 1/100-year floodplain) in flood insurance, zoning and building code policies. In view of climate change, the geographical extent of the 1/100-year flood zone is likely to increase, and insurance and zoning regulations could be made applicable to what is expected to be the future 1/100-year floodplain. Knowing the future 1/100 flood zone offers the possibility of imposing the same requirements that apply to the current 1/100-year flood zone, such as elevation of new buildings and flood-proofing measures, to the future 1/100-year flood zone. Each new development or revitalization programme could then be overlaid with the future 1/100 floodplain maps, and the current NFIP regulation with BFE requirements could be applied to these future flood zones. An alternative policy to mapping the future 1/100 flood zone is to regulate not only the current 1/100 flood zone, but also other flood zones, such as the current 1/500-year flood zone.

Appendix A: Amendments Made to the NFIP Over Time

Table 9.5 *Amendments to the National Flood Insurance Program (NFIP) over time*

Amendment	Problems to be solved	Main changes	Result
Insurance legislation 1969	Limited capacity for detailed flood risk studies and maps	Establish an 'emergency phase' in which the NFIP makes preliminary flood studies and maps of the floodplain Flood insurance is available if the local government agrees to make its best efforts to control building to reduce flood risk	Communities could enter NFIP without detailed flood risk studies
Flood Disaster Protection Act 1973	Low participation of communities in NFIP Low market penetration Insurance incentive for building regulation was inadequate	Governments not in NFIP are ineligible for grants to aid construction in flood-hazard areas Homeowners of communities not in NFIP cannot obtain federal disaster assistance Homeowners in floodplains of communities not in NFIP cannot obtain federally backed mortgages Homeowners in 1/100-year floodplain with federally backed mortgages are required to purchase flood insurance	Participating communities increased from < 3000 to > 18,000 in five years
National Flood Insurance Reform Act 1994	Low market penetration	Strengthen oversight of mandatory flood insurance purchases for homes in 1/100 zone with federal mortgages	Flood insurance demand is observed to be higher for individuals who are subject to the requirements

Table 9.5 *Amendments to the National Flood Insurance Program (NFIP) over time (continued)*

Amendment	Problems to be solved	Main changes	Result
	High losses due to inadequate incentives for mitigation	Limit disaster assistance for households that have suffered flood losses but do not have flood insurance	More than 60% of the NFIP policyholders participated in the CRS in 2000
		Require that households in the 1/100 zone who have received disaster relief should purchase and maintain flood insurance	Empirical evidence shows that the CRS has lowered claims in a sample of communities in Florida
		Set-up of the Community Rating System (CRS) that rewards communities who invest in risk reduction with premium discounts	About US$20 million a year is authorized for mitigation grants
		Provision of grants to states and local governments for actions to reduce flood hazards	
		Extend coverage up to US$20,000 to cover costs of building substantially damaged buildings in compliance with NFIP rules	
Flood Insurance Reform Act 2004	Repetitive losses	Program for mitigation of flood damage to properties affected by severe repetitive loss	Severe repetitive losses continue outpacing FEMA mitigation efforts

Source: Burby and French (1985); Pasterick (1998); Burby (2001); Kriesel and Landry (2004); US Department of Homeland Security (2009); Michel-Kerjan and Kousky (2010)

References

Aerts, J. C. J. H., Major, D. C., Bowman, M. J., Dircke, P., Marfai, M. A., Abidin, H. Z., Ward, P. J., Botzen, W., Bannink, B. A., Nickson, A. and Reeder, T. (2009) *Connecting Delta Cities: Coastal Adaptation, Flood Risk Management and Adaptation to Climate Change*, VU University Press, Amsterdam

Association of State Floodplain Managers (2000) *National Program Review 2000*, ASFPM, Madison

Bingham, K., Charron, M., Kirschner, G., Messick, R. and Sabade, S. (2006) *Assessing the National Flood Insurance Program's Actuarial Soundness*, Deloitte Consulting and American Institutes for Research, Washington, DC

Botzen, W. J. W. and van den Bergh, J. C. J. M. (2008) 'Insurance against climate change and flooding in the Netherlands: Present, future and comparison with other countries', *Risk Analysis*, vol 28, no 2, pp413–426

Botzen, W. J. W. and van den Bergh, J. C. J. M. (2009) 'Managing natural disaster risk in a changing climate', *Environmental Hazards*, vol 8, no 3, pp209–225

Bouwer, L., Bubeck, P. and Aerts, J. C. J. H. (2010) 'Changes in future flood risk due to climate and development in a Dutch polder area', *Global Environmental Change*, vol 20, pp463–471

Browne, M. J. and Hoyt, R. E. (2000) 'The demand for flood insurance: Empirical evidence', *Journal of Risk and Uncertainty*, vol 20, no 3, pp291–306

Burby, R. J. (2001) 'Flood insurance and floodplain management: The US experience', *Environmental Hazards*, vol 3, no 3–4, pp111–122

Burby, R. J. (2006) 'Hurricane Katrina and the paradoxes of government disaster policy: Bringing about wise governmental decisions for hazardous areas', *The Annals of the American Academy of Political and Social Science*, vol 604, no 1, pp171–191

Burby, R. J. and French, S. P. (1985) *Flood Plain Land Use Management: A National Assessment*, Westview Press, Boulder, CO

Crichton, D. (2008) 'Role of insurance in reducing flood risk', *Geneva Papers on Risk and Insurance: Issues and Practice*, vol 33, no 1, pp117–132

de Moel, H., van Alphen, J. and Aerts, J. C. J. H. (2010) 'Flood maps in Europe: Methods, availability and use', *Natural Hazards and Earth System Sciences*, vol 9, pp289–301

Dixon, L., Clancy, N., Seabury, S. A. and Overton, A. (2006) *The National Flood Insurance Program's Market Penetration Rate: Estimates and Policy Implications*, American Institutes for Research, Washington, DC

FEMA (Federal Emergency Management Agency) (2006) *Community Rating System*, FEMA, Washington, DC, pp1–29

FEMA (2008) *Mitigation Grant Programs: Building Stronger and Safer*, FEMA Mitigation Directorate, Washington, DC

FEMA (2009) *Hazard Mitigation Assistance Unified Guidance: Hazard Mitigation Grant Program, Pre-Disaster Mitigation Program, Flood Mitigation Assistance Program, Repetitive Flood Claims Program, Severe Repetitive Loss Program*, FEMA Department of Homeland Security, Washington, DC

FEMA (2010) *Residential Coverage: Policy Rates*, FEMA, www.floodsmart.gov, accessed 30 August 2010

GAO (Government Accountability Office) (2008) *Flood Insurance: FEMA's Rate-Setting Process Warrants Attention*, GAO-09-12, October, US GAO, Washington, DC

Jones, C. P., Coulborne, W. L., Marshall, J. and Rogers, S. M. (2006) *Evaluation of the National Flood Insurance Program's Building Standards*, American Institutes for Research, Washington, DC

King, R. O. (2005) *Federal Flood Insurance: The Repetitive Loss Problem*, CRS Report for Congress, Congressional Research Service, Library of Congress, Washington, DC

Kriesel, W. and Landry, C. (2004) 'Participation in the National Flood Insurance Program: An empirical analysis for coastal properties', *Journal of Risk and Insurance*, vol 71, no 3, pp405–420

Kunreuther, H. C. (1996). 'Mitigating disaster losses through insurance', *Journal of Risk and Uncertainty*, vol 12, no 2–3, pp171–187

Kunreuther, H. C. (2008) 'Reducing losses from catastrophe risks through long-term insurance and mitigation', *Social Research*, vol 75, no 3, pp905–932

Kunreuther, H. C. and Michel-Kerjan, E. O. (2009) *Managing Catastrophes through Insurance: Challenges and Opportunities for Reducing Future Risks*, Working paper 2009-11-30, The Wharton School, University of Pennsylvania, Philadelphia, PA

Kunreuther, H. C. and Roth, R. J. (1998) *Paying the Price: The Status and Role of Insurance against Natural Disasters in the United States*, Joseph Henry Press, Washington, DC

Kunreuther, H. C., Michel-Kerjan, E. O., Doherty, N. A., Grace, M. F., Klein, R. W. and Pauly, M. V. (2009) *At War with the Weather: Managing Large-Scale Risks in a New Era of Catastrophes*, MIT Press, Cambridge, MA

LeBlanc, A. and Linkin, M. (2010) 'Chapter 6: Insurance industry', *Annals of the New York Academy of Sciences – New York City Panel on Climate Change 2010 Report*, vol 1196, pp113–126

Michel-Kerjan, E. O. and Kousky, C. (2010) 'Come rain or shine: Evidence on flood insurance purchase in Florida', *Journal of Risk and Insurance*, vol 77, no 2, pp369–397

Morrissey, W. A. (2006) *FEMA's Flood Hazard Map Modernization Initiative*, CRS Report for Congress, Congressional Research Service, Library of Congress, Washington, DC

National Institute of Building Sciences (2005) *National Hazard Mitigation Saves: An Independent Study to Assess the Future Savings from Mitigation Activities*, Washington, DC

Nicholls, R. (2003) *Case Study on Sea-Level Rise Impacts*, Organisation for Economic Co-operation and Development (OECD), Paris

Nicholls, R. (2009) 'Chapter 5: Adaptation costs for coasts and low-lying settlements', in *Assessing the Costs of Adaptation to Climate Change*, International Institute for Environment and Development and the Grantham Institute for Climate Change, London

NPCC (New York City Panel on Climate Change) (2010) *Climate Change Adaptation in New York City: Building a Risk Management Response*, New York Academy of Sciences, Wiley-Blackwell, New York, NY

Pasterick, E. T. (1998) 'The National Flood Insurance Program', in H. C. Kunreuther and R. J. Roth (eds) *Paying the Price: The Status and Role of Insurance against Natural Disasters in the United States,* Joseph Henry/National Academy Press, Washington, DC

Poussin, J., Botzen, W. J. W. and Aerts, J. C. J. H. (2010) *Stimulating Flood Damage Mitigation through Insurance: An Assessment of the French CatNat System*, Working Manuscript, Institute for Environmental Studies, VU University, Amsterdam

Sarmiento, C. and Miller, T. R. (2006) *Costs and Consequences of Flooding and the Impact of the National Flood Insurance Program*, Pacific Institute of Research and Evaluation, Calverton

Tobin, R. J. and Calfee, C. (2006) *The National Flood Insurance Program's Mandatory Purchase Requirement: Policies, Processes and Stakeholders*, American Institutes for Research, Washington, DC

US Congress (1966a) *A Unified Program for Managing Flood Losses*, House document 465, 89th Congress, 2nd session, US Government Printing Office, Washington, DC

US Congress (1966b) *Insurance and Other Programs for Financial Assistance to Flood Victims*, Committee print, Senate Committee on Banking and Currency, US Government Printing Office, Washington, DC

US Congress (1973) *Flood Disaster Protection Act of 1973*, 93th Congress, 1st session, US Government Printing Office, Washington, DC

US Congress (1994) *National Flood Insurance Reform Act of 1994*, 103th Congress, 2nd session, US Government Printing Office, Washington, DC

US Department of Homeland Security (2009) *FEMA's Implementation of the Flood Insurance Reform Act of 2004*, Department of Homeland Security Report OIG-09-45, Office of Inspector General US, Washington, DC

Wetmore, F., Bernstein, G., Conrad, D., DiVicenti, C., Larson, L., Plasencia, D. and Riggs, R. (2006) *The Evaluation of the National Flood Insurance Program: Final Report*, American Institutes for Research, Washington, DC

10

Navigable Storm Surge Barriers for Coastal Cities: An Overview and Comparison

Piet Dircke, Tom Jongeling and Peter Jansen

10.1 Introduction

Storm surge barriers are flood protection works in rivers or estuaries that can greatly enhance coastal protection for ports and urbanized areas at risk, in particular (e.g. Vrijling, 2001). Since they are often expensive measures (see Table 10.1), such structures are often implemented only after a flood disaster has occurred. For example, the Thames Barrier in the UK and the Delta Works in The Netherlands were developed after the major flood in 1953. Barriers are currently being installed in New Orleans (US) after Hurricane Katrina in 2005. A big advantage of some barriers for port areas such as Rotterdam or New York is that they do not obstruct shipping activities, allowing free navigation under normal conditions (Dircke et al, 2009; Jansen and Dircke, 2009). Table 10.1 provides a global overview of some important or typical storm surge barriers, their year of operation and their costs.

When assessing the possibilities for developing a storm surge barrier in a coastal urban environment, many options are available. Experience thus far with developing and implementing barrier designs suggests that there is not one single perfect gate type, and often a tailor-made design that selects or combines the most favourable aspects of gates and other flood protection measures should be considered in order to find the best solution for a coastal city at risk of flooding (e.g. Jansen and Dircke, 2009). Coastal and, in particular, delta and estuarine cities are often surrounded by dynamic water systems where more than one gate type is required.

Here, a holistic approach is recommended and integrated solutions are required for these systems. This means that although barriers can be a suitable solution for flood protection, for sustainable, economic and environmental arguments, barriers should always be compared to, or combined with, other flood protection or risk-reducing alternatives (e.g. Kabat et al, 2005). These alternatives may include flood walls, dikes or natural coastal systems such as dunes, or even non-structural measures such as building codes for adaptive housing in flood zones.

Table 10.1 *Development costs and construction time of some international storm surge barriers*

Barrier	Type	Country	Construction time (years)	Year of operation	Approximate value in 2010 (million Euros)
Hollandse IJssel Storm Surge Barrier	Lifting gate	The Netherlands	4	1958	98
Oosterschelde Storm Surge Barrier	Lifting gate	The Netherlands		1986	3850
Maeslant Storm Surge Barrier	Floating-sector gate	The Netherlands	8	1997	545
Europoort Barrier and Hartel Barrier	Lifting gate	The Netherlands		1997	253
Ramspol Storm Surge Barrier	Rubber dam	The Netherlands		2002	68
Venice Storm Surge Barrier	Flap gate	Italy		Not yet complete in 2011	4270
New Bedford Hurricane Barrier	Rolling-sector gate	US	4 (Bowman et al, 2008)	1966 (Bowman et al, 2008)	55
Stamford Hurricane Barrier	Flap gate	US	4 (Bowman et al, 2008)	1968 (Bowman et al, 2008)	14.4 (Bowman et al, 2008)
Harvey Canal Flood Protection Barrier, GIWW WCC	Sector gate	US		2008	780
IHNC New Orleans Hurricane Protection Barrier	Sector gate	US		2011	1014
Thames Barrier	Segment gate	UK		1982	2100
Fox Point Hurricane Barrier	Radial gate	US		1966	58
St Petersburg Barrier	Lifting gate and floating-sector gate	Russia		2010	5100
Ems Storm Surge Barrier	Lifting gate and radial gate, plus segment gate	Germany		2002 (PIANC, 2005)	220 (Meinhold, 2005)

Notes: GIWW WCC = Gulf Intracoastal Waterway West Closure Complex; IHNC = Inner Harbor Navigation Canal

In this chapter, we provide a systematic overview, comparison and selection of navigable storm surge barriers that are suitable for coastal cities, to protect cities against the impact of climate change and sea-level rise – where sea-level rise here is interpreted as a main cause for higher future flood levels. The gates presented are proven concepts and most designs have been operating successfully for many years within flood protection systems, particularly in The Netherlands. The available gate types are compared and an overview of the advantages and disadvantages of each gate type is given, as well as a procedure to select the right gate type for a certain situation. For this comparison, different aspects are considered, including type of gate structures, design criteria, risk management, environmental aspects, navigation aspects, technical design details, and aspects of operation and maintenance.

10.2 Possible Functions of a Storm Surge Barrier

The design process for a barrier usually starts with an analysis of the functions of the storm surge barrier, in addition to the basic function of retaining a storm surge. Options capable of providing the desired functions and requirements are selected and further elaborated upon. Key issues in the design are the transfer of hydraulic loads to the foundation, the choice of the mechanical system of the gates in relation to span and operation, and the foundation of the barrier, including bed and bank protections. In addition, aspects such as maintainability, reliability, durability, constructability, morphological impact and environmental impact can play an important role in determining the design. Other requirements and their impact upon the barrier design are listed in Table 10.2.

10.3 Description of Barrier Types

This section presents a description of the six considered barrier types, and discusses their properties and their favourable and unfavourable aspects. Typical examples of best practice for each barrier type are also presented and described. Only gates that are currently operational and were successfully closed at least once under stormy conditions are presented.

10.3.1 Mitre gates

Mitre gates, probably an invention of Leonardo da Vinci, were already common during the 16th and 17th centuries and are often part of shipping locks in canals. Mitre gates are double-leaf gates and the leaves form an angle pointing upstream when the gates are closed. As a result, these movable gates must be strong enough to withstand the water pressure arising from the level difference between adjacent pounds. Typically, the combined lengths of the leaves exceed the lock width by about 10 per cent. When opened,

Table 10.2 *Requirements and their impact upon the storm surge barrier design*

Requirement categories	Requirements for storm surge barrier	Impact upon design
Structure	Risk of failure should be minimized and not exceed the design probability of failure.	Reliable structure design; use of reliable materials; application of reliable driving system; careful operation.
Hydraulics	Amount of storm water overtopping or leaking past the barrier should meet the criteria.	Sufficient height of the barrier; application of seals; construction of breakwaters.
	Strong flow- or wave-induced vibrations and oscillations must be prevented.	Gate with adequate dynamic properties and favourable detailed design in view of dynamic flow forces and feedback phenomena.
	Translatory waves and differential head (caused by closure) should be acceptable.	Selection of suitable closing time and opening time.
Morphology	Designed to reduce/eliminate siltation impacts.	Removal of silt by gate operation; prevention of siltation by an adequate barrier design and layout.
Climate	Limited sensibility for wind loads.	Adequate dynamic properties required in view of wind loads; application of windshields.
Hydrology	Passage of tides and discharge of river flow.	Adequate flow opening.
	Discharge of ice.	Adequate width of flow opening.
	Reverse flow conditions and reverse head.	Adequate structural design and favourable design in view of flow forces and feedback phenomena.
Navigation and transport	Passage of ships.	Width and depth of navigation opening in relation to navigational requirements and ship clearance.
	Connection between both land sides for traffic.	Road and rail bridge.
Environment	Allow for tidal flow in view of water quality and ecology.	Sill level, wet cross-section, effect of density differences.
	Passage of fish.	Wet cross-section, flow velocity, salinity gradient.
Maintenance	Accessible for inspection and maintenance.	Drying possibility of interior.
	Proper choice of materials and techniques for minimum maintenance.	Corrosion protection.
Other	Limited vulnerability to impact of unexpected heavy objects such as barges.	Solid construction and collision impact protection.
	Limited vulnerability to vandalism (or terrorism).	Solid construction, safeguarding of vulnerable parts and other security-related issues.

Table 10.3 *Favourable and unfavourable aspects of mitre gates*

Favourable	*Unfavourable*
Structural aspects, layout and operation	
Unlimited clearance height for shipping	Very small gate span (up to 100 feet)
Little space required	Little or no controlled operation under flow and
Proven concept	waves
Not subjected to wind	
Hydraulic and hydrodynamic aspects	
Horizontal closure	Sensible to vibration as a result of flowing water
Discharge of excess water through gate	Sensible to reverse head
	Sensible to waves

the leaves are housed in lock wall recesses; when closed, after turning through about 60°, they meet in the centre line of the lock (Vrijburcht, 2000). The maximum width of a single gate already built is about 25m (see Table 10.3). A well-known example of a mitre gate is the biggest shipping lock system in The Netherlands: the IJmuiden complex, connecting the port of Amsterdam with the North Sea (see Figure 10.1). This complex has four locks. The biggest 50m-wide lock has straight rolling gates. The other three locks have mitre gates, the biggest of them with a width of more than 20m. The locks were built during the early part of the 20th century and were renovated between 1985 and 1995.

Figure 10.1 *Example of a mitre gate in The Netherlands*

Source: Vrijburcht (2000)

10.3.2 Vertical lifting gates

Vertical lifting gates are widely used and have a satisfactory record of operation. Because of their widespread use, much experience is available on construction techniques and on functioning and behaviour under flow and wave conditions. The hoisting towers and the tower foundation are usually constructed within cofferdams, while concrete sills may be floated in and immersed on a gravel base or pile foundation that has been constructed underwater. The bed adjacent to the sill and the towers may be riprap protected. Lifting gates may not always be acceptable from an aesthetic point of view. Table 10.4 lists the pros and cons of vertical lifting gates.

Table 10.4 *Favourable and unfavourable aspects of vertical lifting gates*

Favourable	Unfavourable
Structural aspects, layout and operation	
Large gate span (up to 300 feet)	Limited clearance height for shipping
Little space required	Raised gate subject to wind load
Controlled operation under flow and wave	Water depth $\leftrightarrow$ gate height
Raised gate accessible for maintenance	Wheel gates weak spot, wearing
Proven concept	Smooth slide required $\leftrightarrow$ growth underwater
Hydraulic and hydrodynamic aspects	
Vertical closure	Sensitivity to vibrations
Discharge of excess water	Small stiffness during operation
Overflow and reverse flow acceptable	Subject to down-pull flow forces and wave loads
Underside free of sill	
Limited vertical flow forces and wave loads	

An example of a vertical lifting gate is the Krimpen Barrier, which was constructed in the River Hollandsche IJssel (see Figure 10.2). It is located where the river joins the River Rhine, about 30km away from the outflow of the Rhine into the North Sea. This barrier is part of the sea defence system of The Netherlands. For security reasons, a double-barrier system has been constructed, with two gates in tandem. A small leakage gap is kept open between the gate underside and sill. This leakage gap may sometimes cause flow-induced vibrations (de Jong and Jongeling, 1995). A shipping lock has been constructed beside the barrier. The concrete sills and the abutments have, in subsequent stages, been constructed within cofferdams.

Another example of a vertical lifting gate is the Hartelkanaal Storm Surge Barrier in The Netherlands. The Hartelkanaal is a shipping canal for ocean-going vessels, providing access from the North Sea to Europoort, an industrial area near Rotterdam (see Figure 10.3). The Hartelkanaal Storm Surge Barrier is part of the Dutch sea defence system and consists of two gates, one gate with a span of 321.5 feet (98m) and a

Figure 10.2 *Krimpen Storm Surge Barrier: The flood protection gates are in a raised position*

Source: Vrijburcht (2000)

Figure 10.3 *Hartelkanaal Storm Surge Barrier: The flood protection gates are partly in a raised position*

Source: Vrijburcht (2000)

smaller gate with a span of 161.7 feet (49.3m). The sliding gates are driven by hydraulic cylinders with a long piston, which are hinged to the side towers. The gates are lens shaped; the retaining plate at the high-water side is through vertical beams connected to the arched truss girders. When not in use, the clearance height between the mean water level and the gate underside amounts to 45.9 feet (14m). In closed position, the gates are about 0.65 feet (0.2m) above the concrete sill, leaving a leakage gap open. A floating beam across the canal protects the barrier from ship collisions. The shipping lock beside the barrier is used when the barrier is closed. During extreme storm conditions, the water level on the sea side may rise above the crest of the gates; this will result in overflow of the gates, but also in a limitation of the horizontal hydraulic load while the gates still reduce the storm surge. A reduced gate height was chosen not only to limit the hydraulic load, but also to limit the height of the towers, the weight of the gates and the length of the hydraulic cylinders.

A third example of a large-scale vertical lifting gate barrier is the Eastern Scheldt Storm Surge Barrier in The Netherlands (see Figure 10.4). The Eastern Scheldt (Oosterschelde) Storm Surge Barrier is The Netherlands' major storm surge barrier. It closes off the Eastern Scheldt Estuary from the North Sea in the case of a storm surge, but remains open to maintain the environmental values of the estuary (Bakker and Vink,

Figure 10.4 *Eastern Scheldt Storm Surge Barrier: View of flood tide and closed barrier*

Source: Vrijburcht (2000)

1994; Nienhuis et al, 1994). The gates of the barrier are closed when a storm surge is anticipated. The barrier has been built using pre-constructed elements that include concrete piers, concrete sill beams, concrete upper beams, concrete road bridge elements and steel lifting gates. The sill and adjoining bed protection have been constructed of several layers of riprap in combination with asphalt layers. There are 62 flow openings, each flow opening with a breadth of 131.2 feet (40m). The height of the gates varies between 19.4 feet (5.9m) and 39.0 feet (11.9m), depending upon their location in the three main gullies. The frame of the gate consists of a space truss; the arched elements of the retaining plate are vertically attached to the space truss. When the gates are lowered, they close off the flow opening between upper beams and sill beams. The sliding gates are lifted with the help of hydraulic cylinders; the horizontal loads are transferred to the supports in the gate recesses in the piers. Because the lifting gates are applied in combination with fixed concrete sill beams and upper beams, they do not allow for shipping, but a separate shipping lock has been provided. The barrier can be closed within approximately one hour. The open space truss of the gates faces the sea and may be subjected to heavy wave loads (design wave conditions: wave height up to 19 feet (5.8m) and wave period up to 10 seconds). In the initial design, plate girders were applied; but these girders were replaced by an open truss in order to reduce the wave (shock) loads.

10.3.3 Flap gates

Flap gates are not visible when the flood protection barrier is not in use. The gates are stored in a bottom recess and with one end hinged to the sill; the free end emerges above the water surface when the barrier is put in operation. Two gate types are discussed: gates driven by hydraulic cylinders and pneumatic gates operated by air injection into flotation tanks. Flap gates are not widely applied in flood protection schemes. This is probably due to the fact that inspection, maintenance and replacement are difficult, while the issue of silting of the gate recess may be an additional drawback (see Table 10.5). Nevertheless, the flap gate barrier design also has some prominent advantages, such as the invisibility of the barrier, the distributed load transfer to the foundation, and the unlimited breadth of the flow opening. Concrete caissons, possibly also with the flap gates pre-installed, can be floated to the site and immersed on a prepared base of compacted soil with a gravel top layer or a pile foundation. The bed adjacent to the caissons may be riprap protected.

An example of a flap gate is the Stamford Hurricane Barrier in Connecticut, US (see Figure 10.5). The East Branch Barrier at Stamford comprises a 2840 foot (866m)-long earth-and-rock dike with a top elevation of more than 17 feet (5m). The structure features a 90 foot (27m)-wide channel opening protected with a single steel flap gate. The hollow steel gate rests on the bottom of the 18 foot (5.5m)-deep channel and is raised to close the opening. The gate is operated by hydraulic cylinders, which are housed in the vertical side walls. The gate is lifted in 20 minutes.

Table 10.5 *Favourable and unfavourable aspects of flap gates*

Favourable	Unfavourable
Structural aspects, layout and operation	
No limitation of the span	Natural frequencies low; small stiffness, great mass
Separate flaps; reduced failure risk	
No clearance height	Pneumatic: not fully controlled
Little space required	Hydraulic: concentration cylinders
Suitable for deep waters	Underwater: corrosion, growth
Controlled operation flow and wave	Hinges may wear out in sand
Not subjected to wind	Maintenance difficult
Invisible when not in use	
Hydraulic and hydrodynamic aspects	
No strong confinement of horizontal flow (gates used alternately)	Sensitivity to vibrations
Vertical closure, single flap	Small stiffness during operation
Excess water through one flap or by lowering the gate crest	Subject to down-pull flow forces and wave loads

Figure 10.5 *Stamford Hurricane Barrier: Aerial view*

10.3.4 Horizontally moving or rotating gates (slide gates, sector gates and floating-sector gates)

Flood protection gates that move over a carriageway or slideway on a sill on the river bed have some important disadvantages, which may be the reason that they have not yet been applied in major flood protection schemes. These disadvantages are the need for deep side chambers in which the gates are housed when not in use, and the risk of malfunctioning when silting occurs on the sill (see Table 10.6). When the span is limited, gates that rotate about a vertical axis can be kept free of the sill; this may reduce the risk of malfunctioning. In order to limit the forces on the slideways and in the hinges, it may be necessary to install flotation tanks in the gates. The tanks themselves, however, may not evoke dynamic wave or flow forces on the gates in a vertical direction. The side chambers and abutments with gate supports can be built within cofferdams, while the sills (concrete caissons) may be floated to the site and immersed on a gravel base or pile foundation that has been constructed underwater or built within a cofferdam.

An example of horizontally moving or rotating gates is the New Bedford Hurricane Barrier in Massachusetts, US (see Figure 10.6). This hurricane barrier comprises a 9100 foot (2774m)-long dam with a crest level of more than 20 feet (6m); the 150 foot (46m)-wide navigation opening can be closed with two sector gates, which are housed in side chambers in the abutments. The 60 foot (18m)-high gates roll with steel wheels on a concrete sill. The gates can be closed in 12 minutes. The structure is part of a larger 3.5 mile (5.6km)-long rock barrier that protects to 26 feet (8m) above high tide. It is the largest stone structure in the eastern US.

Table 10.6 *Favourable and unfavourable aspects of moving or rotating gates*

Favourable	Unfavourable
Structural aspects, layout and operation	
Large span feasible	Large space and deep excavation required for chambers
No clearance height limitation	
Not subjected to wind	Flat and smooth slideway required
Suitable for deep waters	Silting may hamper operation
Immediately ready for operation	Sector gates: load transfer to hinges, maintenance, corrosion, growth
Free of sill; reduced load on sill	
Stable structure; no load concentration	Straight gates: sluice gates may be required; time for dock filling
Dry docks: maintenance, no collision	
Hydraulic and hydrodynamic aspects	
Limited differential head and horizontal flow contraction in last stage of closure	Sector gates: ship collision; siltation in open chambers
Excess water: through sluice opening	
Suitable for reverse head and flow	
Not sensitive to flow vibrations	

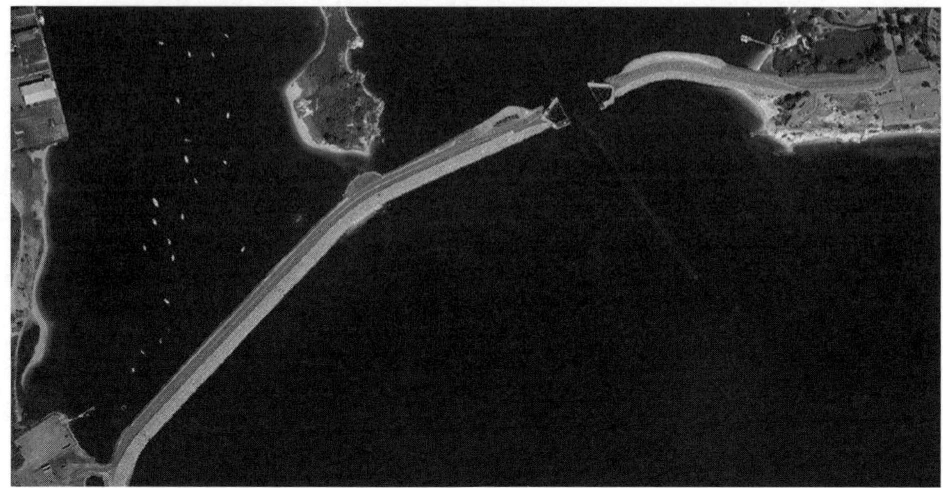

Figure 10.6 *New Bedford Hurricane Barrier: Aerial view*

Source: Whalingcity (2010)

Another example of horizontally moving or rotating gates is the flood protection barrier in the Harvey Canal near New Orleans, Louisiana, US (see Figure 10.7). The flood protection barrier in the Harvey Canal near New Orleans was recently constructed and was closed successfully by the US Army Corps of Engineers (USACE) for the first time during Hurricane Gustav on 1 September 2008. The barrier protects against hurricane storm surges from the Gulf of Mexico and is part of the larger West Bank project. The barrier comprises two sector gates which enclose a 125 foot (38m)-wide channel. The channel is 16 feet (4.9m) deep, and the gates protect to an elevation of +11.5 feet (+3.5m). The gate can be closed in ten minutes.

10.3.5 Floating-sector gates

Floating-sector gates have some important advantages: the gates can be stored in a relatively inexpensive, shallow dry dock in the abutments, which enables easy maintenance, and the gates can be immersed on the sill when the sill is covered with silt (see Table 10.7). A significant disadvantage of the floating gates is the sensitivity to flow-induced oscillations and dynamic wave loads. The dry docks and the abutments with ball hinges can be built within cofferdams. The sill can be built underwater. The bed adjacent to the sill may be riprap protected.

An example of a floating-sector gate is the Rotterdam Storm Surge Barrier (Maeslantkering), The Netherlands (see Figure 10.8). This well-known landmark barrier protecting the city of Rotterdam was ready for operation in 1997 and had its first and until now only successful closure under storm conditions ten years later, in

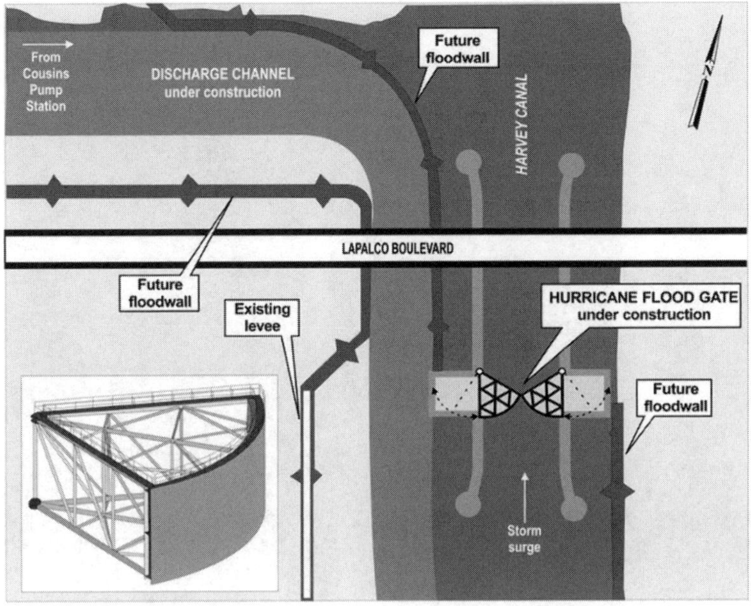

Figure 10.7 *Harvey Canal Flood Protection Barrier*

Source: USACE (2010)

Table 10.7 *Favourable and unfavourable aspects of floating-sector gates*

Favourable	*Unfavourable*
Structural aspects, layout and operation	
Large gate span possible	Large space required
No clearance height limitation	Operation complicated; in-flowing water may not be controlled
Shallow dry dock: easy inspection and maintenance and collision protection	A negative differential head may cause problems (pull-up forces ball hinges)
Can be immersed if sill is covered with silt	Objects on sill can cause damage
No flatness of sill required	Load concentration; forces on hinges
	Mobilization time: filling of dry docks
Hydraulic and hydrodynamic aspects	
Vertical closure of flow opening (no strong horizontal flow contraction)	Sensitivity to flow-induced oscillations
Separate sluice openings may be applied to reduce differential head and discharge excess water	Sensitive to dynamic wave forces
	Limited resistance to negative differential head

Figure 10.8 *Rotterdam Maeslant Storm Surge Barrier: The gates are floated into the river*

Source: Vrijburcht (2000)

November 2007. The barrier consists of two floating-sector gates, each with a radius of 807 feet (246m) and arch length of 682 feet (208m). The gate arm is connected to a single ball hinge on the abutments.

When not in use, the gates are housed in relatively shallow side docks with a high elevated floor, which can be closed with the help of gates and pumped dry. This enables inspection and maintenance 'in the dry'. Each gate is driven by a locomotive engine that is connected to the abutment by means of a rod that can move along a vertical pile. The locomotive itself runs on a rack and pinion rail on top of the gate. After the gates have been floated into the river, the flotation boxes are filled with water, and the gates sink down onto the sill. The gates stand on fender blocks that are affixed to the under-side of the gate at a centre-to-centre distance of 19.7 feet (6m). The sill consists of large rectangular concrete blocks, 10.5 feet (3.2m) thick, on a gravel base. The gravel base has been constructed in a trench that was cleared of silt and filled with sand; the gravel top layer was flattened within a tolerance range of +/– 0.16 feet (+/– 0.05m), using floating equipment. Subsequently, the sill blocks were placed with the help of a lifting crane on a barge. The horizontal hydraulic load on the gates is transferred to the ball hinges. The ball hinges are constructed on top of huge concrete gravity caissons, which are filled with sand. The hinges are elevated well above the mean water level. The gates can be floated into the river and immersed within 1.5 hours; additionally, mobilization and filling of the dry docks with water requires about one hour. Floating-sector gates are sensitive to flow-induced oscillations when not well designed (Jongeling et al, 1995).

10.3.6 Vertically rotating gates (segment and radial gates)

Two types of vertically rotating gates are applied in storm surge barrier schemes: segment gates with circular side disks that are stored in a bottom recess, and conventional radial gates that are rotated above the water level and leave space for smaller ships to pass underneath. When closed gate bodies are used, the torsion stiffness is high, enabling the construction of gates with a long span and supporting gate arms only at the two sides. Segment gates in bottom recesses may be vulnerable to silting. The abutments with driving systems can be built within cofferdams. The sill can either be built within a cofferdam or prefabricated and floated to the site. The bed adjacent to the sill may be riprap protected (see Table 10.8). Vertically rotating gates are used in the London Thames Storm Surge Barrier (UK) and the Ems Storm Surge Barrier (Germany).

A well-known example of a vertically rotating gate is the London Thames Storm Surge Barrier in the UK (see Figure 10.9). This storm surge barrier in the River Thames comprises four main navigable openings with a breadth of 200 feet (61m) and two smaller navigable openings with a breadth of 103 feet (31.5m), which are closed off by means of segment gates; four non-navigable openings with a breadth of 103 feet (31.5m) are closed by radial gates. The segment gates have a radius of 40.0 feet (12.2m) and are stored in a bottom recess in the concrete sill when not in use. The gate body is connected to circular side disks, also with a radius of 40.0 feet (12.2m), which rotate in

Table 10.8 *Favourable and unfavourable aspects of vertically rotating gates*

Favourable	Unfavourable
Structural aspects, layout and operation	
Large gate span feasible	Load transfer and concentration
Immediately ready for operation	Segment gate: high sill tolerance demands; vulnerable to
Controlled operation flow and wave	silting, objects and corrosion
Little space required	Segment gate: access and maintenance
Not subjected to wind	Radial gate: limited clearance height
Segment gate: no clearance limitation	
Inspection and maintenance	
Hydraulic and hydrodynamic aspects	
Limited horizontal flow contraction	Segment gates: sensitive to oscillation in the case of
Excess water through gate	overflow
Suitable for reverse head and flow	Open gates subject to down-pull forces and wave loads
Radial gate kept free of sill	

Figure 10.9 *London Thames Storm Surge Barrier: Downriver view*

Source: EA (2010)

a vertical plane about central pivot bearings. The side disks are partly filled with cast iron to counterbalance the weight of the gate body and are driven by hydraulic jacks.

The hollow gate body is filled with water when the gate is immersed. When the gate body is lifted, the internal water flows out through check valves near the toe of the body; air flows in simultaneously through openings in the side disks. Conversely, when the gate body is lowered, water flows into the hollow gate through openings in the side disks, and air is simultaneously blown off through the air vents in the side disks. This water-filling system is chosen to minimize silt from entering into the gate interior. The segment gates can be rotated 180° so that the gate body is fully lifted above the water and is accessible for inspection and maintenance. The accessibility in the narrow interior of the gate body, however, is difficult. The radial gates in the non-navigable openings have more or less a similar closed gate body as the segment gates, but the two gate arms, one at either side, are designed as non-circular conventional gate arms. When not in use, these gates are lifted above the water level. The full barrier can be closed within one hour.

Another example of a vertically rotating gate is the Ems Storm Surge Barrier in Germany (see Figure 10.10). The Ems Storm Surge Barrier protects against storm surges from the North Sea; but it is also used to temporarily dam up the water in the river to enable large ships to sail from the upstream shipyard to the sea. The barrier has two navigable and five non-navigable openings. The main navigable opening with a breadth of 196.8 feet (60m) has been provided with a segment gate that is stored in

Figure 10.10 *Ems Storm Surge Barrier: Artist's impression of the aerial view*

Source: Meinhold (2005)

a recess in the sill when not in use. The segment gate, having a closed gate body, has circular side disks with a radius of 39.4 feet (12m). For inspection and maintenance purposes, the segment gate can be rotated above the water level. The secondary navigable opening with a breadth of 164 feet (50m) has been provided with a radial gate, with two gate arms, that is rotated above the water when not in use, leaving a clearance height for shipping of 24.1 feet (7.35m). Both gates are driven by hydraulic cylinders. The five remaining non-navigable flow openings with a breadth of 205.1 to 164 feet (62.5m to 50m) have been provided with vertical lifting gates. These gates are combined with a fixed upper structure. The barrier can be closed in 30 minutes.

A final example of a vertically rotating gate is the Fox Point Hurricane Barrier, Providence, Rhode Island, in the US (see Figure 10.11). This barrier was constructed between 1960 and 1966; the 3000 foot (914m)-long hurricane barrier at Providence features three 40 foot (12m)-wide navigable openings that are provided with Tainter gates, each measuring 40 feet by 40 feet (12m by 12m) and weighing 53 tons. The structure is founded on a concrete mat 8 feet (2.4m) thick and 61.5 feet (19m) wide. Because the channel is 15 feet (4.6m) deep, protection is provided up to an elevation of 25 feet (7.6m). Clearance under the raised gate is 25 feet (7.6m). Obviously, only small boats may pass through the gates. The gates are operated by electric hoists, requiring 30 minutes to close. The 700 foot (213m)-long concrete structure also contains a five-pump, 7000 cubic feet per second (200m³ per second) pump station to discharge water trapped behind the closed gates.

Figure 10.11 *Fox Point Hurricane Barrier*

Source: www.rhode-island-pictures.com/index.html

10.3.7 Inflatable rubber dam

The Ramspol Barrier, in operation since 2002, is the only major flood protection barrier in the world based on inflatable rubber dams. Nevertheless, rubber dams are widely used in the world, but for other purposes, mainly in river engineering and water control applications, and for the creation of water reservoirs. Rubber dams have not yet been constructed in deep water, possibly because the fabrication of reinforced rubber sheet with large dimensions is difficult. Rubber dams have some marked advantages, such as the invisibility of the barrier, the distributed load transfer to the foundation, and the unlimited breadth of the flow opening (similar to the flap gates barrier) (see Table 10.9). For construction of rubber dams, a temporarily cofferdam is required.

An example of an inflatable rubber dam is the Ramspol Storm Surge Barrier, The Netherlands (see Figure 10.12) (van der Horst and Rovekamp, 1999). The inflatable Ramspol Storm Surge Barrier has been constructed to protect the hinterland from storm surges from the IJssel Lake. The barrier comprises three identical rubber dams, one of them spanning a shipping channel. The dams have a length of about 246 feet (75m) measured along the crest, and their inflated height amounts to about 27 feet (8.2m) above the sill. When the barrier is not in use, the rubber sheets are stored in bottom recesses. When the water level exceeds a certain alarm level, the barrier is closed and the dams are filled with water and air. The air is blown in through openings in the abutments; simultaneously, water from the upstream side flows freely into the dam

Table 10.9 *Favourable and unfavourable aspects of inflatable rubber dams*

Favourable	Unfavourable
Structural aspects, layout and operation	
No limitation of span	Flexible structure, low frequencies, small stiffness, great mass
No clearance height limitation	
Not subjected to wind	Internal pressure determines stability
Little space required	Control of storage and immersion of rubber sheet
Direct transfer of hydraulic load	Not suitable for deep water
Invisible when not in use	Difficult inspection, maintenance and replacement of rubber sheet
No need for hinges and driving system	
	Vulnerable to vandalism
Hydraulic and hydrodynamic aspects	
Vertical closure of the flow opening	Ships or objects collision
Not sensitive to silting of sill	Strong flow contraction in last stage
	Considerable response to wave loads
	No spill of excess water; overflow vibrations

Figure 10.12 *Ramspol Storm Surge Barrier*

Source: Tom Jongeling

through pipes in the base. When the dams are deflated, the air escapes through vents in the abutments and the water is pumped out. With the help of guiding rollers, the sheet can be retracted into the recess. The dams can be erected and deflated in flowing water (design flow velocity up to 10.8 feet per second, or 3.3m per second), and are resistant to strong wave loads. The barrier can be closed in about one hour.

10.4 Comparison and Selection of Gate Types

Table 10.10 shows criteria that can be used to select a suitable storm surge barrier for a particular site. The primary function of a storm surge barrier is to withstand the storm surge. All considered types of barriers are capable of impeding a storm surge. However, requirements such as high reliability, or the ability to be operational at high wind or water flow velocities, may be the reason that certain types of gates (such as mitre gates or rolling horizontal gates) prove less favourable. If large ships have to pass the barrier without any hindrance or delay, as will be required for a busy entrance channel to a large harbour, design options such as a rubber dam or mitre gates, may be ruled out.

The first step is to look at the critical functions and their requirements. These requirements can rule out certain types of barriers. During the design process, the ability of the design to meet the critical requirements should be verified and checked by specialists. The second step is to look at criteria such as costs, maintenance, durability and landscape. Given the requirements and environmental conditions, each type of gate will have a certain score based on these criteria. By comparing the scores of the different design options, the best design can be selected. Constructing a score table is a specialist's job. Choosing between costs, maintenance, durability and landscape can be difficult. Public opinion can influence the final decision. A third step is to determine the

Table 10.10 *Comparison of gate type characteristic*

	Mitre	Vertical lift	Flap	Horizontal	Vertical rotate	Rubber
Span > 30m	–	+	+	+	+	+
Span > 100m	–	–	+	+	–	–
Water depth > 10m	+	+	+	+	+	–
Impact upon landscape	+	–	+	+	–/+	+
Maintenance	+	+	–	0	+	0
Currents and waves	–	+	0	0/+	0/+	0
Closure time	+	+	+	+	+	0/–
Space required	+	+	+	–	+	+
Colliding ships	0/–	+	+	0/–	+	0
Reliability	–/+	+	0/+	–/+	+	0
Clearance height	+	–	+	+	–/+	+

Note: – not favourable up to not feasible; 0/– below average/vulnerable; 0 average/possible; 0/+ above average; + favourable/proven technology; –/+ score depends on design choices and conditions.

wet cross-section of the barrier in view of tidal and river flows, as well as navigation. This requirement will determine the size of the barrier, but not its type. A large flow cross-section can also be provided with several smaller openings. If a barrier is to be combined with a (rail)road bridge, design options with limited spans become more favourable. A traffic connection is always possible without a physical combination with a barrier. This function will seldom determine the type of barrier. Table 10.10 compares the most important favourable and unfavourable aspects of the described gate types.

In general, *mitre gates* are primarily used as gates in shipping locks. At shipping locks, waves and current are limited, while the navigation width seldom exceeds 50m. In such a case, mitre gates have proven to be very cost effective. If conditions are more severe, such as strong tidal currents and high waves, mitre gates are less attractive or even not possible. Mitre gates have never been built for spans exceeding 50m but have recently been proposed for bigger spans up to 70m.

Vertical lifting gates are reliable and financially attractive for barriers if there is no height requirement. If ships need to pass underneath, vertical lifting gates are less attractive. However, up to a free clearance of 30m, vertical lifting gates can be very competitive. Vertical lifting gates are also very visible. At attractive urban or natural locations, this visibility may not be appreciated, although it can also be considered as a new landmark for a region. The span of vertical lifting gates is limited to about 100m. If passing ships require a bigger width, vertical lifting gates are less suitable.

If a visible barrier is not allowed, *flap gates* are a logical solution. Flap gates enable an almost unlimited barrier width. An obvious disadvantage of flap gates is that they are located underwater. This makes them more difficult for inspection and maintenance, and more vulnerable for sediments. These aspects can even threaten the operational reliability of the barrier.

If a big span is required, *horizontally rotating sector gates* (floating, sliding or rolling) can be the right answer. A span of up to 350m is possible, allowing even the biggest ships to pass. Horizontal-sector gates provide unlimited vertical clearance and do not strongly influence the open landscape view. Horizontal-sector gates require a lot of space for side docks in which the gates are placed when not in use. This makes this type of barrier more expensive. The horizontal-sector gate is typically a gate type that will be selected for the protection of a big port area.

Vertically rotating gates share many advantages and disadvantages of the vertical lifting gates. Since the operating mechanism needs no towers, the impact upon the landscape can be limited. Vertical rotating gates that are stored in a bottom recess below the water level share many advantages and disadvantages with flap gates. Contrary to flap gates, the span is limited; but inspection and maintenance are easier because the gate can fully rotate above the water surface.

Inflatable barriers or *rubber dams* are still not a widespread proven technology for storm surge barriers. Many rubber dams have been built, but only once as a storm surge barrier. If dimensions are limited, rubber dams can prove to be a very interesting option, requiring little maintenance.

10.5 Conclusions

The comparison and selection of suitable gate types discussed in this chapter is obviously not complete. Major issues such as costs, construction time, environmental issues and many other aspects must also be taken into consideration when designing a flood protection system for a coastal city. There is no simple decision for finding the best gate possible because an optimal design depends upon specific requirements and characteristics. A systems engineering approach is recommended, consisting of several interactive steps in which, first, the functions and requirements are determined, the design options are considered, and finally a comparison of design options is made in a trade-off matrix in order to select the best option (e.g. Rijkswaterstaat et al, 2008). After determining the best design option, the components of the selected barrier type are chosen and the procedure starts again (functions, requirements, options, trade-offs) in order to select the best design option for every component.

For bigger flood protection systems, a combination of different gate types can be the optimal solution. Such solutions are under construction in St Petersburg and in New Orleans, and have been successfully constructed in the Ems near Hamburg. In these schemes, many flow openings with lifting gates have been provided, in addition to some wide navigation openings with unlimited clearance for passing ships.

Finally, one should bear in mind that there is never one perfect solution. For the well-known Rotterdam Storm Surge Barrier (Maeslantkering), six completely different designs were proposed by contractors, who were to come up with different competitive solutions. All of these six designs fulfilled the design criteria that were given for the competition.

Acknowledgements

This chapter is based on an a general overview of navigable flood gate structures that was originally written as part of the *Inner Harbor Navigation Canal Flood Gates Conceptual Study* for the USACE New Orleans District, carried out by Deltares and ARCADIS (USACE, 2006). Parts of this overview were also used in a global overview of navigable storm surge barriers to select suitable gate types for New York at an American Society of Civil Engineers (ASCE) seminar held on 30 to 31 March 2009 in New York (Hill, 2011). Some of the information that was used for this overview, including many photographs, came from Rijkswaterstaat, The Netherlands.

References

Bakker, C. and Vink, M. (1994) 'Zooplankton biomass in the Oosterschelde (SW Netherlands) before, during and after the construction of a storm-surge barrier', *Hydrobiologia*, vol 282–283, no 1, pp127–143

Bowman, M., Hill, D., Buonaiuto, F., Colle, B., Flood, R., Wilson, R., Hunter, R. and Wang, J. (2008) 'Threats and responses associated with rapid climate change in Metropolitan New York', in M. McCracken, F. Moore, J. C. Topping, Jr. (eds) *Sudden and Disruptive Climate Change: Exploring the Real Risks and How We Can Avoid Them*, Earthscan, London pp119–142

de Jong, R. J. and Jongeling, T. H. G. (1995) 'In-flow vibrations in hydraulic structures', in *Proceedings of Sixth International Conference on Flow-Induced Vibration*, 10–12 April 1995, pp667–673

Dircke, P. T. M., Jongeling, T. H. G. and Jansen, P. L. M. (2009) 'A global overview of navigable storm surge barriers: Suitable gate types for New York from a Dutch perspective', in *ASCE Proceedings of the 2009 Seminar Against the Deluge: Storm Surge Barriers to Protect New York City*, 30–31 March 2009, pp93–102

EA (UK Environment Agency) (2010) *The Thames Barrier*, www.environment-agency.gov.uk/homeandleisure/floods/38353.aspx

Hill, D. (ed) (2011) *Against the Deluge: Storm Surge Barriers to Protect New York City*, Conference Proceedings, Polytechnic Institute of New York University, Brooklyn, NY, 30–31 March 2009, Annals of the New York Academy of Sciences, New York, NY

Jansen, P. L. M. and Dircke, P. T. M. (2009) 'Verrazano Narrows Storm Surge Barrier: A Dutch vision', in *ASCE Proceedings of the 2009 Seminar Against the Deluge: Storm Surge Barriers to Protect New York City*, 30–31 March 2009, pp119–127

Jongeling, T. H. G., Bakker, A. D. and Nederend, J. M. (1995) 'Oscillation of a sector-gate barrier caused by fluid resonance', in *Proceedings of Sixth International Conference on Flow-Induced Vibration*, 10–12 April 1995, pp131–138

Kabat, P., Vellinga, P., Aerts, J., Veraart, J. and van Vierssen, W. (2005) 'Climate proofing The Netherlands', *Nature*, vol 438, pp283–284

Meinhold, W. (2005) *Working Group 26 PIANC: Project Review Ems-Barrier*, Federal State Lower Saxony (Niedersachsen), Germany

Nienhuis, P. H., Smaal, A. C. and Knoester, M. (1994) 'The Oosterschelde estuary: An evaluation of changes at the ecosystem level induced by civil-engineering works', *Hydrobiologia*, vol 282–283, no 1, pp575–592

PIANC (2005) *Design of Movable Weirs and Storm Surge Barriers*, PIANC report, InCom Working Group 26 reduced version, 1 June 2005, www-new.anast.ulg.ac.be/doc/WG26_1.pdf

Rijkswaterstaat, Bouwend Nederland, ProRail and ONRI (2008) *Guideline Systems Engineering for Public Works and Water Management*, Rijkswaterstaat, www.leidraadse.nl/downloads/Guideline_systems_engineering_public_works_and_watermanagement_in_the_netherlands.pdf

USACE (US Army Corps of Engineers) (2006) *Inner Harbor Navigation Canal Flood Gates Conceptual Study*, USACE New Orleans, ARCADIS (Lawton, Dircke, Bandy) and Deltares (Jongeling)

USACE (2008) *Stamford Hurricane Dam*, www.nae.usace.army.mil/recreati/nrb/NRBhome.htm, accessed December 2010

USACE (2010) *Harvey Canal Sector Gate*, www.mvn.usace.army.mil/images/hurr_fldgate_graphic_Harvey_lrg.jpg, accessed December 2010

van der Horst, A. Q. C. and Rovekamp, N. (1999) 'Design of the storm surge Barrier Ramspol', in *STUFIB Conference Proceedings*, Osaka, Japan

Vrijburcht, A. (2000) *Hydraulische aspecten van stroomsluizen en hoogwaterkeringen*, TUDelft, Delft, The Netherlands

Vrijling, J. K. (2001) 'Probabilistic design of water defense systems in The Netherlands', *Reliability Engineering & System Safety*, vol 74, no 3, pp337–344

Whalingcity (2010) *New Bedford Hurricane Barrier Construction*, www.whalingcity.net/picture_hurricane_barrier.html, accessed December 2010

11

Dealing with Uncertainty through (Participatory) Backcasting

Susan van 't Klooster, Pieter Pauw and Jeroen Aerts

11.1 Introduction

Complex phenomena such as climate change are associated with large uncertainties and the risks can be substantial. As climate change and other systems are not fully understood, uncertainties are considerable and risks cannot be fully limited. Due to this level of ignorance, experts, decision-makers, stakeholders and the public who are increasingly confronted with climate-related risks must contend with uncertainty.

In this chapter we illustrate how stakeholders can be involved in the design of adaptation measures under uncertainty. We demonstrate the use of participatory backcasting – an interactive policy tool – in combination with interactive spatial planning. First, we give a brief introduction to backcasting. In the next sections we share our experience with the application of backcasting in two Dutch climate change adaptation projects to reduce future flood risk, addressing the following aspects:

- case introduction;
- formulating context-specific design principles;
- description of the two exercises.

We conclude with some final remarks about lessons learned.

11.2 A Brief Introduction to the Backcasting Approach

Backcasting can be understood as a (participatory) tool for policy analysis. Backcasting, literally 'looking back from the future' (Quist, 2007, p11),[1] is a scenario approach in which alternative desired (or even undesired) futures are taken as a starting point. After defining (un)desirable futures, the next step is to 'move backward to the present situation' (Quist, 2007) in order to identify what actions must be taken to attain (or avoid) these futures. By providing a framework for dealing with uncertainty, backcasting promises to strengthen the rationale basis for decision-making. Backcasting is often considered a powerful tool in interacting with stakeholders (e.g. Quist, 2007).

Backcasting was developed (and is mainly applied) in the field of climate change mitigation. It originates from the 1970s, in response to the oil crisis. In these exercises, a low energy demand in society and the development of renewable energy technology were taken as a starting point (see Quist, 2007). Backcasting has been applied for the exploration and initiation of system innovations towards sustainability (see Quist, 2007, for a comprehensive historical overview). Backcasting has not been used exclusively in energy studies (Robinson, 1982; Anderson, 2001; for further references, see Quist, 2007), but also in other climate-related domains such as freshwater availability (e.g. the SCENES project: van Vliet and Kok, 2010), mobility (e.g. Miola, 2008), hydrogen technology, nutrition and land use (e.g. Bakkes, 2009) (cf. Quist, 2007; Banister et al, 2000a, 2000b).

Backcasting is, however, a relatively new approach for the development of climate change adaptation strategies under uncertainty. In the field of climate change adaptation, it has not been extensively conducted or at least hardly evaluated and reported.

11.2.1 Backcasting as an alternative to forecasting

Backcasting is often portrayed as an alternative to 'forecasting' (e.g. Robinson, 1990; Quist, 2007). Forecasting can be understood as the attempt to predict the future. This predictive style of assessing the future is usually associated with trend extrapolation and the establishment of time series and point estimates. Forecasting aims to produce very specific statements about the value of a particular indicator at a specified moment in the future, such as a country's gross domestic product (GDP) in 5 years' time or demographics in 50 years' time. This approach to the future is very data intensive and involves quantitative analysis and modelling as a dominant means to predict the future. The main assumption employed in forecasting is the idea of continuity. It is assumed that the future will be shaped by processes, mechanisms and factors that shaped the past. It is furthermore assumed that those processes, mechanisms and factors are sufficiently known and understood to support prediction (see also van Asselt et al, 2010).

When the future under consideration is imbued with uncertainty and complexity, and processes, mechanisms and factors that might determine the future are not or are only partially known, predicting the future is impossible. In these cases, it makes much more sense to develop scenarios of how the future may develop. A scenario is not a prediction, but a description of a hypothetical situation – a story about a possible future.[2] Most scenarios take the present as the starting point and coherently describe various 'hypothetical sequences of events' (Kahn, cited in Aligica, 2004, p75). By identifying so-called drivers of change, different narrative storylines are constructed which are then fleshed out quantitatively (using models) or qualitatively. This mapping of a 'possibility space' (Berkhout and Hertin, 2002; Berkhout et al, 2002) is aimed at the systematic assessment of the implications of various evaluations of uncertainty and, ultimately, the identification of robust strategies. Notwithstanding ambitions to provide an alternative for the predictive style of assessing the future, these scenarios are often criticized. A general point of critique is that they provide a too narrow and bounded variety of possible futures and that they have a conventional 'business-as-usual' (trend-based) character. Scenario development, in general, is considered too probability-focused and discontinuity-aversive,[3] ignoring 'wild cards' and low-possibility futures, ignoring uncertain, surprising events and non-linear developments (cf. Brooks, 1986; Robinson, 1990; Ayres, 2000; Marien, 2002).

In response to this criticism, 'backcasting' has been put forward as an alternative scenario approach for the exploration of alternative futures and providing as much information as possible on uncertainties with which the future is imbued. Backcasting and forecasting can be understood as two different scientific traditions representing different philosophical views and perspectives (see Table 11.1).

Table 11.1 *Forecasting versus backcasting*

	Forecasting	**Backcasting**
Philosophical views	Causality; determinism	Causality and teleology; partial indeterminacy
Perspective	Dominant trends; likely futures; how to adapt to trends	Societal problem in need of solution; desirable futures; scope for human choice
Methods	Various (econometric) models	Partial and conditional extrapolations highlighting polarities and limits

Source: adapted from Dreborg (1996)

Furthermore, the advantage of backcasting over more conventional methods to guide adaptation strategies, such as the commonly used cost–benefit analysis, is that it offers a framework for systematic assessment of externalities, uncertainty and the interplay of human beings and nature (i.e. the factors complicating the assessment of costs and benefits in the field of sustainability science).

11.2.2 The backcasting approach

Instead of taking the present as the starting point, backcasting starts with the articulation of one or more desired (or undesired) future(s). A backcasting scenario process generally follows four different steps (cf. Quist, 2007). The first step involves a creative process of defining (a) desired (or undesired) future(s). The next step is working backwards from that future(s) to identify strategies, measures and policies needed to realize (or avoid) that future(s). This creative phase is then followed by the third step, in which the underlying assumptions are evaluated in terms of feasibility and consequences

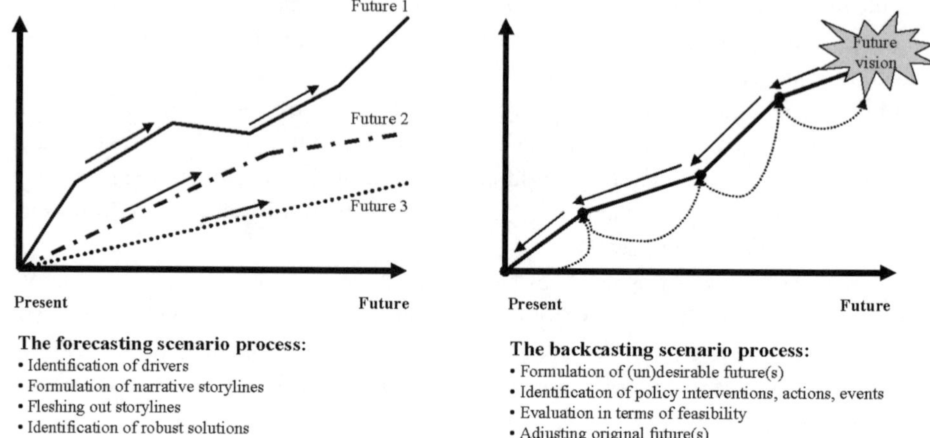

Figure 11.1 *Backcasting versus forecasting scenario process*

by reflecting on the implications of the long-term perspective for short-term policy-making. After the identification of policy interventions, actions and events, the original future vision is generally adjusted (see also Figure 11.1). Backcasting is an iterative process aimed at resolving internal inconsistencies and mitigating undesirable impacts and side-effects revealed in the course of the analysis (Quist, 2007).

In the backcasting methodology, dealing with uncertainty is inscribed in several ways. Strategies for dealing with uncertainty include:

- *Put normative futures at the heart of the exercise.* This is done by starting the exercise with the identification and construction of future vision, which can represent either a desired future or an undesired future. Backcasting is often a 'single vision exercise'. The backcasting methodology prescribes that these single visions should embrace a realm of ideas, wishes and interests and that they become 'shared multi-actor constructions' (Quist, 2007). This also implies that the constructed visions should allow stakeholders to adjust the constructed future vision when new insights or conditions occur (Quist, 2007).
- *'Get the system in the room'.* To ensure that different value frames are included in the analysis, backcasting aims to get 'the whole system in the room' (Weisbord, 1993) by involving people from different relevant backgrounds and positions.
- *Articulate and confront value frames.* Backcasting explicitly aims at guiding interactivity and the articulation and confrontation of different value frames. It is assumed that including and confronting different frames of reference does not only enhance creativity and out-of-the-box thinking, it also enhances the understanding of the assumptions on which visions, policy choices and preferences are based and broadens the scope of solutions to be considered.

- *Aim towards a future vision that embraces a realm of relevant ideas, wishes and interests.* Dreborg (1996) suggests that a backcasting study should not only describe the kind of 'value-related considerations', but also provide images of the future that are based on 'different concepts and values'. This way, a backcasting project enables groups with different views to utilize the results and to gain a deeper understanding of the issue at stake (Quist, 2007).

- *Set a long time horizon.* Backcasting exercises generally have their starting point in the distant future. The assumption is that having a long time horizon shifts the focus away from present conditions and therefore stimulates 'out-of-the-box' thinking. It is argued that this enables a discussion remote from daily concerns and circumstances and permits radical change. As people are generally conservative to change, a long time horizon would be needed to encourage the participants to think more in terms of radical changes and discontinuities (cf. Dreborg, 1996; Kok et al, 2006).[4]

- *Start from a (un)desirable future.* Starting from a (un)desirable future is considered as a way of avoiding the tendency of retaining the goal of reliable prediction (Robinson, 1990). Working backwards is assumed not only to facilitate breaking through the evolutionary paradigm – the gradual incremental unfolding of the future – but also to treat surprise random events and changed conditions.

- *Postpone feasibility/plausibility discussions.* By starting with (un)desired future visions remote from daily concerns and circumstances, participants are implicitly instructed to defer judgement of ideas until later. The suggestion is that postponing feasibility/plausibility discussions facilitates 'out-of-the-box' thinking and enhances creativity.

- *Aim for sequential decision-making.* Finally, backcasting prescribes sequential decision-making. The notion of adaptive management is central to backcasting. It is a way of dealing with uncertainty by 'seeking to maximize flexibility', 'keeping many options open' and 'avoiding lock-in' (Quist, 2007). With regards to uncertainty, sequential decision-making is a way of exploiting improved understanding as time goes by and of considering new conditions, processes, developments and (unexpected) events. In our backcasting experiment, we did not investigate the follow-up, spin-off and implementation of the backcasting exercises described and analysed in this chapter.[5] We will therefore not evaluate the issue of sequential decision-making in the remainder of this chapter.

Over the years, backcasting has been applied regularly for the development of climate mitigation strategies. For climate change adaptation, backcasting is a relatively new approach that has not been extensively conducted, or, at least, hardly evaluated and described.[6] Applying backcasting in the field of climate adaptation requires further methodological development and specification, taking into account context-specific characteristics. In order to set up a climate adaptation backcasting experiment in a Dutch context, we therefore tried to identify some main characteristics of the climate adaptation discussions and to formulate additional design principles.

11.3 Case Introduction: Backcasting Climate Adaptation in The Netherlands and Its Province of Groningen

Climate change will have a profound effect upon flood risks in The Netherlands (e.g. Bresser et al, 2005). A large part of the country's population lives in flood-prone areas and a substantial share of the national income is earned in the same low-lying locations (Aerts et al, 2007). The sea level at the Dutch coast is expected to rise between 39cm and 85cm by the year 2100 (KNMI, 2006). The consequences of sea-level rise in The Netherlands are exacerbated by subsidence of land, which will locally lead to a lowering of the ground level of 1m by 2100 (van Dorland and Jansen, 2007). In addition, Middelkoop et al (2001) estimate that peak flows of the River Rhine (one of the three major rivers in The Netherlands) may increase by about 5 to 8 per cent by the year 2050. Without additional innovative investments, current technical measures such as dikes will not meet the legally binding Dutch safety standards.

Traditionally, Dutch water management can be characterized by an engineering approach in which technocratic solutions, such as raising dikes, are the dominant strategy to maintain the safety targets. Alternative safety measures were hardly considered. During past decades, and especially after the near floods of 1993 and 1995, it has been increasingly recognized that alternative strategies, such as insurance and spatial planning actions (e.g. flood zoning and building codes), as well as coordinated action, are required (e.g. Middelkoop et al, 2004; Kabat et al, 2005). New water management strategies are to be developed by a more 'interactive' and 'area-oriented' regional approach. In The Netherlands, there are many recent initiatives for the development of alternative strategies to dealing with the challenges of climate change, such as the formation of the National Programme on Spatial Planning and Adaptation to Climate Change (ARK), in which the national government collaborates with regional and local governments to jointly develop an adaptation strategy, and a number of major research programmes, such as Climate Changes Spatial Planning (KvR) and Living with Water. The two case studies central to this chapter are also part of this 'new turn' in Dutch water management. The first case study focused on climate-proofing The Netherlands as a whole and was part of the Safety First project – a national-scale project that investigated how long-term changes in climate, land use, governance and socio-economic trends will affect flood safety in The Netherlands (Aerts et al, 2008). The backcasting workshops for the second project in Groningen was part of a regional project on climate-proofing this Dutch province under the national Climate Changes Spatial Planning programme (Roggema, 2009).

11.4 Formulating Context-Specific Design Principles: Backcasting Climate Change Adaptation in The Netherlands

The Intergovernmental Panel on Climate Change (IPCC, 2007) has defined adaptation as an adjustment in ecological, social or economic systems in response to actual or expected climatic stimuli and their effects or impacts. Adaptation can be both a response to adverse effects or vulnerabilities and a capitalization of opportunities that arise due to climate change. In the low-lying delta country of The Netherlands, climate change adaptation in flood management is mainly framed as an adjustment to increasing flood risk under long-term uncertainties. Three different characteristics of the Dutch adaptation debate stand out:

1 *Different interpretation frames with regard to climatic uncertainty.* In The Netherlands, different interpretation frames are used by policy-makers, professionals and stakeholders with regard to the rate and magnitude of climate change. The regional climate scenarios of the Royal Netherlands Meteorological Institute (KNMI) are the norm; however – especially after the release of Al Gore's *An Inconvenient Truth* – a significant number of people also feel that bigger changes must be anticipated (e.g. Haasnoot et al, 2009). In a workshop setting, different interpretation frames need to be acknowledged and constructively used in order to avoid paralysing discussions among the participants about the plausibility of claims with regard to our future climate.

2 *A dominant technocratic perspective on Dutch water management.* The management of water resources is currently undergoing a paradigm shift from a command-and-control approach towards a more integrated and participatory approach (e.g. Pahl-Wostl et al, 2007a, 2007b). Although it has been broadly recognized that the traditional Dutch 'hydrological and engineering approach' needs to be replaced by alternative, more adaptive approaches (e.g. Kabat et al, 2005), the key players in Dutch water management[7] are still 'wrestling' with old commitments, structures and cultures. In addition, linear, predictive probability-based accounts of the future (which are in line with the command-and-control approach) still constitute the dominant way of reasoning (e.g. Middelkoop et al, 2004; Pahl-Wostl et al, 2007a, 2007b).

3 *Combining technocratic and spatial adaptation strategies.* In The Netherlands, water and spatial planning have traditionally been separate policy domains, with different cultures, ambitions, approaches and methods. In view of climate change and the need to adapt, it is increasingly acknowledged that this separation of policy domains cannot be maintained and that new and innovative strategies need to be developed by integrating both domains. The challenge is, therefore, to facilitate this integration in a workshop setting.

In order to deal with these specific challenges, we formulated three additional design principles with regards to dealing with uncertainty. First, backcasting generally uses a reference scenario (in our case, we used the business-as-usual scenario) as a comparative framework against which the backcasted scenarios can be understood. The risk of using such a reference scenario is that it may instigate debates on the likelihood or implications of the reference scenario (cf. Robinson, 1990). As different interpretation frames coexist with regards to the future climate condition, anticipating potential 'paralysing' discussions about the plausibility of climate-related claims is relevant. We proposed two sets of assumptions about future climatic conditions as starting points for the development of future visions and the context within which the scenario analysis could proceed. One set included extreme assumptions with regards to sea-level rise, precipitation and temperature change, whereas the other set included even more extreme assumptions. The assumptions of extreme climatic conditions were also meant to break through the technocratic evolutionary perspective and to encourage participants to think in terms of radical changes and discontinuities. Third, we decided to support the backcasting exercise with the use of maps (cf. Tress and Tress, 2003) to further stimulate and facilitate reflection and deliberation on participants' assumptions and to facilitate the integration of 'water' and 'spatial planning'. Visualization through a map is a powerful tool when communicating with stakeholders in order to grasp the possible impacts of alternative developments and to collaboratively search for innovative adaptation strategies (cf. Shaw et al, 2009).

Table 11.2, which formed the basis of our backcasting experiment, summarizes the design principles involved.

Table 11.2 *Strategies for dealing with uncertainty inscribed in the backcasting methodology, as well as additional design principles*

Strategies inscribed in the backcasting methodology

- Put normative futures at the heart of the exercise.
- 'Get the system in the room.'
- Articulate and confront value frames.
- Aim towards a future vision that embraces a realm of relevant ideas, wishes and interests.
- Set a long time horizon.
- Start from a (un)desirable future.
- Postpone feasibility/plausibility discussions.
- Aim for sequential decision-making.

Additional design principles

- Use a different set of assumptions (about distant future (e.g. 2100) climate conditions) as a starting point for the discussion.
- Use maps/visualization techniques.
- Formulate extreme assumptions (about distant future climate conditions).

11.5 Description of the Two Backcasting Exercises

Backcasting exercises may differ from each other in a number of ways: the emphasis placed on articulating the vision *vis-à-vis* analysing pathways to achieving this vision; the role played by technical or financial implications; and the place given to stakeholder involvement (cf. Quist, 2007). Our backcasting exercises were set up as a methodological/analytical experiment for which we adopted a 'light' version of backcasting in the sense that we did not engage in extensive modelling (cf. Robinson et al, 2006), we involved a relatively small group of people, and follow-up and spin-off in terms of an actual implementation of the backcasting results (and, therefore, a thorough evaluation of costs and benefits) was not a primary goal. Although we worked with a relatively small group (13 to 20 people), we aimed at involving a broad range of participants. Organizing the groups was a precise and conscious effort in creating well-balanced groups in which people with different ideas, ways of thinking and backgrounds could interact. Participants ranged from scientists, representatives of (local) governments and water boards, water engineers, landscape architects/designers and non-governmental organizations (NGOs) representing the Dutch or Groningen 'water world' and the 'world of spatial planning'.[8] We also invited some so-called 'remarkable people' (van der Heijden, 1996) (i.e. experts who were not the 'usual suspects' from both worlds). The idea was to stimulate fresh and potentially innovative ideas, prevent group thinking by questioning traditional ways of thinking, increase interaction amongst participants, and expand the rationality of the group. In the Safety First case study, 20 (day one) and 18 (day two) people participated. In the Groningen case study, 17 (day one) and 13 (day two) people participated. In both cases the group was divided over two scenarios. Most of the participants of the first workshop were also present at the second. The goal of the first meeting was to formulate future visions of a climate-proof country/province and to draw new maps of what the participants desired as the way in which The Netherlands/Groningen would look by 2100, based on projections of sea-level rise, and increasing temperatures, precipitation and number of storms. The process was initiated by a plenary discussion on what 'climate proof' actually means and on which socio-economic and spatial indicators were to be included in the definition. Second, desired ingredients of the future visions of The Netherlands/Groningen were drawn up individually and discussed on a plenary basis in order to make participants familiar with the matter and to ensure that the assignment was well understood. Third, the participants were divided into two groups who discussed the general outlook and important elements of the future vision/map. These 'elements' were not fixed, as participants were able to change their minds about what is desirable during the exercise after seeing the outcomes of specific choices made during the backcasting exercise. We considered the backcasting exercise, in other words, as an iterative process in which the desired future vision is a product of the process of trying to reach it.

Both backcasting exercises followed a similar step-wise procedure (see also Table 11.3). We organized two workshops per case study and invited the same people to both workshops. During both workshop days, poster-sized geographical and hypsographic maps about current land-use patterns, water systems and elevation were provided to support the development of future visions and policy interventions. Participants were also expected to develop their own maps. In each group a geographic information systems (GIS) expert/landscape architect participated to assist the participants by explaining the maps provided to them and helping them to visualize their ideas. During the first backcasting meetings, participants mapped their own visions of a climate-proof The Netherlands and Groningen in 2100. Eight colours of clay and different markers were used for visualization on a transparent overlay of the geographic map. As the process advanced, more and more detail was brought in. For the second meeting we used a big paper timeline (1m × 2m) from the present to the year 2100 on which the participants could plot intermediate objectives, required interventions, threats, opportunities and challenges.

The Safety First backcasting exercise resulted in two maps of The Netherlands in 2100. One map was based on extreme and one on even more extreme assumptions about the future climate condition. The extreme assumptions were based on the extreme 'W+' assumptions provided by the Royal Netherlands Meteorological Institute (KNMI), as developed in 2006. The even more extreme assumptions were provided to the participants by the research team, inspired by Al Gore's *An Inconvenient Truth* (see also Table 11.4). The Groningen exercise started with the same climate assumptions. However, this exercise resulted in three visions of what the province should look like in 2100.

All maps show ideas on how to cope with sea-level rise, increased river discharge, more extreme precipitation patterns and other assumed climate-related impacts. Plates 9 and 10 summarize the future visions that were depicted in each of the maps.

During the second workshop day, the participants developed strategies to realize their visions. After a plenary discussion on the maps produced (to refresh the memory), the group split into two again to analyse how the envisaged future(s) could be achieved by looking back from these futures and identifying what would be needed to attain them. As a first step, intermediate objectives were set using a timeline and (if these objectives had a spatial component) the map as documenting material. Second, the required interventions were set. These could be political, social, technological, environmental, etc. Again, the interventions were visualized in the timeline and the map. Third, threats and opportunities were identified and the most important challenges in reaching the desired outcome per 2100 formulated. In this step, the participants also investigated the feasibility of the developed measures and visions. Finally, the intermediate objectives, required interventions, main challenges and final aim were jointly presented in a plenary session.

Table 11.3 *Goals, steps and output of the two backcasting exercises*

Day 1			Day 2		
Goal: Define and visualize (desired) future visions			*Goal: Identify strategies, policies and programmes, and evaluate them in terms of feasibility and consequences using timelines*		
Step 1: Plenary discussion	Step 2: Discussion in two groups	Step 3: Plenary discussion	Step 1: Plenary discussion	Step 2: Discussion in two groups	Step 3: Plenary discussion
Create shared understanding of 'climate-proofing'	Discuss the (extreme and even more extreme) climate assumptions	Present the future visions/maps	Discuss the future visions/maps: do the participants still identify themselves with it? Are adjustments required?	Set intermediate objectives	Present the timelines
Identify general ingredients of the future vision	Develop future visions and visualize these	Evaluate the exercise		Identify required interventions	Evaluate exercise
				Identify threats and opportunities	
				Identify challenges	
Output: Descriptions of (desired) future visions and their visualizations			*Output: Timeline with intermediate objectives, required interventions, threats and opportunities and challenges*		

Table 11.4 *Assumptions regarding future climate change
in The Netherlands and Groningen exercise*

		Extreme (W+ KNMI)	More extreme (Al Gore: An Inconvenient Truth)
Precipitation	Summer	−38%	−45%
	Winter	+28%	+40%
Temperature	Summer	+5.6°C	+6°C
	Winter	+4.6°C	+6°C
Sea-level rise		0.85m	5m

11.6 Lessons Learned

For climate change adaptation discussions, such as adaptation against flood and flood-related risks through water management and spatial planning, backcasting is a relatively new approach. This section outlines our most salient observations with regards to the application of the backcasting approach for dealing with uncertainty in climate change adaptation. To this end, we discuss each of the strategies for dealing with uncertainty inscribed in the backcasting methodology and additional design principles, as identified in Table 11.2.

Putting normative futures at the heart of the exercise. In our backcasting exercise, the participants felt comfortable discussing socially and environmentally desirable futures. In addition, this normativity-centred approach seemed to facilitate the exploration of implications of alternative developments and the value frames that underlie them. During the exercise, assumptions on which visions, policy choices and preferences were based were regularly questioned, explained and made explicit (see 'Articulation and confrontation of different value frames' below).

Putting the climate change adaptation discussion in a different normative context re-framed the discussion and seemed to enable the participants to spot policy interventions that they otherwise would possibly have missed. During the exercise and in the evaluation afterwards, the majority of the participants mentioned that the exercise resulted in changed perspectives and innovative ideas on climate-proofing The Netherlands and Groningen. This seems to indicate that putting normative futures at the heart of the exercise helps to broaden the scope of climate-proofing solutions (cf. Dreborg, 1996).

'Getting the system in the room'. The backcasting methodology places a strong emphasis on the inclusion of diverging values. Adaptation discussions lay at the intersection of many policy domains (e.g. water management, spatial planning, environment, economic affairs), disciplines (they have an economic, environmental, socio-cultural or institutional/political dimension), scale levels (municipal, provincial, national or European), institutions/organizations (governments, NGOs or private companies) and

different types of actors (policy-makers, scientists or the wider public). There is, in other words, not a single organization or subsystem that can constitute the 'heart' of the backcasting exercise. As a well-defined and clear-cut playing field is absent in a climate change adaptation context, 'getting the whole system in the room' and thereby including the relevant value frames in the analysis is not self-evident.

Articulation and confrontation of different value frames. Backcasting prescribes that systematic assessment of different normative interpretations should be part of the entire exercise. In both The Netherlands and Groningen exercises, different and opposing adaptation approaches could be discerned (i.e. defensive versus offensive and technological versus socio-economic strategies). To illustrate the differences, Table 11.5 provides some examples of proposed interventions.

Table 11.5 *Diverging strategies*

	Defensive	*Offensive*
Technological	• Building a super-dike around the cities of Rotterdam and Amsterdam and the 'Green Heart' (Netherlands exercise) • Compartmentalized inundation of the lower-lying part of the province (Groningen exercise)	• Building a new island in the North Sea (Netherlands exercise) • Creating new dunes, thereby moving the coast line northwards to the current location of the Frisian Islands in the North Sea (Groningen exercise)
Socio-economic	• Steered migration to higher grounds (areas 5m above sea level) (Netherlands exercise) • Moving housing, businesses and other functions to areas 5m above sea level (Groningen exercise)	• Enhancing the adaptive capacity of local communities (Netherlands exercise) • Stimulating/subsidizing new adaptive agricultural economies (water- and/or salt-'resistant' crops/aquaculture, recreation) (Groningen exercise)

Despite differences in approaching adaptation, during the exercises the varying strategies turned out to be complementary and integrative. For example, the idea of compartmentalized inundation (technological strategy) gave rise to a discussion about how and under what conditions this technological measure could enhance the economic and social viability of the region (socio-economic strategy). Similarly, discussing the conditions and effects of steered migration (defensive strategy) raised the issue of what to do with people who refuse to move away and the alternative of enhancing the adaptive capacity of local communities (offensive strategy). The shared ambition of finding no-regret options for climate-proofing The Netherlands and Groningen seemed to emphasize functional relationships, to facilitate combining and integrating different type of strategies, and to crystallize consensus about preferences.

Using a different set of assumptions and embracing different value frames. Backcasting is generally a single vision exercise that ideally leads to a vision that embraces a realm of ideas, wishes and interests (Quist, 2007). Due to uncertainty surrounding the future climate condition and the coexistence of different interpretation frames, we decided to work with two sets of assumptions regarding precipitation, temperature and sea-level rise, and divided the group accordingly. We realized that this could be at cross-purposes with the aim of creating so-called 'shared multi-actor constructions' (Quist, 2007). Observations in practice, however, show the opposite. Although both groups worked simultaneously with different climate assumptions, the upshot of the exercises turned out to be surprisingly similar, even in the Groningen exercise, where one group proposed two different visions on the same climatic condition (i.e. the earlier mentioned defensive and offensive strategies). Despite the different starting points, both backcasting exercises led to complementary future visions. This was also brought up by the participants themselves during the plenary presentations of the constructed future visions. They argued that the future visions could virtually be united without the occurrence of significant tensions or contradictions. Our backcasting experiment seems to indicate that even when different development paths are followed, a strong vision can be constructed that embraces a realm of relevant ideas, wishes and interests, and that starting from different assumptions regarding the future climate condition is not necessarily an impediment to constructing a single vision. Further research is needed to investigate if this also holds in situations where more diverse and competing frames are included in the analysis.

Using maps/visualization techniques. Due to the strong spatial nature of climate change adaptation, and in order to further stimulate and facilitate reflection and deliberation on assumptions of the participants, we used several maps during the exercises. The maps proved very useful in several ways as they enabled communication between people from different backgrounds. In science and technology studies, objects that productively stimulate interaction across professional boundaries are referred to as 'boundary objects' or focal points around which knowing-in-practice may arise (Star and Griesemer, 1989). First, the maps provided a common point of reference in the discussions. For example, discussing and determining the size and shape of the inundated areas in Groningen would have been impossible without zooming in on the actual spatial specifics of a situation. This is because, from a practical point of view, one needs to see the altitude of different parts of the province, but also because inundation is a measure that causes negative feelings and a sense of uncontrollability if it cannot be confined by a drawn dike.

Second, the maps enabled participants to translate abstract ideas about desirability into a more concrete visual level. In addition, the possibility of three-dimensional (3D) construction stimulated participants to make multifunctional designs, such as the idea of building houses in, and roads on, new dikes. Although participants were initially a bit hesitant, the use of clay in particular seemed to stimulate creativity as it enabled participants to simply mould and remould, move and remove their designs. Third, the maps enabled the participants to collaborate and systematically evaluate the spatial consequences of proposed visions, choices and preferences in an integrative way and, as such, to develop sufficiently shared meanings to move forward.

Setting a long time horizon. In both The Netherlands and Groningen exercises, we used a time horizon until 2100. Discussing strategies, actions and policies remote from daily concerns and circumstances turned out to be a difficult task for most participants. Despite several moderating strategies, the participants continued to feel more comfortable discussing and exploring the short- and medium-term future. This is, for example, illustrated by the fact that out of 61 intermediate objectives on the three timelines of the Groningen groups, 56 were placed before 2050.

Starting from a (un)desirable future. Starting with a creative process of defining desired (or undesired) future visions and then working backwards from that future (instead from historically and/or currently dominant trends) is assumed to break through the evolutionary paradigm – not an easy task. In particular, when short- and, to a certain extent also, medium-term planning measures were discussed, the participants began to see drawbacks and impossibilities, especially in financial and procedural terms. The pitfall here is that the degree to which participants consider proposed measures, plans and programmes for the future 'relevant' is increasingly referred to in terms of the degree to which these propositions can be realized under current conditions and circumstances. As a result, especially innovative ideas were easily disqualified and dismissed.

Postponing feasibility/plausibility discussions. During both exercises, we observed that desirability claims became easily questioned in terms of feasibility/plausibility. This tendency of considering proposed measures against the degree to which they could be realized under current conditions and circumstances seemed to hamper an open and creative process of creating future visions that are remote from daily concerns and circumstances, and a systematic assessment of new patterns, structures, mechanisms and possibilities did not come off the ground. At the same time, this intertwining of desirability claims, on the one hand, and plausibility/feasibility claims, on the other, is probably unavoidable. The separation between desirability and feasibility inscribed in the backcasting methodology is an analytic distinction that can be delayed or postponed, but not, in practice, maintained.

Formulating extreme assumptions. Not surprisingly, more extreme assumptions on the effects of climate change led to more extreme alterations of present-day protection and water management. Extreme assumptions encouraged the participants to reflect more on changed conditions and, subsequently, to propose new, yet unknown, spatial infrastructure, such as building new roads and other infrastructure at 5m above sea level and the construction of a super-dike. Our observation of the backcasting exercises seem to indicate that the participants (in both exercises) had difficulties in handling different climate assumptions in a balanced way. During the discussions, the participants exclusively focused on the potential effects of sea-level rise and possible responses. As a result, potential effects of changed and more extreme precipitation and temperature patterns were not systematically scrutinized. This focus on sea-level rise can be explained by the fact that this effect was considered straightforward, visible, linear and relatively stable, whereas the effects of temperature rise or changes in precipitation were not, and were therefore more difficult to handle. Another possible explanation is that for the other changes, the concrete impacts are not immediately apparent. For example, it is difficult to visualize what an increase of winter precipitation of + 40 per cent means. In addition, the concrete impacts of these changes are probably not immediately apparent. By prioritizing and reducing the number of possible impacts, the participants found a way to move forward.

11.7 Discussion: Backcasting Climate Change Adaptation

There is a growing awareness that systems we once thought behaved in a (relatively) predictable manner are characterized by uncertainty, value diversity and discontinuity. To cope with the resulting unpredictability, we need methods and approaches that can facilitate the improvement of our capacity to cope with and adapt to changed conditions, and possibly change them into improved and more desirable pathways.

In this chapter, we analysed to what extent backcasting can improve our capacity to adapt to climate change. In general, backcasting turns out to be a promising approach for developing adaptation strategies under uncertainty. It contributes to strengthening the rational basis for decision-making by:

- improving the understanding of the complex nature of climate change, decision-making and the assessment of impacts and vulnerabilities; and
- assisting in making informed and integrated decisions on practical adaptation options, measures and programmes.

11.7.1 Improving understanding

Backcasting facilitates a reframing of the issue, thereby enabling participants to collectively search for integrated solutions. It seems to be a powerful tool for providing information on uncertainties with which the future is imbued and, in particular, for articulating and confronting value frames and the construction of future visions. The backcasting methodology provides a framework for the systematic development of future visions that embrace a realm of ideas, wishes and interests.

In dealing with multiple (uncertain) climate effects and impacts, discussing strategies, actions and policies remote from daily concerns and circumstances and breaking through the evolutionary paradigm, some challenges remain. Although the use of extreme assumptions encourages participants to reflect more on changed conditions, the discussions habitually focused on the short and medium term, and linear extrapolative arguments and – occasionally – the discarding of novel ideas based on historically grounded accounts were not uncommon.

Although backcasting provides a relatively simple framework for the exploration of the future, backcasting exercises are very complicated and challenging thought experiments. In particular, translating future visions to intermediate goals, identifying policy interventions, actions and events, and formulating potential obstacles and innovative solutions, thereby taking account of a changing environment, is not an easy task. Backcasting exercises require time and huge intellectual power, and their success is not only the upshot of a well-thought-through backcasting process (e.g. using the right techniques and creating the right conditions), but is also strongly dependent upon the willingness and capacity of participants to stretch their ways of thinking.

11.7.2 Assisting in making informed decisions

Putting the climate change adaptation discussion in the context of a higher-level objective of climate-proofing enabled the participants to spot policy interventions that they otherwise could have missed. However, a remaining challenge with regard to using backcasting in the domain of climate adaptation is the question of how to link the analysis to action. The application of backcasting in the domain of climate mitigation is often explicitly targeted towards the actual coordination and implementation of strategies, policies, programmes and actions. The assumption is that for future visions to become effective, their actual creation does not suffice. As a result, backcasting exercises are often aimed towards the induction of bottom-up processes driven by stakeholders that lead to follow-up and spin-off (cf. Quist, 2007). The management of future visions in the context of climate change adaptation deserves special attention. Adaptation discussions lie at the intersection of many policy domains, disciplines, spatial-scale levels, institutions/organizations and different types of actors. There is, in other words, not a single organization or subsystem that can constitute the 'heart' of the backcasting exercise. Since a well-defined and clear-cut playing field is absent in the field of climate change adaptation, 'getting the whole system in the room' is not self-evident.

For backcasting exercises at a national (or even international) level, it may be even more difficult to initiate bottom-up processes driven by stakeholders that lead to follow-up and spin-off. On the other hand, maps and other visualization techniques in the field of climate change adaptation could play a positive role in motivating behavioural change and capacity-building of the participants after the actual backcasting exercise, thereby contributing to the realization of the created future vision(s). Applying backcasting in the field of climate change adaptation challenges us to rethink the ability to shape the future. Due to uncertainty, the ability to influence the future should not be overestimated. Our backcasting experiment shows that the strength of backcasting in this domain lies mainly in offering a framework for strengthening the rational basis for decision-making rather than directly manoeuvring the system towards more desirable directions.

Acknowledgements

We would like to thank the workshop participants for their time, intellectual effort and valuable ideas. Our backcasting experiment also greatly benefited from the methodological expertise and professional moderation of Nicole Rijkens-Klomp, Michael van Lieshout (Pantopicon) and Marleen van de Kerkhof (IVM). The backcasting exercises were part of two bigger research projects: Safety First (funded by the Directorate-General of Water, BSIK Living with Water and BSIK Climate for Space) and the Hotspot Groningen (under the national Climate Changes Spatial Planning programme). We would like to thank Rob Roggema of Groningen Province and our project team member at the Safety First project for their inspiring cooperation. A final word of thanks to all reviewers, in particular Nicole Rijkens-Klomp and Marleen van de Kerkhof for their fruitful suggestions.

Notes

1 Lovins (1976) introduced the 'backward-looking-analysis' to explore the long-term energy policy for the US; Elmore (1980) developed the 'backward mapping' approach, and Robinson (1982) eventually coined the term 'backcasting' as an alternative to traditional planning and forecasting methods.
2 In the context of foresight, scenarios are also referred to as thought experiments and 'what if' analysis.
3 With the term 'discontinuities', we refer to temporary or permanent, and sometimes unexpected, trend breaks in society caused by interactions between events and long-term processes (cf. van Notten, 2005).
4 The temporal scope of the analysis is also determined by the analytical realm under consideration. For example, climate-related studies generally involve very long-term processes and effects. A long time horizon allows for the treatment of long-term global change phenomena.
5 See Quist (2007) for an analysis of the impact of several backcasting experiments. He describes several enabling and constraining factors (internal and external).
6 Extensive analysis of future studies has shown that this does not exclusively hold for backcasting studies, but for futures studies in general (van Asselt, et al, 2010). In cases where methodological accounts are provided, they are usually (very) short descriptions of some main steps or are confined to a simple scheme.
7 Traditional key players in Dutch water management are Rijkswaterstaat and the 26 water boards (*Waterschappen*) governing the regional water issues.
8 In The Netherlands, water and spatial planning have traditionally been two more or less separate domains with both their own culture, approaches and methods.

References

Aerts, J. C. J. H., Kolen, B., van de Most, H., Kok, M., van 't Klooster, S. A., Satijn, B. and Leusink, A. (2007) *Waterveiligheid en klimaatbestendigheid in breder perspectief*, Adaptation Program Space and Climate, Institute for Environmental Studies, Amsterdam, The Netherlands

Aerts, J. C. J. H., Sprong, T. and Bannink, B. (eds) (2008) *Aandacht voor Veiligheid [Living with Water]*, Adaptation Program Climate for Space, DG Water, Institute for Environmental Studies, Amsterdam, The Netherlands

Aligica, P. D. (2004) 'The challenge of the future and the institutionalization of interdisciplinarity: Notes on Herman Kahn's legacy', *Futures*, vol 36, no 1, pp67–83

Anderson, K. L. (2001) 'Reconciling the electricity industry with sustainable development: Backcasting – a strategic alternative', *Futures*, vol 33, pp607–623

Ayres, R. U. (2000) 'On forecasting discontinuities', *Technological Forecasting and Social Change*, vol 65, no 1, pp81–97

Banister, D., Stead, D., Steen, P., Åkerman, J., Dreborg, K., Nijkamp, P. and Schleicher- Tappeser, R. (2000a) *European Transport Policy and Sustainable Mobility*, Spon Press, London

Banister, D., Dreborg, K., Hedberg, L., Hunhammar, S., Steen, P. and Åkerman, J. (2000b) 'Transport policy scenarios for the EU in 2020: Images of the future', *Innovation*, vol 13, pp27–45

Bakkes, J. A. (2009) *Getting into the Right Lane for 2050: A Primer for EU Debate*, Netherlands Environmental Assessment Agency/Stockholm Resilience Centre, The Hague, Stockholm

Berkhout, F. and Hertin, J. (2002) 'Foresight futures scenarios: Developing and applying a participative strategic planning tool', *Greener Management International*, vol 37, pp37–52

Berkhout, F., Hertin, J. and Jordan, A. (2002) 'Socio-economic futures in climate change impact assessment: Using scenarios as "learning machines"', *Global Environmental Change*, vol 12, no 2, pp83–95

Bresser, A. H. M., Berk, M. M., van den Born, G. J., van Bree, L., van Gaalen, F. W., Ligtvoet, W., van Minnen, J. G. and Witmer, M. C. H. (2005) *The Effects of Climate Change in the Netherlands*, Netherlands Environmental Assessment Agency, MNP report no 773001037, Bilthoven, The Netherlands

Brooks, H. (1986) 'The typology of surprises in technology, institutions, and development', in W. C. Clark and R. E. Munn (eds) *Sustainable Development of the Biosphere*, Cambridge University Press, Cambridge, UK

Dreborg, K. H. (1996) 'The essence of backcasting', *Futures*, vol 28, no 9, pp813–828

Elmore, R. (1980) 'Backward mapping: Implementation research and policy decisions', *Political Science Quarterly*, vol 94, pp601–616

Haasnoot, M., Middelkoop, H., van Beek, E. and van Deursen, W. P. A (2009) 'A method to develop sustainable water management strategies for an uncertain future', *Sustainable Development*, doi:10.1002/sd438

IPCC (Intergovernmental Panel on Climate Change) (2007) *Climate Change 2007: Synthesis Report, Contribution of Working Groups I, II and III to the Fourth Assessment Report of the Intergovernmental Panel on Climate Change*, Core Writing Team: R. K. Pachauri and A. Reisinger (eds), IPCC, Geneva, Switzerland

Kabat, P., Vellinga, P., Aerts, J., Veraart, J. and Van Viersen, W. (2005) 'Climate proofing The Netherlands', *Nature*, vol 438, pp283–284

KNMI (Royal Netherlands Meteorological Institute) (2006) *KNMI Climate Change Scenarios 2006*, KNMI scientific report WR 2006-02, de Bilt, The Netherlands

Kok, K., Patel, M., Rothman, D. S. and Quaranta, G. (2006) 'Multi-scale narratives from an IA perspective: Part II. Participatory local scenario development', *Futures*, vol 38, pp285–311

Lovins, A. B. (1976) 'Energy strategy: The road not taken?', *Foreign Affairs*, vol 55, no 1, pp63–96

Marien, M. (2002) 'Futures studies in the 21st century: A reality-based view', *Futures*, vol 34, no 3/4, pp262–281

Middelkoop, H., Kwadijk, J. C. J., Daamen, J., Gellens, D., Grabs, W., Lang, H., Parmet, B. W. A. H., Schader, B., Schulla, J. and Wilke, K. (2001) 'Impact of climate change on hydrological regimes and water resources management in the Rhine basin', *Climatic Change*, vol 49, pp105–128

Middelkoop, H., van Asselt, M. B. A., van 't Klooster, S. A., van Deursen, W. P. A., Kwadijk, J. C. J. and Buiteveld, H. (2004) 'Perspectives on flood management in the Rhine and Meuse rivers', *River Research and Applications*, vol 20, pp327–342

Miola, A (2008) *Backcasting Approach for Sustainable Mobility*, European Commission, Joint Research Centre, Official Publications of the European Communities, Luxembourg

Pahl-Wostl, C., Craps, M., Dewulf, A., Mostert, E., Tabara, D. and Taillieu, T. (2007a) 'Social learning and water resources management', *Ecology and Society*, vol 12, no 2, p5, www.ecologyandsociety.org/vol12/iss2/art5

Pahl-Wostl, C., Sendzimir, J., Jeffrey, P., Aerts, J., Berkamp, G. and Cross, K. (2007b) 'Managing change towards adaptive water management through social learning', *Ecology and Society*, vol 12, no 2, p30, www.ecologyandsociety.org/vol12/iss2

Quist, J. (2007) *Backcasting for a Sustainable Future: The Impact after 10 Years*, PhD thesis, Eburon Academic Publishers, Delft, The Netherlands

Robinson, J. (1982) 'Energy backcasting: A proposed method for policy analysis', *Energy Policy*, vol 10, pp337–344

Robinson, J. (1990) 'Futures under glass: A recipe for people who hate to predict', *Futures*, vol 22, pp820–843

Robinson, J., Carmichael, J., Van Wynsberghe, R., Tansey, J., Journeay, M. and Rogers, L. (2006) 'Sustainability as a problem of design: Interactive science in the Georgia Basin – special issue on interactive sustainability', *The Integrated Assessment Journal*, vol 6, no 4, pp165–192

Roggema, R. (2009) *Hotspot Klimaatbestendig Groningen*, Final Report, Climate for Space, Province of Groningen, The Netherlands

Shaw, A., Sheppard, S., Burch, S., Flanders, D., Wiek, A., Carmichael, J., Robinson, J. and Cohen, S. (2009) 'Making local futures tangible: Synthesizing, downscaling, and visualizing climate change scenarios for participatory capacity building', *Global Environmental Change*, vol 19, pp447–463

Star, S. L. and Griesemer, J. R. (1989) 'Institutional ecology, "translations" and boundary objects: Amateurs and professionals in Berkeley's Museum of Vertebrate Zoology, 1907–39', *Social Studies of Science*, vol 19, pp387–420

Tress, B. and Tress, G. (2003) 'Scenario visualisation for participatory landscape planning: A study from Denmark', *Landscape and Urban Planning*, vol 64, pp161–178

van Asselt, M. B. A., van 't Klooster, S. A., van Notten, P. W. and Smits, L. A. (2010) *Foresight in Action: Developing Policy-Oriented Scenarios*, Earthscan, London van der Heijden, K. (1996) *Scenarios: The Art of Strategic Conversation*, Wiley & Sons, London

van Dorland, R. and Jansen, B. (2007) *Het IPCC-Rapport en de betekenis voor Nederland,* PCCC, De Bilt/Wageningen, The Netherlands

van Notten, P. (2005) *Writing on the Wall: Scenario Development in Times of Discontinuity*, Boca Raton, FL, Dissertation.com

van Vliet, M. and Kok, K. (2010) 'Backcasting as multi-scale governance tool: On the development of local robust actions and their implications for higher scales', in *Conference Program and Book of Abstracts, Scaling and Governance Conference 2010 on Towards a New Knowledge for Scale Sensitive Governance of Complex Systems,* Wageningen, The Netherlands, 11–12 November 2010, Wageningen, The Netherlands

Weisbord, M. R. (1993) *Discovering Common Ground*, Berret-Koehler Publishers, San Francisco, CA

12

Governance of Climate Change in Coastal Cities: The Example of Hong Kong

Maria Francesch-Huidobro

12.1 Introduction

Coastal cities are sites of economic activity and centres of human and technological development for the entire world economy and their respective nations. Their location ensures accessibility to assets within and outside their borders, while increasing their vulnerability to current and potential climate hazards such as cyclones, intense precipitation, flooding, sea-level rise, and coastal erosion and sedimentation. City authorities are confronted with the duty to protect their cities from such hazards; thus, their governing capacity for applying mitigating and adaptive measures to climate changes is constantly challenged.

Until the end of the 1990s, city climate change governance was led by individuals working in municipal bureaus, inter-bureau working groups and municipal city networks (Bulkeley and Betsill, 2003). During the last decade, city climate change governance has been driven by a multiplicity of stakeholders from the public, private and people sectors often participating in transnational alliances that serve as forums for scientific, technological and governance exchanges. Such alliances mostly focus on activities such as policy learning amongst their members, the transfer of best practices, and the representation of their members' interests at national, regional and international platforms (Alber and Kern, 2009; Kern and Bulkeley, 2009). While it is evident that cities can benefit from engaging in these activities, they can only profit from them through the good governance of the multilevel web of interdependent actors at different territorial levels and across the public non-profit divide, as well as by learning from other cities in order to find policy solutions to their problems (Stren and Cameron, 2005; Pemberton, 2009).

Coastal cities' authorities have added challenges to the governance of climate change. It is often the case that flood-prone, densely populated coastal cities' capacity to mitigate and adapt to the consequences of climate change is affected by their ability to manage a shared body of knowledge on which responsible and affected parties base their discussions about possible solutions and create consensus and cooperation (*best-practice transfer*). Coastal cities' ability to lessen the impact of and adapt to climatic changes may also be affected by the learning process through which responsible and interested parties review their strategies and interpretations, incorporating additional information within their frames (*policy transfer*). Finally, cities' mitigation and adaptability may be affected by their capacity to manage the dynamics of sub-national and transnational climate change politics that are structured by localized and transnational forces of political competition, executive–legislative relations and interest group representation, all of which contain a multilevel dimension (*governance capacity*). It is around the two themes of 'policy learning' and 'multilevel governance capacity' that the discussion that follows is framed. The case of Hong Kong is used as an empirical account.

The chapter is organized as follows: in section 12.2 the context of governance in Hong Kong, the parameters by which climate change is monitored and the risks that it poses to people and assets are discussed. Special attention is given to the risk from flooding that has been and still is a problem in Hong Kong. Section 12.3 analyses the governance and policy solutions that Hong Kong is beginning to put into place to manage the challenges posed by climatic trends. Special attention is given to discussing the role of government, private firms and non-governmental organizations (NGOs) in incipient collaborative government arrangements. Section 12.4 discusses the challenges of governing capacity that Hong Kong and, possibly, other coastal cities encounter.

12.2 Governance Context and Climate Challenges of Hong Kong

Hong Kong, situated at the mouth of the delta formed by the Xijiang (West), Beijiang (North) Dongjiang (East) and Zhujiang (Pearl) rivers as they enter the South China Sea, is a Special Administrative Region (SAR) of the People's Republic of China. China is a signatory of both the United Nations Framework Convention on Climate Change (UNFCCC) and the Kyoto Protocol, and it has notified the United Nations to extend the conventions to the Hong Kong SAR from May 2003 onwards. Classified as a developing country, China is not required to meet specific reduction targets prescribed in the protocol under the principle of 'common but differentiated responsibility'. Hong Kong is, therefore, not required to meet specific reduction targets either. Although the Hong Kong government's *laissez-faire*, non-interventionist approach and economic priority constrain its ability to conceive decisive policies for the environment, Hong Kong is nevertheless monitoring the negative effects that climate change is having upon the city and its population and has taken several mitigating and adaptive measures.

Moreover, in October 2007, Hong Kong joined the C40 Large Cities Climate Leadership Group and participated in 2008 in the C40 Tokyo conference on the theme of Adaptation Measures for Sustainable Low Carbon Cities. In December 2009, Secretary for the Environment Edward Yau participated in the third C40 meeting in Seoul, chairing a plenary session on 'Engaging stakeholders in the fight against climate change'. Discussions at the conference revolved around the theme 'Cities' achievements and challenges in the fight against climate change'.

In January 2010, the Hong Kong Business Coalition on Climate Action, a non-profit organization, called 'for determined actions by the HKSAR Government and close collaborations with businesses to reduce carbon and other greenhouse gas emissions, through setting concrete targets and benchmarking against best practices of other metropolitan cities in developed countries, with a view to demonstrate Hong Kong's leadership as a world city', adding that 'to implement policies and programmes efficiently requires strong leadership and close cooperation among all relevant bureaux and departments' (Business Environment Council, 2010). As is so often the case in Asian countries, the process of coordinating private and public interests rests with the Hong Kong bureaucracy that is expected to assume a central role (Cheung and Scott, 2003; Painter, 2005, p335; Ng, 2007).

12.2.1 Climate parameters

Climate change in Hong Kong can be attributed to both global warming as a result of an enhanced greenhouse effect and to urbanization. Changes in essential meteorological parameters and potential impacts recorded by the Hong Kong Observatory (HKO) are revealing.

The HKO suggests that because of Hong Kong's population density of 6480 persons per square kilometre, the trend of *temperature* increase is faster than that of the rest of the world (Hong Kong Observatory, 2009). The annual mean temperature recorded at the Hong Kong Observatory Headquarters since 1887 indicates a steady rise (see Figure 12.1). During the post-war years from 1947 to 2007, the average rise amounted to 0.17°C per decade. The warming at the Hong Kong Observatory Headquarters has been significantly faster in the period from 1989 to 2007, at a rate of 0.34°C per decade. Compared with the 1980 to 1999 average of 23.1°C, the annual mean temperature in Hong Kong during the decade of 2090 to 2099 is expected to rise by 4.8°C (Hong Kong Observatory, 2009).

Reliant on imported energy and water, the coastal city is susceptible to climate stressors in the surrounding regions, particularly in the Pearl River Delta (PRD). These include intensive rainfall events and sea-level rise. The *rainfall* over different regions of Hong Kong has had a generally rising trend, increasing by 46mm every decade since 1946 (Hong Kong Observatory, 2009).

An analysis of the *sea-level* records of Victoria Harbour during the last 50 years from the tide gauge stations at North Point (1954 to 1985) and Quarry Bay (since 1986) shows that the mean sea level in Victoria Harbour has risen 0.13m from 1954 to 2007, at an average rate of 2.4mm increase per year (Hong Kong Observatory, 2009). The trend was similar to that over the South China Sea as measured from satellites using

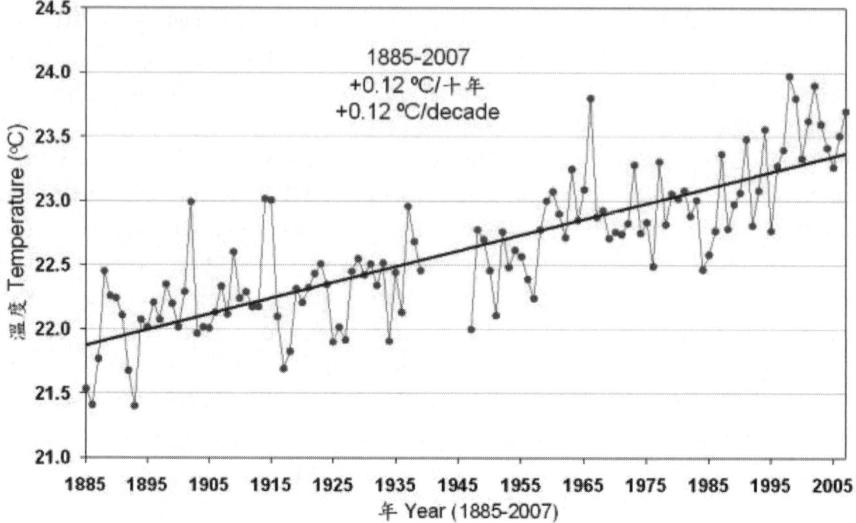

Figure 12.1 *Annual mean temperature recorded at the Hong Kong Observatory Headquarters, 1885–2007*

Note: Data are not available from 1940 to 1946.
Source: Hong Kong Observatory (2009)

remote sensing techniques from 1993 to 2005 (see Figure 12.2). A major impact of the mean sea-level rise in Hong Kong is an increase in sea flooding from storm surges caused by tropical cyclones. With a rise of 0.18m or 0.59m in mean sea level and assuming no changes in storm characteristics, the extreme sea levels in Victoria Harbour for various return periods would become higher, as illustrated in Table 12.1.

12.2.2 Climate risks

As coastal cities feel the irreversible effects of climate change, adaptation is becoming a policy must. The challenges that Hong Kong faces in terms of adaptability are related to the promotion of better management of flood risk and to the innovation of delta technology. In this regard, Hong Kong appears to be highly vulnerable to the consequences of climate change due to a number of factors. First, the city heavily depends upon reclaimed land for its infrastructural development. Key developments, including its financial district, its airport and its future cultural district in West Kowloon, are built on reclaimed land, which is only a few metres above sea level. The current mean sea level is about 1.4m above chart datum (see Figure 12.2). The trend during the period of 1954 to 2007 is of an increase of 2.4mm per year, or a total of 0.13m during the same period (see Table 12.1). The calculation is made as 2.4mm divided by 1000 equals 0.0024 times 53 years (1954 to 2007) equals 0.1272m.

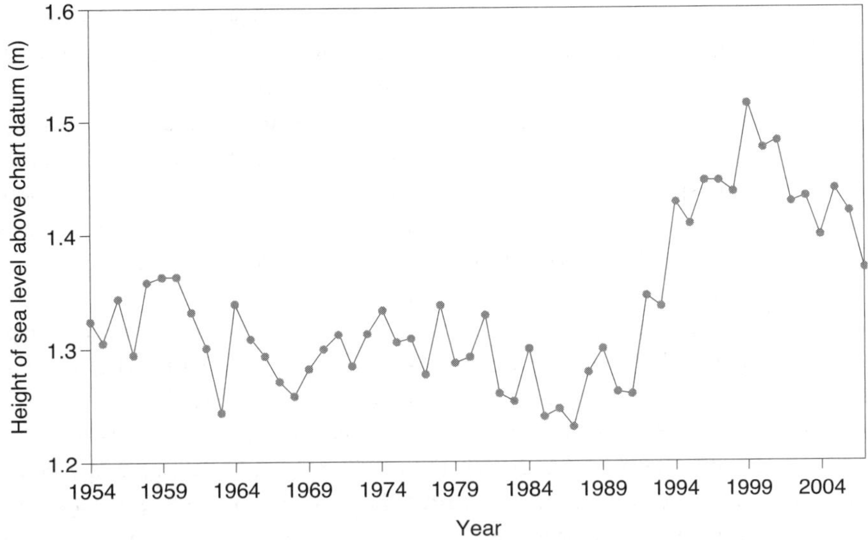

Figure 12.2 *Annual mean sea level at North Point/Quarry Bay, 1954–2007*

Source: Hong Kong Observatory (2009)

The city is also located in a region prone to extreme weather conditions such as typhoons, sudden and extreme rainfall, and vector-borne diseases; but its drainage infrastructure has been challenged during recent years by several incidences of extreme weather. For example, on 7 June 2008, the hourly record of rain was 145.5mm, the highest since records began in 1887. This caused heavy flooding in low-lying areas such as Hong Kong West and parts of the New Territories. A study by the Hong Kong Observatory found that 'the extreme 1 to 3-hourly rainfall amounts and the extreme daily minimum and maximum temperatures at the Hong Kong Observatory Headquarters exhibited

Table 12.1 *Sea-level trends*

Return period (years)	Extreme sea level above principal datum (m)		
	Current mean sea level	A rise of 0.18m in mean sea level	A rise of 0.59m in mean sea level
2	2.8	2.9	3.3
5	3.0	3.1	3.5
10	3.1	3.3	3.7
20	3.2	3.4	3.8
50	3.4	3.5	4.0
100	3.5	3.7	4.1
200	3.6	3.8	4.2

Source: Hong Kong Observatory (2009)

statistically significant long term rising trends' (Mok and Wong, 2008). Greenpeace is undertaking a study on the economic cost of the 7 June 2008 flood that saw 632 flood spots across the territory (interview with Greenpeace, 27 April 2009).

Hong Kong relies on imports for the supply of water, food and energy; thus, it will suffer from the consequences of the disruption of trade due to flooding or other climatic crises in affected exporting regions. The city's economic activity also depends upon the mobility of its population through an integrated mass transit railway system vulnerable to failure. Both its people and its infrastructure are highly concentrated, with a concentration of asset value. In a crisis, a lot will fall together.

Nevertheless, Hong Kong also has a number of advantages and significant adaptive capacity: a highly reliable infrastructure; expertise in coping with extreme weather conditions and crises (i.e. severe acute respiratory syndrome (SARS), avian flu and so on); relatively high gross domestic product (GDP) per capita of US$42,700 (PPP); and a reputation for innovation (CIA, 2010). These advantages, however, cannot belie the uncertainties that the city confronts. The IPCC has been unable to forecast a figure for sea-level rise if it were to factor in the melting of ice in the Antarctic and Greenland. It is also unclear how the fact that some of Hong Kong's key suppliers of goods which are poorly covered against risk may increase Hong Kong's vulnerability to climate change. Finally, it is unclear whether Hong Kong possesses enough capacity to design its adaptive infrastructure to the latest scientific data. Adaptation options should include actions which help to prevent (reduce the probability of an impact from occurring), prepare (act to understand the risk and develop an effective response), respond (act to reduce the consequences of damage) and recover (act to return to normal but with a higher level of resilience).

The physical parameters that can be identified as requiring attention with regards to adaptation capacity include rising temperatures; increasing inter-annual variation in rainfall; increasing severity and/or frequency of extreme weather events; rising sea levels; and an unrecorded parameter by the Hong Kong Observatory – namely, increasing ocean acidity. The sectors in which this adaptive capacity may be observed are biodiversity and nature conservation, built environment and infrastructure, energy supply, financial services, food resources, human health and water resources (see Table 12.2). It is perhaps the risk of flooding that causes greater concern to Hong Kong's public administrators.

12.2.3 Flooding and flood protection

Typhoons and floods have been a recurring phenomenon predominantly determined by intense precipitation from tropical cyclones and monsoons passing through Hong Kong, and seriously flooding the rural low-lying areas and natural floodplains in the northern part of the territory and older urban areas (Fung et al, 2004) (see Table 12.3). Residential blocks on higher ground, slopes and embankments are also vulnerable to flooding. Risks increase when fish ponds and agricultural land are filled and built

Table 12.2 *Adaptation parameters and impact upon water resources in Hong Kong*

Impacts of climate change	Rising temperatures	Increasing incidence of extreme weather events	Increasing inter-annual variation of rainfall (i.e. drought/too much precipitation)	Rising sea levels	Increasing ocean acidity
Adaptation measures	Increase consumer demand Increase evaporation rates and reduction of raw water yields Reduction in water quality from erosion, salinity, etc.	Reduced water yield due to overflowing of reservoirs/rivers (Donjiang riparian cities) Damage to reservoir structures and surrounding areas Danger to personnel Reputational damage Decreased water quality due to increased debris and turbidity Increased industrial pollution along the Donjiang (East) River	Reduced water yield due to decreased supplies in reservoirs and the Donjiang Increase in health, reputational and cost damages due to decreased supplies	Increased salinity of the Pearl River Delta (PRD) with decreased availability from current sources and need to find additional sources Increased risk of flooding and consequential industrial contamination	Increasing corrosion rates of subsurface infrastructure

Source: modified from Environmental Resources Management presentation, December 2008

up, eliminating a natural buffer against floods (Drainage Service Department, 2009). During recent years, Hong Kong has invested a lot in measures to prevent present and future flooding.

While climate change adaptation is somewhat new to Hong Kong policy-makers, flood protection has always been a concern from the time that Hong Kong Island came under British rule in 1841. Although the first drainage system was constructed mainly to improve sanitation and health, infrastructure soon developed, with drainage (flood

Table 12.3 *Some of Hong Kong's historical typhoons and extreme weather events*

Year	Event	Description
1906	Unnamed typhoon	Sudden gales and storm surges of the typhoon caused vast damage along the sea front in Hong Kong and resulted in heavy loss of life. The death toll of about 10,000 (90% boat people) was a shockingly high figure for the community of less than 450,000 people at that time.
1937	Great Hong Kong Typhoon	The villages along the coast of Tolo Harbour were severely flooded by the storm surge of the Great Hong Kong Typhoon. Over 10,000 lives were lost, mostly fishermen who were living in their boats. The surge was about 3.8m.
1962	Typhoon Wanda	Typhoon Wanda coincided with a high tide, causing a tidal wave reaching as much as 7m high. The effect was disastrous, with over 130 deaths, 72,000 people left homeless and vast damage to infrastructure, the fishing fleet and farm land.
1971	Typhoon Rose	On 17 August alone, 288.1mm of rain was recorded, resulting in flooding and landslides. 100 people were killed and 5644 left homeless. A total of 24 buildings were devastated and 653 huts destroyed, 110 cases of road blockage were recorded and 30,000 telephones rendered out of order.
2001	Typhoon Utor	The surge of 1.1m on 6 July 2001 coincided with an astronomical high tide. Sea levels reached 3.4m at Quarry Bay. This is the highest sea level recorded in the Victoria Harbour since Typhoon Wanda in 1962 (4m). Storm surges caused severe flooding in numerous places. Some villagers sustained serious economic loss.
2008	Heavy rainfall event	With a total monthly precipitation of 1346.1mm, June 2008 had the heaviest rainfall in Hong Kong history. A Black Rainstorm Signal was issued on 7 June, when 301mm fell within a day; 162 floods, 622 landslides and 2 casualties were recorded. According to a preliminary and conservative cost estimate, the damage of the heavy rainfall on 7 June alone was 55 million Euros (US$75 million).

Source: Drainage Service Department (2009); Pauw and Francesch-Huidobro (2010)

protection) systems and sewer systems separated. With an ever-increasing population and industrial activity during the 1960s and 1970s, newly built gullies, culverts and streams could not reduce the chance of flooding. As a result, in 1987, the director of civil engineering became responsible for coordinating all aspects of flood control in Hong Kong, leading to the drainage and flood control strategy adopted in 1990. Simultaneously, the Drainage Services Department was established, and since then investment in flood protection increased dramatically (Drainage Services Department, 2009).

Hong Kong's painful experiences with storm surges promoted the adoption of stringent design standards for coastal infrastructure (see Figures 12.4 and 12.5). Land reclamation is today always raised 4m above sea level to cater for sea-level rise. Physical and computational models are widely applied for the design of bridges, coastal infrastructure and drainage. The drainage system that copes with intense (cyclone) precipitation has a recurrence interval of flooding prevention of 1 in 50 to 200 years in urban areas (see Table 12.4). Continuing efforts are made to separate the sewage system from storm-water drainage for health and safety reasons. Important measures include the use of dry-weather flow interceptors that collect polluted water during dry times, but allow storm-water overflow. Other technological measures to reduce the flood risk include flood storage tanks (e.g. under sports grounds), raising entrances of buildings and using inflatable dams (see Figures 12.3 to 12.5), pumping stations and drainage tunnel schemes.

Figure 12.3 *Inflatable dam at Kam Tin River*

Source: Drainage Service Department (2009)

Table 12.4 *Average recurrence interval of flooding prevention
in Hong Kong drainage systems*

Types of drainage	Recurrence interval flooding prevention (years)
Urban drainage trunk systems	200
Urban drainage branch systems	50
Main rural catchment drainage channels	50
Village drainage	10
Intensively used agricultural land	2–5

Note: These figures are based on type of land use, economic growth, socio-economic needs, consequences of flooding, and cost–benefit analysis of flood mitigation measures.
Source: Fung et al (2004, p100)

Furthermore, a world-class storm warning system has been installed to monitor and forecast stormy weather. In combination with warning services, the building design of infrastructure and increasing public awareness have greatly reduced the threat posed by tropical cyclones with respect to loss of lives, which has decreased during the last 25 years (see Figure 12.6). At this juncture, we have a better understanding about the context of governance, the parameters by which climate change is monitored, and the

Figure 12.4 *The flood-vulnerable subway system in Hong Kong has lifted entrances
to prevent flooding*

Source: Pauw and Francesch-Huidobro (2010)

Figure 12.5 *Nathan Road at its junction with Jordan Road flooded in 1997*

Source: Drainage Service Department (2009)

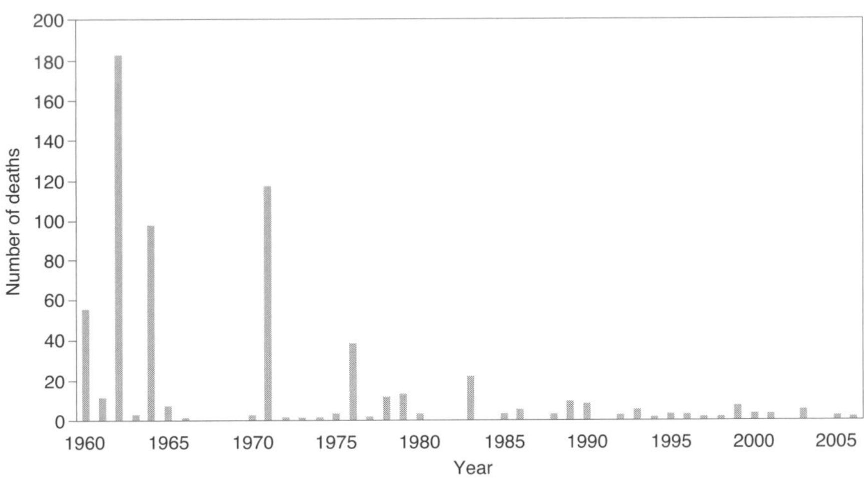

Figure 12.6 *The total number of deaths and missing people reported during tropical cyclone events in Hong Kong from 1960 to 2006*

Note: A clear declining trend in death toll can be observed.
Source: Lee and Wong (2007)

risks that it poses to people and assets. How is this process managed in the increasingly intertwined web of private and public actors?

12.3 The Dynamics of Governing Climate Change in Hong Kong

Given current climatic trends, an increasing ability to manage a shared body of scientific and technological knowledge, and a greater institutional self-assertion, Hong Kong is putting in place its own measures and models of collaboration to expand its climate mitigating and adaptive capacity. Noteworthy was the C40 workshop in November 2010 on low carbon living organized by the Environmental Campaign Committee (ECC) and the think tank Civic Exchange, with the support of the Environment Bureau which provided a platform for exchanges among local stakeholders and representatives of other C40 cities (Legislative Council, Panel on Environmental Affairs, 2010). Figure 12.7 diagrammatically presents the various Hong Kong stakeholders and how they relate to international and national actors, and to each other. Their contributions and interactions are discussed in the following sections.

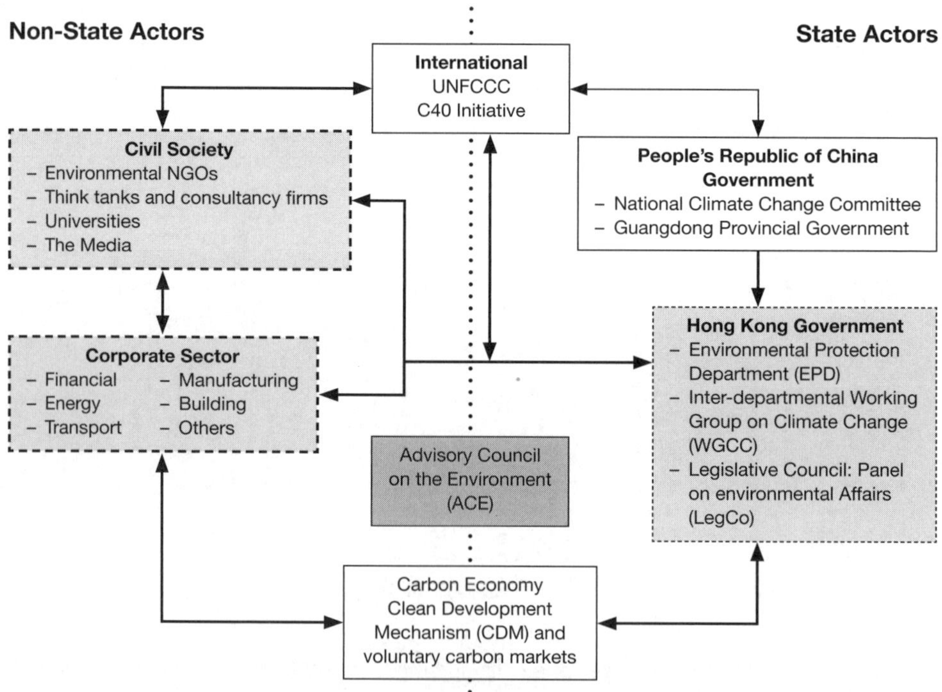

Figure 12.7 *Climate change governance in Hong Kong*

Source: Chu and Schroeder (2010)

12.3.1 Government

From the chronology of climate change-related initiatives, one appreciates that private firms and civil society groups, rather than government, were the first movers of climate change policy since 2006 (see Table 12.5). From 2007, the government began paying attention to climate change during Legislative Council debates of the Chief Executive Annual Policy Addresses of 2007 and 2008 (Policy Address, 2007, 2008). In January 2008, the Inter-departmental Working Group on Climate Change (IWGCC) was set up, commissioning shortly after an 18-month consultancy to evaluate Hong Kong's mitigation and adaptation strategies. The report is yet to be released, as we confirmed during a discussion with several members of the IWGCC in July 2010 (interview, 13 July 2010). In 2009, the government established a Hong Kong $450 million fund for energy-saving measures mostly related to promoting the use of electric vehicles (EVs).

Nevertheless, the Hong Kong government has not set emissions reduction targets or action plans to meet them. China's climate action plan spelled out in the China National Climate Change Programme of 2007 and the White Paper on *China's Policies and Actions in Addressing Climate Change* of October 2008 offer much more concrete commitments to reduce greenhouse gases (GHGs) by 2020 (National Development and Reform Commission, 2007). The lead that the government has taken from the local think tank Civic Exchange and the Environmental Campaign Committee is, nevertheless, hopeful: as mentioned earlier, a workshop on low carbon living with the support of the Environmental Fund Campaign (Legislative Council, Panel on Environmental Affairs, 2010) is being convened. The workshop is the first platform opened to Hong Kong stakeholders to come together and meet stakeholders from other cities to share experience on two strategies to curb GHG emissions to which the Hong Kong government has pledged some commitment: the energy efficiency of buildings and promoting the use of electric vehicles.

12.3.2 Non-governmental organizations (NGOs)

During the last decade, a new breed of civil society has emerged in Hong Kong that is transforming itself from simply being a set of institutions delivering services to being agents of advocacy (Chan and Chan, 2007). As presented in Figure 12.7, civil society groups have organized campaigns to raise awareness and public interest ahead of any government initiative. A study conducted by Nielsen Global Online Survey found that between 2007 and 2008, people's awareness about climate change had jumped 14 per cent, partly associated with media coverage on the IPCC reports and the documentary version of *An Inconvenient Truth* (Nielsen Company and ECI, 2007). However, civil society efforts are offset by the fact that climate change in Hong Kong is frequently perceived as something inevitable, linked to air pollution coming from the north across the border and therefore not in Hong Kong's control – this despite the fact that electricity generation and transportation account for about 62 and 16 per cent of GHG emissions, respectively (EPD, 2009). A draft report by Greenpeace states that China proposes a total energy reduction of 62 per cent of 2007 levels if Hong Kong is to meet the IPCC target by 2020 (Greenpeace China, 2010).

12.3.3 Private firms

Largely left unregulated when it comes to environmental legislation, the private sector in Hong Kong, mostly 50 large corporations with a total market share of 76 per cent of the Hong Kong Stock Exchange Main Board as of 2008, has adopted international standards such as ISO 14001 and codes of conduct in their formulation of strategies that address opportunities and threats due to climate change. Chu and Schroeder (2010) reported in a working paper on corporate governance of climate change in Hong Kong that four of the top ten listed companies they interviewed assume a leadership role in climate change initiatives, though they can hardly be considered representative as 98 per cent of businesses in Hong Kong are small- to medium-sized enterprises of no more than 100 employees who perceive that voluntary environmental compliance would reduce profits and competitiveness. Nevertheless, the private sector took the lead in setting up climate change strategies, as can be seen in Table 12.5.

Earlier, a city's ability to mitigate and adapt to climate change was hypothesized as a function of the ability of interested parties to manage a body of knowledge from which to discuss solutions, and to create consensus and cooperation; the learning process through which actors review their strategies and interpretations as they go along; and their ability to manage the dynamics of multilevel sub-national and transnational politics. What challenges does Hong Kong encounter in its capacity to govern climate change?

12.4 Assessing Hong Kong's Governing Capacity for Climate Change

The coastal city of Hong Kong has equipped itself with knowledge on climate change science and technology and, thus, has made a few mitigating and adaptive commitments, although there are strong vested interests that determine the pace of emissions reduction, such as the power companies, the transport sector and estate developers. Power companies, especially China Light and Power, have recently been proactive in utilizing renewable energy and setting up energy-efficient projects in China (Shenzhen), but have been rather sluggish in adhering to the polluter pays principle. The public transport sector (especially buses) has not successfully made the switch to Euro V engines. As for estate developers, the Building Energy Efficiency Code released during the fourth quarter of 2009 prescribes voluntary compliance. Weak regulations on the height of buildings is also a concern, as they create a wall effect that, besides blocking

Table 12.5 *Chronology of climate change actions in Hong Kong*

Year	Government	Corporate sector	Civil society
November 2006			Study on The Impacts of Climate Change in Hong Kong and the PRD (Civic Exchange)
December 2006			Hong Kong Climate Change Conference (University of Hong Kong)
2007		Hong Kong Shanghai Banking Corporation (HSBC) five-year Climate Partnership	
May 2007		International Climate Change Conference in 2007 (Analogue Group)	
September 2007	Adopted Asia Pacific Economic Council's 25% carbon-intensity reduction target of 2005 levels by 2030		
November 2007	Joined the C40 Cities Network	Climate Change Business Forum formed; feasibility study for carbon trading in Hong Kong (Hong Kong Stock Exchange)	
December 2007		Carbon offset programmes (Cathay Pacific and Dragon Air)	
January 2008	Created Inter-departmental Working Group on Climate Change (IWGCC)		
March 2008	Consultation on mandatory building energy codes		
April 2008			Lung Fu Shan Environmental Education Centre created (EPD and University of Hong Kong)
June 2008	Emissions Trading Platform launched to permit generation of Clean Development Mechanism (CDM) credits under National Development and Reform Commission of China	Two carbon trading workshops held (co-organized by British Consulate and Civic Exchange); Carbon Ventures Department created (China Light and Power Holdings – CLPH)	Return on Investment Sustainability Conference (*The Economist*); Greater Pearl River Delta Low Carbon Economy and Development Forum (The Climate Group (TCG) and National Development Resource Council); Hong Kong Climate Change Business Forum created

Table 12.5 *Chronology of climate change actions in Hong Kong (continued)*

Year	Government	Corporate sector	Civil society
July 2008	Partnership with 37 volunteered businesses to audit 100 buildings using the Guidelines to Account for and Report on GHG Emissions and Removals for Buildings		First Asian Youth Summit on Climate Change (University of Hong Kong)
October 2008	A 20% profit tax deduction for five years on environment protection installations, including renewable energy facilities mainly ancillary to buildings	Climate Change Conference: Making Business Sense of the Low Carbon Economy (HSBC)	Low-Carbon Office Operations Programme (LOOP) partnered with four business pilots to measure and reduce office buildings emissions (WWF); The Climate Conference 2008 (TCG, HSBC and Swire Pacific)
March 2009			Earth/One-Hour Lights Off Day in Hong Kong (WWF)
April/May 2009	Hong Kong $450 million funding for buildings owners to implement energy saving projects; Environmental Protection Department introduces electric vehicles	CLPH partnered with Wilson Parking and Link Management to put up ten pilot electric vehicle charging stations in car parks	Climate Change Business Forum (CCBF) launches How to Reduce GHG Emissions Guide
June 2009			Conference on China and Global Climate Change (Lingnan University)
July/August 2009		Hong Kong Stock Exchange releases a consultancy paper on establishing a carbon trading Certified Emissions Reduction (CER) futures exchange	Workshop on Climate Projections and Economics of Climate Change (British Consulate-General Hong Kong and Hong Kong Observatory); panel discussion on carbon trading (CCBF); consultancy report on climate change governance in Hong Kong (CCBF and Reset)
September/ October 2009			International Climate Change Conference 2009 (TCG, Hong Kong Climate Change Business Forum and HSBC)
December 2009			The 2009 Climate Leaders' Summit (TCG)
November 2010	C40 Low Carbon Living and Climate Change Workshop	C40 Low Carbon Living and Climate Change Workshop	C40 Low Carbon Living and Climate Change Workshop

Source: Chu and Schroeder (2010)

views, traps pollutants and increases the heat island effect. Thus, arguably, many ideas about climate change and policy responses have entered into the discourse about climate change in the city; but little has actually been taken seriously by parties who are instrumental in successfully implementing mitigation strategies.

At the government level, in his 2008–2009 Policy Address, the chief executive outlined the commitments to enhance Hong Kong's quality of life, moving the city to a low carbon economy, involving specifically legislation of mandatory compliance of the Building Energy Codes in order to enhance the energy efficiency of buildings, currently accounting for 89 per cent of total power consumption (Policy Address, 2008, 2009, p98). An increase in the use of clean fuels was also promised as the government is able to tap into a stable supply of nuclear energy, natural gas and liquefied natural gas through the memorandum of understanding signed with the National Energy Administration on 28 August 2008, under which gas-fired energy, which currently accounts for 28 per cent of the total, would increase to 50 per cent of the total.

In April and May 2009 the government took a few proactive steps, attributed to its responsiveness to the Green New Deal call of the United Nations and the policy announcements made by the chief executive in his last policy address. In his trip to Beijing, Secretary for the Environment Edward Yau visited the Ministry of Environmental Protection and met Vice-Minister of Environmental Protection Zhou Jian to exchange views on how to collaborate in transforming the Pearl River Delta Region into a green and quality-living area. Mr Yau also called on the administrator of the national energy administration, Zhang Guobao, to discuss the implementation of the memorandum of understanding on energy cooperation and on the vice-minister of land and resources, Wang Min, to brief him on the preparation works for setting up a national geological park in Hong Kong (Policy Address 2008–2009, p104). Yau's visit to China was followed by visits to Japan, Canada and the US, where the secretary for the environment discussed issues as diverse as the promotion of electric cars, green investment, the development of geological parks and research on 'smart cities'. This agenda is more ambitious than the policies Edward Yau had earlier initiated, involving only a levy on plastic bags implemented in July 2009, a review of the air quality objectives with due regard for the World Health Organization (WHO) guidelines, and a revision of car engines' standards.

Thus, even though policy officials and private actors demonstrated a substantial grasp of, and familiarity with, the climate change challenges, and are aware of the corresponding mitigation and adaptation strategies, much has yet to be institutionalized and legitimized as norms and codes of conduct. The Inter-departmental Working Group on Climate Change (IWGCC), the highest-level authority for policy coordination, has yet to offer any comprehensive strategy (interviews with Environmental Protection Department, 26 May 2009 and 13 July 2010). As illustrated when discussing risks and responses to risk, responses are often initiated by the government in a sectoral manner, defined narrowly according to the bureaucratic set-up and failing to adopt a holistic and integrated approach.

Indeed, Hong Kong has yet to successfully manage the knowledge that it has acquired on mitigation and adaption measures. Knowledge management within the government remains hierarchical. The problems are defined, and knowledge is disseminated according to bureaucratic division of labour, a model well fitted for routine policy issues but inappropriate for climate change, a 'wicked problem' plagued by much uncertainty and long-term implications that require a holistic response. This is clearly illustrated by the fact of having multiple government entities, each dealing separately with a contributing sector to global warming (e.g. transport, housing and electricity supply), when all contribute to energy consumption and greenhouse gas emissions. Such a state of affairs hampers the possibility of having a broader strategy with the whole city as the subject of the scheme. As a coordinating platform, the IWGCC, too, fails to solicit an encompassing handling of the problem. After all, established under the Environmental Protection Department (EPD), the group is neither able, nor given the authority, to initiate a radical departure from the existing sectoral approach, and does not have the ability to persuade other responsible bureaux/departments to reformulate their agendas and measures. Currently, only the Cross-Boundary International Group under the EPD is responsible for climate change issues – something that betrays the lack of political attention by the government about the transnational nature of the problem (interviews, Environmental Protection Department, 26 May 2009).

These barriers to effective governance suggest a lack of priority and political will by the government to address climate change challenges, in addition to the significant disjunction between discourse and practice of the city's climate change strategy. Much has been learned by actors; but the strategies are yet to be updated; political capacity to manage the multilevel politics, similarly, remains under-supplied. Authorizing the secretary for the environment, as head of the policy formulation portfolio on environmental protection, to coordinate climate change-related policies might strengthen the coordinating and decision-making efforts.

The public and people sectors are important actors in putting Hong Kong on a low carbon development track. The people sector, mostly led by private firms, has found that there are opportunities for investment and strengthened social responsibility. If and when the government comes up with regulations that effectively sanction excessive emissions and incentivizes initiators, Hong Kong will be on a firmer track to becoming a low carbon economy. The administrative-led and *laissez-faire* governance style in Hong Kong has meant that policies are formulated under the shadow of a few big and many small dominating businesses. Environmental protection, such as curbing human-generated GHG emissions and slowing climate change, remain secondary to economic growth, although the rhetoric of government and leading businesses is beginning to be translated into actions, especially in the building and transportation sectors.

The case of the coastal city of Hong Kong illustrates the interdependence among government, businesses and civil society in shaping climate change governance in Hong Kong. Despite the detailed recording of climate parameters, the evidence of risks posed to people and property by global warming and climate change, and the urgency of putting in place mitigating and adapting strategies, there is still a small-scale, inconsistent and voluntary string of actions by a few in the private sector with little echo in the industry at large or in government.

This chapter therefore suggests wider questioning on the institutional capacity of cities to make decisions in the absence of a commitment both from the local and national governments, especially when the international Kyoto negotiations and the subsequent Copenhagen Accord have yet to produce a clear, just and accountable system of global climate change responsibility. These deficiencies may underpin the reason why the Hong Kong government is constrained in its capacity, not in terms of resources or knowledge, but in terms of political will, to take the climate matter in hand.

In the final analysis, effective climate governance requires enhancing linkages among government, business and civil society that ought to collaborate in nurturing both stable financial and long-term human resources for low carbon development. The rise of multilevel governance offers two challenges for public administrators. First is the allocation of roles to politicians and civil servants in steering public policy within multilevel regimes: simply put, who governs what? The second challenge is the management of interfaces between levels of government and the multitude of organizations involved in these mediations. How can policies be implemented in a reasonably uniform way when such leeway is left to cities' governments and to sectors within them? It is in the light of these questions that this chapter provides an overview of climate change governance challenges in Hong Kong. The objective here was not to definitely evaluate the suitability of the measures that Hong Kong has put into place to solve its climate change-related problems, but to apply a set of data in order to better understand the process by which policy is established, on the one hand, and how policy is governed, on the other.

Acknowledgements

Research for this chapter was funded by a City University of Hong Kong Start Up grant: 7200153(SA). The chapter is a substantially revised version of Francesch-Huidobro's articles 'Risks, policy learning and challenges of climate governance in Hong Kong' and 'Institutional deficit and legitimacy: Challenges to the governance of emerging responses to climate change in Hong Kong' (May 2011). Parts of this chapter are based on Pauw and Francesch-Huidobro (2010); Pieter Pauw is thanked for his contribution.

References

Alber, G. and Kern, K. (2009) 'Governing climate change in cities: Modes of urban governance in multi-level systems', in OECD (ed) *OECD Conference Proceedings*, OECD, Paris

Bulkeley, H. and Betsill, M. (2003) *Cities and Climate Change: Urban Sustainability and Global Environmental Governance*, Routledge, London

Business Environment Council Communiqué (2010)

Chan, E. and Chan, J. (2007) 'The first ten years of the HKSAR: Civil society comes of age', *Asia Pacific Journal of public Administration*, vol 29, no 1, pp77–99

Cheung, A. B. L. and Scott, I. (2003) *Governance and Public Sector Reform in Asia: Paradigm Shifts or Business As Usual?*, Routledge–Curzon, London and New York

Chu, S. Y. and Schroeder, H. (2010) 'Private governance of climate change in Hong Kong: An analysis of drivers and barriers to corporate action', *Asian Studies Review*, vol 34, pp278–308

CIA (2010) 'Hong Kong GDP per capita (2009 est)', *The World Factbook*, https://www.cia.gov/library/publications/the-world-factbook/geos/hk.html, accessed 21 August 2010

Drainage Services Department (2009) *Sewerage and Flood Protection: Drainage Services 1841–2008*, Government of Hong Kong Special Administrative Region, Hong Kong

EPD (Environmental Protection Department) (2009) *Inventories of Greenhouse Gas Emissions in Hong Kong*, Updated March 2009, EPD, Hong Kong

Fung Wing Yee, Lam Ka Se and Hung Wing Tat (2004) *Provision of Service for Characterizing the Climate Change Impact in Hong Kong*, Environmental Protection Department, Hong Kong

Greenpeace China (2010) *Climate Change Bill: Economic Costs of Heavy Rainstorm in Hong Kong*, Greenpeace China report

Hong Kong Observatory (2009) *Climate of Hong Kong*, www.hko.gov.hk/climate_change/climate_change_e.htm

Interviews, Environmental Protection Department, 26 May 2009, 13 July 2010

Interviews, Greenpeace, China, 27 April 2009

Interviews, Inter-departmental Working Group on Climate Change, 13 July 2010

Kern, K. and Bulkeley, H. (2009) 'Cities, Europeanization and multi-level governance: Governing climate change through transnational municipal networks', *Journal of Common Market Studies*, vol 47, no 2, pp309–332

Lee, T. C. and Wong, C. F. (2007) 'Historical storm surges and storm surge forecasting in Hong Kong', Hong Kong Observatory, Paper for the JCOMM Scientific and Technical Symposium on Storm Surges (SSS) in Seoul

Legislative Council, Panel on Environmental Affairs (2010) *C40 Workshop to Be Held in Hong Kong in 2010*, 21 January

Mok, H. Y. and Wong, M. C. (2008) 'Trends in Hong Kong climate parameters relevant to engineering design', Presented to the HKIE Civil Engineering Conference, January 2009, HKO Reprint no 832, www.hko.gov.hk/publica/reprint/r832.pdf

National Development and Reform Commission, People's Republic of China (2007) *China's National Climate Change Programme*, www.ccchina.gov.cn/WebSite/CCChina/UpFile/File188.pdf, accessed 21 August 2010

Ng, M. K. (2007) 'Sustainable development and governance in East Asia world cities', *Journal of Comparative Policy Analysis*, vol 9, no 4, pp321–325

Nielsen Company and ECI (Environmental Change Institute) (2007) *Global Nielsen Survey: Consumers Look to Governments to Act on Climate Change*, Press Release, University of Oxford, UK, pp1–5

Painter, M. (2005) 'Transforming the administrative state: Reform in Hong Kong and the future of the developmental state', *Public Administration Review*, vol 65, no 3, pp335–346

Pauw, P. and Francesch-Huidobro, M. (2010) 'Hong Kong', in P. Dircke, J. Aaerts and A. Molenaar (eds) *Connecting Delta Cities: Sharing Knowledge and Working on Adaptation to Climate Change*, City of Rotterdam, Rotterdam

Pemberton, S. (2009) 'Editorial', *Policy & Politics*, vol 37, no 3, pp313–316

Policy Address (2007) Government of Hong Kong Special Administrative Region, Hong Kong

Policy Address (2008) Government of Hong Kong Special Administrative Region, Hong Kong

Policy Address (2009) Government of Hong Kong Special Administrative Region, Hong Kong

Stren, R. and Cameron, R. (2005) 'Metropolitan governance reform: An introduction', *Public Administration and Development*, vol 25, pp275–284

13

Climate Adaptation in New York City

David C. Major, Vivien Gornitz, Radley Horton,
Daniel Bader, William Solecki and
Cynthia Rosenzweig

13.1 Introduction

New York City, the financial, cultural and communications capital of the US, is at risk from many aspects of climate change, including rising sea level, storm surge, rising temperatures, heat waves, inland flooding and more frequent droughts in upland watersheds. Leadership by New York City in dealing with climate change has been, and will be, an important element in the regional success of coping with such impacts. Relative (local) sea-level rise in the New York area is greater than worldwide sea-level rise primarily due to on-going glacial isostatic readjustments (Peltier, 2004; Horton and Rosenzweig, 2010). New York City has enormous assets and a large population at risk from sea-level rise and storm surge. However, only some parts of the city are located at or close to sea level; elevations in other areas of the city are significantly higher, ranging up to 409.8 feet (124.9m) at Todt Hill on Staten Island.

New York City and regional agencies have been concerned about climate change, its impacts, and adaptation and mitigation for many years. The first comprehensive study, the Metro East Coast report, was part of the First National Assessment of Climate Variability and Change (Rosenzweig and Solecki, 2001). Subsequently, the New York City Department of Environmental Protection (NYCDEP) established a Climate Change Task Force in 2004 (Rosenzweig et al, 2007a) to review both vulnerabilities and adaptation possibilities, and has since established a programme for dealing with climate change in the agency's operations and infrastructure investments (NYCDEP, 2008). As part of the city's long-term sustainability planning (New York City, 2007), Mayor Michael Bloomberg in 2008 convened the New York City Panel on Climate Change (NPCC). This panel of experts, chaired by Cynthia Rosenzweig and William Solecki, was charged with advising on issues related to climate change and adaptation (NPCC, 2010; Rosenzweig et al, 2011). The NPCC has assisted the New York City Climate Change Adaptation Task Force, also established by the city in 2008, in developing a coordinated adaptation plan for the city. The task force consists of over 40 public- and private-sector stakeholders. In the wider region, the NYSERDA ClimAID Team (2011) assessment of New York State and an economic study of climate impacts in New Jersey (Solecki et al, forthcoming) address issues of coastal flooding.

This chapter focuses on the most important challenge facing the city and the region: the combination of rising sea level, storm surge and the resulting increased frequency of coastal flooding. The chapter first addresses the city's geographical and demographic characteristics (section 13.2), historical sea-level rise and coastal flooding in New York City (section 13.3) and climate scenarios (section 13.4). Next, the chapter focuses on climate adaptation options and planning for future adaptation in New York City (sections 13.5 and 13.6).

13.2 The City and the Region

New York City is the largest city, with the largest metropolitan region, in the US. Its population has been growing: the 2009 estimated population was 8.4 million, up 5 per cent from 2000 (US Bureau of the Census, undated). Located on a large natural harbour on the Atlantic coast of the north-eastern US, the city consists of five boroughs: the Bronx, Brooklyn, Manhattan, Queens and Staten Island. The total land area is 305 square miles (790 square kilometres). The metropolitan area is defined in several ways. The official statistical entity called the New York–Northern New Jersey–Long Island Metropolitan Statistical Area (MSA) had a 2008 estimated population of 19 million, reflecting growth of 3.7 per cent since 2000. A wider region, defined by the US Office of Management and Budget based on commuting patterns, is the New York–Newark–Bridgeport, New York–New Jersey–Connecticut–Pennsylvania Combined Statistical Area (CSA). The CSA had an estimated population of 22,232,494 as of 2009, a growth of 4 per cent since 2000 (US Bureau of the Census, undated).

13.3 Historical Sea-Level Rise and Coastal Flooding in New York City

A historical overview of sea-level rise and coastal flooding in the New York City area, given in this section, provides the background for the following discussions of climate scenarios and adaptation possibilities.

13.3.1 Historical sea-level rise

Prior to the Industrial Revolution, sea level rose along the East Coast of the US at rates of 0.34 to 0.43 inches (0.87cm to 1.1cm) per decade, primarily because of regional subsidence as the Earth's crust slowly readjusts to the melting of the ice sheets since the end of the last ice age (glacial isostatic adjustments). Within the last 100 to 150 years, however, as global temperatures have increased, regional sea level has been rising more rapidly (1mm to 2mm per year faster) than over the last 1000 years. Currently, rates of sea-level rise in the New York metropolitan area range between 0.86 and 1.5 inches (2.2cm to 3.8cm) per decade, with a long-term rate since 1900 averaging 1.2 inches (3cm) per decade. The local (or relative) sea-level rise rates, measured by tide gauges, include both the effects of recent global warming and regional subsidence. Most of the observed current climate-related rise in sea level over the past century can be attributed to expansion of the oceans as they warm, although melting of land-based ice may become the dominant contributor to sea-level rise from the 1990s (Cazenave and Llovel, 2010). It should be noted that, while the New York metropolitan area faces increased risks of coastal flooding due to climate change, the highest relative sea-level rise on the East Coast does not occur in the New York region, but rather appears to be around the small city of Norfolk, Virginia (Kaufman, 2010), built on marshland and experiencing large amounts of subsidence in addition to sea-level rise. Norfolk is already taking preventative action against more frequent and damaging floods.

13.3.2 Types of floods in New York City

New York City does not have substantial flood risk from upland riverine flooding as does the city of Rotterdam, located in the Rhine Delta (although other areas in the New York metropolitan region are at risk from such floods). Rather, the main flooding risks are, first, coastal flooding from storms and, second, inland flooding from intense rainfall. Hurricanes and nor'easters are the two types of storms that have historically caused the most damage in the New York metropolitan region (e.g. Hill, 1996). Hurricanes occur during mid summer and early autumn, whereas nor'easters are most prevalent during early winter to early spring. Although hurricane surges can be much higher, they are typically shorter in duration (on the order of hours), in contrast to nor'easters that can last several days and, hence, are likely to span several tidal cycles, including high tide. Therefore, strong nor'easters can produce flooding comparable to a weak hurricane. Depending upon the storm track, the near right-angle bend in the coastline between New Jersey and New York tends to funnel surge water towards the New York City estuary, potentially amplifying the flooding.

New York City and the surrounding area have experienced numerous hurricanes, some of which have produced serious damage and loss of life. Future rising sea levels associated with global warming will produce higher floods associated with hurricanes and other coastal storms, and the interval between floods of a given elevation could drop sharply (Horton and Rosenzweig, 2010, p177). The only hurricane whose eye hit what is now New York City directly occurred on 3 September 1821, making landfall at Jamaica Bay. Moving northward rapidly, the storm (rated a category 1–2 on the Saffir–Simpson scale) produced a storm surge of 13 feet (4m) in only one hour, causing widespread flooding of lower Manhattan as far north as Canal Street. However, the storm caused few deaths because the population density was low at the time (Rosenzweig et al, 2007b).

In the metropolitan area, the Long Island Express, or Great Hurricane of 1938 (category 3 at landfall in Long Island), hit central Long Island and crossed Long Island Sound into southern New England on 21 September 1938, killing nearly 700 people and injuring thousands more. The storm, striking with little advance warning (this in the days before weather satellites), raised a wall of water 25 to 35 feet (7.6m to 10.7m) high, sweeping away barrier dunes and buildings. The surge levels were extra high, in part due to the fact that the storm made landfall close to high tide at full moon and perigee (closest approach of the Moon to Earth in its monthly orbit) at the autumnal equinox. Because of the tremendous population growth and development that have occurred since then, damages from such a storm today would be enormous. Hurricane Donna, in September 1960, generated the highest water level at The Battery tide gauge within the last half century, flooding large parts of lower Manhattan (Major et al, 2010).

The worst nor'easter in New York City during the 46-year period of detailed tidal records was on 11 December 1992, which hit New York City with hurricane-force wind gusts of up to 90 miles per hour, causing tidewaters to rise 7.7 feet (2.34m) above normal. The nor'easter crippled transportation, businesses and schools, produced numerous power failures, and flooded wide areas. Many other hurricanes and nor'easters have hit the region (Major et al, 2010). Plate 11 shows current estimated inundation zones of parts of New York City from hurricanes of different magnitudes. Extending more widely into the region, the Coastal Resilience Project of The Nature Conservancy is engaged in mapping flood zones on the north and south shores of Long Island and the Connecticut shore of Long Island Sound using sea-level rise scenarios created with NPCC methods, together with estimated coastal storm heights from tide gauges and models (Ferdaña et al, 2010).

Inland flooding in New York City, as opposed to floods from storm surge or inland wind damage from hurricanes and nor'easters, can occur from intense rainfall events. There is some expectation that these will increase with climate change (Horton and Rosenzweig, 2010, p165). This type of flooding results in local inundation and combined sewer overflows, both of which are serious issues in a highly developed city faced with tight environmental regulation. The New York City Department of Environmental Protection (NYCDEP) has been investigating such intense rainfall with a view to updating intensity–duration–frequency curves (NYCDEP, 2008). The worst recent storm of this nature was the 7 August 2007 storm (MTA, 2007; Rosenzweig et al, 2007b). This downpour dropped 1.7 inches (43.2mm) in one hour, with a daily total of 2.8 inches (71.1mm) in Central Park (Barron, 2007), and caused system-wide transportation delays and other damages.

Dangers of future flooding lie with the increased damages likely with rising sea level, and the less certain but widely expected increase in the intensity of storms. The Columbia University Earth Institute's Center for Climate Systems Research has developed progressively more detailed downscaled Global Climate Model (GCM) scenarios for the city and region since the original work reported in Rosenzweig and Solecki (2001), most recently in Horton and Rosenzweig (2010). These are described in the next section.

13.4 Climate Scenarios

13.4.1 Methodology for temperature and precipitation scenarios

The NPCC (2010) study uses 16 Global Climate Model outputs driven with 3 Intergovernmental Panel on Climate Change (IPCC) greenhouse gas emissions scenarios (A2, A1B and B1) (IPCC, 2000). The outputs for the model grid box(es) covering New York City are used as the basis for scenarios for three decadal time slices (2020s, 2050s, 2080s), relevant to different levels of planning. The combination of 16 GCMs and 3 emissions scenarios produces a 48 (16 × 3)-member matrix of outputs for temperature and precipitation (Horton and Rosenzweig, 2010). For each time period and climate variable, the suite of outputs constitutes a 'model-based' probability function. The future model results are compared to model results for the 1971 to 2000 baseline period. For example, mean temperature change projections are calculated as the difference between each model's future simulation and the same model's baseline simulation; mean precipitation is based on the ratio of a given model's future precipitation to the same model's baseline precipitation (expressed as a percentage change). The relevant changes are then applied to the actual baseline values for 1971 to 2000 to produce the scenarios. The central ranges of outputs (i.e. the mid 67 per cent) for the scenarios are shown in Table 13. 1. The central range of values for temperature increases is + 4.0°F to 7.5°F (2.2°C to 4.2°C) and the central range of values for precipitation is +5 to 10 per cent for the 2080s.

13.4.2 Sea-level rise scenarios methodology

The methodology for calculating future scenarios of sea-level rise for the New York City region is adapted from the IPCC methodology (IPCC, 2007). This modified IPCC methodology includes two global (global thermal expansion and melt water from glaciers, ice caps and ice sheets) and two local (local land subsidence and local water surface elevation) components. The local components are discussed in IPCC (2007) but are not included in the global IPCC calculations. Inasmuch as only 7 of the 16 AR4 models used for temperature and precipitation scenarios have the necessary outputs for sea-level rise scenarios, the model-based probability function contains a matrix of 7 × 3 = 21 cells. The central range of results is shown in Table 13.1; the central range for sea-level rise using this method is +12 to 23 inches (31 to 59cm) for the 2080s.

Within the scientific community, the possibility that the IPCC estimates of future sea-level rise may substantially underestimate ocean levels has been extensively discussed (Rahmstorf, 2007; Horton et al, 2008). For this reason, an alternative 'rapid ice-melt' scenario was developed, based on palaeoclimate studies. Starting around 20,000 years ago, global sea level rose 394 feet (120m) and reached nearly present-day levels around 8000 to 7000 years ago. The average rate of sea-level rise during this approximately 10,000- to 12,000-year period was 0.39 to 0.47 inches per year

Table 13.1 *Baseline climate and mean annual changes*

	Baseline 1971–2000	*2020s*	*2050s*	*2080s*
Air temperature Central range*	55°F	+ 1.5°F –3.0°F	+ 3.0°F –5.0°F	+ 4.0°F –7.5°F
Precipitation Central range*	46.5 inches** (118cm)	+ 0–5%	+ 0–10%	+ 5–10%
Sea-level rise** Central range*	n/a	+ 2–5 inches (5–13cm)	+ 7–12 inches (18–30cm)	+ 12–23 inches (30–58cm)
Rapid ice-melt scenario***	n/a	~ 5–10 inches (13–25cm)	~ 19–29 inches (48–74cm)	~ 41–55 inches (104–140cm)

Notes: This table is based on 16 Global Climate Models (7 GCMs for sea-level rise) and 3 emissions scenarios. Baseline is 1971 to 2000 for temperature and precipitation, and 2000 to 2004 for sea-level rise. Data derive from the National Weather Service (NWS) and the National Oceanic and Atmospheric Administration (NOAA). Temperature data are from Central Park; precipitation data are the mean of the Central Park and La Guardia Airport values; and sea-level data are from The Battery at the southern tip of Manhattan (the only location in NYC for which comprehensive historical sea-level rise data are available).
* Central range = middle 67% of values from model-based probabilities; temperature ranges are rounded to the nearest half-degree, precipitation to the nearest 5%, and sea-level rise to the nearest inch.
** The model-based sea-level rise projections may represent the range of possible outcomes less completely than the temperature and precipitation projections.
*** 'Rapid ice-melt scenario' is based on acceleration of recent rates of ice melt in the Greenland and West Antarctic ice sheets and palaeoclimate studies.
n/a = not available.
Source: Horton and Rosenzweig (2010)

(1 to 1.2cm per year). This average rate of ice melt was added to the other (unchanged) three components to provide an upper-bound estimate of sea-level rise (for additional description of the method, see Horton and Rosenzweig, 2010, Annex C.) The central range of forecasts for scenarios using this method is also shown in Table 13.1; the values are approximately 41 to 55 inches (105 to 141cm) for the 2080s.

A third approach to sea-level rise is the semi-empirical method of Rahmstorf and Horton (Rahmstorf, 2007; Horton et al, 2008; Vermeer and Rahmstorf, 2009), which is based on a historical relationship between observed sea-level rise and global surface temperature. The results of this method for the 2080s, along with the other two methods, are shown in Figure 13.1. The semi-empirical method yields somewhat higher estimated values than those based on the IPCC-derived methodology used by the NPCC (2010), but lower than the results for the rapid ice-melt scenarios.

The overall conclusion of this work is that New York City will, with a very high degree of certainty, face sea-level rise of at least 1 to 2 feet (30.5 to 58cm) during the 21st century and, with somewhat less certainty, significantly higher sea levels.

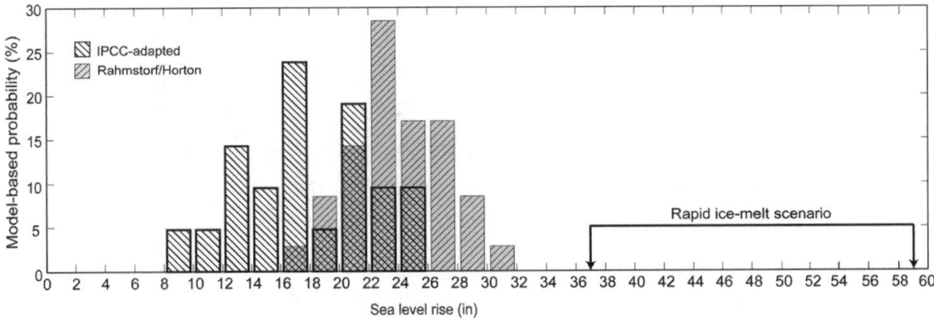

Figure 13.1 *IPCC, rapid ice-melt and Rahmstorf/Horton sea-level rise scenarios for the 2080s*

Source: Horton and Rosenzweig (2010)

13.4.3 Extreme events methods

The largest impacts of extreme temperature and precipitation (with the exception of drought) tend to occur on daily rather than monthly timescales. Because monthly output from climate models is considered more reliable than daily output, a hybrid projection technique was employed in Horton and Rosenzweig (2010). Modelled changes in monthly temperature and precipitation are based on the same methods described for the annual data; monthly changes over time in each of the 16 GCMs and 3 emissions scenarios were then applied to the observed daily Central Park climate data from 1971 to 2000 to generate 48 years of daily time series. The results are shown in the first two sections of Table 13.2. This is a simplified approach to the projection of extreme events since it does not allow for possible changes in variability through time. However, while changes in variability for most climate hazards are highly uncertain, this approach can assist long-term planners to prepare for extreme events by providing a framework for decision. The results for coastal flooding are shown in the last section of Table 13.2. The coastal flooding results illustrate the central dilemma facing New York City in a manner that is quite accessible to policy-makers: the changing frequency of floods of various recurrence intervals. For example, the current 1/10-year flood is expected to recur during the 2080s every one to three years.

13.5 Climate Adaptation for New York City

13.5.1 Adaptation assessment framework

Within the institutional structure described at the beginning of this chapter, the NPCC (Major and O'Grady, 2010) identified an eight-step process to help stakeholders identify at-risk infrastructure and to develop adaptation strategies to address those risks (see

Table 13.2 *Quantitative changes in extreme events*

	Extreme event	Baseline (1971–2000)	2020s	2050s	2080s
Heat waves and cold events	Number of days per year with maximum temperature exceeding:				
	90°F (32°C)	14	23– 29	29–45	37– 64
	100°F (38°C)	0.4*	0.6–1	1–4	2– 9
	Number of heat waves per year**	2	3–4	4–6	5– 8
	Average duration (in days)	4	4–5	5	5–7
	Number of days per year with minimum temperature below 32°F (0°C)	72	53–61	45–54	36–49
Intense precipitation and droughts	Number of days per year with rainfall exceeding:				
	1 inch (2.5cm)	13	13–14	13–15	14–16
	2 inches (5.1cm)	3	3–4	3–4	4
	4 inches (10.2cm)	0.3	0.2–0.4	0.3–0.4	0.3–0.5
	Drought to occur, on average***	~ once every 100 years	~ once every 100 years	~ once every 50–100 years	~ once every 8–100 years

Coastal floods and storms****				
1/10- year flood to reoccur, on average	~ once every 10 years	~ once every 8–10 years	~ once every 3–6 years	~ once every 1–3 years
flood heights in feet (m) associated with 1/10-year flood	6.3 (1.9)	6.5–6.8 (2.0–2.1)	7.0–7.3 (2.1–2.2)	7.4–8.2 (2.2–2.5)
1/100-year flood to reoccur, on average	~ once every 100 years	~ once every 65–80 years	~ once every 35–55 years	~ once every 15–35 years
flood heights in feet (m) associated with 1/100-year flood	8.6 (2.6)	8.8–9.0 (2.7)	9.2–9.6 (2.8–2.9)	9.6–10.5 (2.9–3.2)
1/500-year flood to reoccur, on average	~ once every 500 years	~ once every 380–450 years	~ once every 250–330 years	~ once every 120–250 years
flood heights in feet (m) associated with 1/500-year flood	10.7 (3.3)	10.9–11.2 (3.3–3.4)	11.4–11.7 (3.5–3.6)	11.8–12.6 (3.6–3.8)

Notes: Extreme events are characterized by higher uncertainty than mean annual changes. The central range (middle 67% of values from model-based probabilities) across the GCMs and greenhouse gas emissions scenarios is shown.

* Decimal places shown for values less than 1 (and for all flood heights), although this does not indicate higher accuracy/certainty. More generally, the high precision and narrow range shown here are due to the fact that these results are model-based. Due to multiple uncertainties, actual values and range are not known to the level of precision shown in this table.

** Defined as three or more consecutive days with maximum temperature exceeding 90°F (32°C).

*** Based on minima of the Palmer Drought Severity Index (PDSI) over any 12 consecutive months.

**** Does not include the rapid ice-melt scenario.

Source: Horton and Rosenzweig (2010)

also Rosenzweig and Solecki, 2001; Rosenzweig et al, 2007a; MTA, 2007; NYCDEP, 2008). These adaptation assessment steps are designed to be incorporated within the risk management, maintenance and operations, and capital planning processes of the agencies and organizations that manage and operate critical infrastructure (see Figure 13.2). The set of steps is general enough to be useful for a range of jurisdictions and infrastructure sectors, yet specific enough to serve as the template for developing and implementing any given sector's adaptation efforts. They can be modified to assist in climate change adaptation in other urban areas, with region-specific adjustments related to climate risk information, critical infrastructure and protection levels (it is of interest that a 2009 National Research Council report includes a discussion of the New York City climate change planning procedures as an appendix; see NRC, 2009, Appendix A).

The eight adaptation assessment steps are as follows:

1   Identify current and future climate hazards.
2   Conduct an inventory of infrastructure and assets.
3   Characterize risk of climate change on infrastructure.
4   Develop initial adaptation strategies.
5   Identify opportunities for coordination.
6   Link strategies to capital and rehabilitation cycles.
7   Prepare and implement adaptation plans.
8   Monitor and reassess.

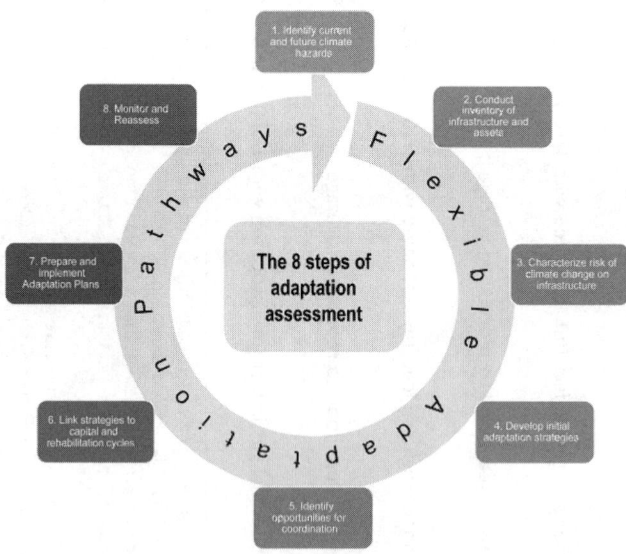

Figure 13.2 *Steps in the adaptation process*

The eight steps reflect the fundamental nature of adaptation planning, which is, in principle, a highly complex, multidimensional optimization problem. They are designed to enable stakeholders to adapt to climate change in practice, while at the same time staying within a suitable conceptual framework (see also Chapter 11). They are intended to be broadly applicable to climate change adaptation in many jurisdictions. The extent to which each of these steps has been implemented to date in New York City varies – there are fully developed climate scenarios (step 1), and detailed inventory questionnaires (NPCC, 2010) have been completed for step 2, while steps 3 to 8 are in progress or are in the planning stage. Among the most interesting applications of adaptation planning within this assessment approach to date are the use of a risk management framework (Yohe and Leichenko, 2010), flexible adaptation pathways (Major and O'Grady, 2010, p243), a three-dimensional climate change risk assessment approach (see Figure 13.3) and an adaptation prioritization matrix (Major and O'Grady, 2010, p257).

13.5.2 Types of adaptation

A variety of adaptation options is described in NRC (2010) for all elements of climate change, including sea-level rise and storm surge. Many adaptations for current climate variability, either in place or contemplated, are also highly relevant for sea-level rise and storm surge. Among the most important and discussed for New York City are the following:

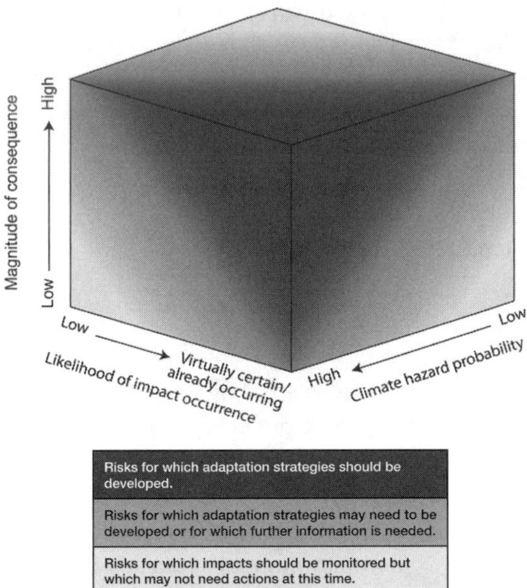

Figure 13.3 *Three-dimensional climate change risk assessment*

Source: Major and O'Grady (2010)

- *Emergency evacuation.* The New York City Office of Emergency Management (NYCOEM) has contingency plans for a wide variety of emergencies, including flooding. Guidance on these for coastal storms is provided on the NYCOEM website (www.nyc.gov/html/oem/html/about/planning_coastal_storm.shtml). Most people in New York City live within a relatively short distance of higher ground, so that with adequate warning and good management, emergency evacuations can be effective in many coastal storm situations.

- *Land-use changes.* Areas at risk can be reserved for parks and other low-density uses that can survive flooding with minimal damage. Land-use planning and environmental regulation generally are highly developed in New York City (Sussman and Major, 2010) and can be used effectively as part of overall adaptation to sea-level rise and storm surge (it should be noted, however, that there is significant new construction in waterfront areas within the current 100-year flood zone).

- *Flood walls.* Much of New York City's coastline is already hardened, and flood walls can be raised for critical facilities, such as wastewater treatment plants (WWTPs) and transportation facilities.

- Raising critical elements of infrastructure. This is planned now for one WWTP, the Rockaway WWTP, at a cost of US$30 million (Walsh, 2009).

- *'Soft' adjustments.* These include a range of adaptations to coastal flooding, including beach nourishment, dune raising and replanting, and restoration of saltmarshes. A 2010 exhibit at the Museum of Modern Art, New York City (Rising Currents) showcased the work of five architecture teams in developing innovative 'soft' adjustments to New York Harbor for rising sea levels (Nordenson et al, 2010). Elements of the architects' strategies to deal with sea-level rise, which were based on the NPCC sea-level rise scenarios described above and are aimed at creating a broader, softer, lower-gradient land-sea fringe, include:
  - new wetlands;
  - piers and slips;
  - offshore windmills;
  - oyster beds;
  - artificial islands;
  - subway car reefs;
  - offshore piers (these adaptations are shown in overview in Plate 12).

- *Barriers.* The most ambitious plan to protect parts of New York City from damaging storm surges in the context of sea-level rise is a set of storm surge barriers (see Figure 13.4). In one formulation, these would consist of barriers on the upper East River, the Arthur Kill (the waterway between Staten Island and New Jersey) and the Narrows (the entrance to the inner bay), or alternatively a larger gateway system (ASCE, 2009). Bowman et al (2005) have studied the hydrologic feasibility of such barriers. Preliminary cost estimates for the New York Harbor barriers are $1.5 billion for the upper East River site, US$1.1 billion for the Arthur Kill, US$6.5 billion

Figure 13.4 *Conceptual design of a surge barrier at the Verrazano-Narrows Bridge, New York City*

Source: Dircke et al (2009)

for the Narrows barrier, and US$5.9 billion for the gateway barrier system (ASCE, 2009). According to Aerts et al (2009, p75):

These options are at present only conceptual, and would require very extensive study of feasibility, costs, and environmental and social impacts before being regarded as appropriate for implementation. New York City has high ground in all of the boroughs and could protect against some levels of surge with a combination of local measures (such as flood walls) and evacuation plans; and barriers would not protect against the substantial inland damages from wind and rain that often accompany hurricanes in the New York City region.

Aside from environmental concerns and costs, the proposed barriers would not protect all vulnerable low-lying city neighbourhoods; in particular, parts of Staten Island, Brooklyn and Queens lie outside the protected zones, as do other nearby sections of the metropolitan area.

13.5.3 Information on costs and benefits

Relatively little information has been developed on the costs and benefits of adaptation to sea-level rise and storm surge in New York City. For a geographically diverse range of adaptations to current hazards, the Multihazard Mitigation Council gives an estimate of US$4 in benefits for every US$1 spent on hazard mitigation (Multihazard Mitigation Council, 2005a, 2005b). Jacob et al (2011) provide a careful but still approximate estimate of the costs (economic losses and physical damages from transportation outages) to the New York metropolitan area of a storm with an approximate recurrence interval of 100 years, comparing the current costs to costs with 2 and 4 feet (0.6 and 1.2m) of sea-level rise. The results are that the additional costs due to 2 feet of sea-level rise are US$12 billion in 2010 dollars, and the incremental costs for an additional 2 feet of sea level rise are US$14 billion in $2010 dollars. The total cost of sea-level rise to the city during this century would include damages from all storms, not just the one studied in the example, and this suggests very high levels of costs. Leichenko et al (2011) make a rough extrapolation of incremental insured property damages from flooding in all New York State coastal counties due to climate change, with an estimate of US$157 million to $202 million in additional annual damages without adaptation. Figures such as these are useful primarily in considering orders of magnitude of damages, suggesting future damages in the billions of dollars per year from climate change in New York City if no protective measures are taken.

No overall costs for adaptation to coastal flooding have yet been made. Table 13.3 gives cost figures for adaptation to current climate variation in the metro New York City area, which can provide some of the basis for estimating the costs of adaptations to future sea-level rise and storm surge.

Table 13.3 *Estimated costs for some adaptations in New York City and other regional locations*

Adaptation	Climate (current or future) and/or other variables	Location and year of proposal or project	Estimated cost
Reconnecting a saltmarsh	Adapt to development	Long Island Sound (Connecticut shoreline) (2003)	Total cost US$60,000–$141,000 for 10 acres
Wetlands restoration	Sea level, storm surge	Jamaica Bay–Elders West (2010)	US$10 million for 40 acres
Wetlands restoration	Sea level, storm surge	Soundview (2010)	US$5 million for 4 acres
Sea wall repair	Sea level, storm surge	Roosevelt Island (2001)	US$6,222,000
Beach nourishment	Sea level, storm surge	Coney Island (1995)	US$9 million
Beach nourishment	Sea level, storm surge	Westhampton Beach (1996)	US$30.7 million
Storm surge barriers	Sea level, storm surge	New York Harbor (2009)	US$9.1 billion for three-barrier system

Source: Leichenko et al (2011)

13.6 Adaptation: Future prospects

This chapter shows that many of the elements for effective adaptation planning for sea-level rise and storm surge have begun to be implemented in New York City. However, much remains to be done to ensure appropriate adaptation to climate change over the next decades and throughout the century.

13.6.1 Planning for adaptation

Many needed continuing and additional elements of planning for efficient and effective adaptation to climate change can be identified now. These include:

- Prepare complete high-resolution LIDAR (light detection and ranging) elevation datasets for New York City. This mapping is in process and will replace much coarser topographic data.
- Provide a detailed inventory of as-built infrastructure within, say, 10 to 20 feet (3 to 6m) of current sea level, calibrated to the current geodetic datum, NAVD88. This will be essential for planning.
- Review all large-scale infrastructure projects currently in the planning stage to make appropriate adjustments for adaptation to climate change.

- Prepare a set of plans for adaptation, using flexible adaptation pathways (Major and O'Grady, 2010, p243) for the short, medium and long terms. These will need to be reviewed on a regular basis as new information on sea-level rise and storm surge accumulates.
- Develop detailed benefit–cost estimates for different available adaptation options to assist in the development of a climate adaptation plan for various time periods.
- Continue to utilize the latest climate change inputs as a central element in New York City planning for climate adaptation. These will be particularly important in monitoring and reassessment.
- Foster strong political leadership in climate change adaptation; this has been present in New York City during the last decade, but often absent elsewhere. In New York City, such leadership means, in particular, a strong on-going mayoral commitment to climate change adaptation.

13.6.2 Future prospects

New York City has developed an effective approach to climate change adaptation that encompasses a number of best practices, including high-level proactive political leadership; links to larger sustainability activities; and involvement of multiple layers of government and a wide range of public- and private-sector stakeholders and experts. It is hoped that the work of the NPCC and other stakeholders, within the larger context of the city's efforts, will contribute to the implementation of an effective programme of adaptation to climate change both immediately and over the long term.

By mid century, the city will know much more about the outcomes of two outside possibilities that would be relevant to the need for and success of adaptation (Major, 2011). These are:

1 Policy-makers will confront and seriously reduce the problem of greenhouse gas emissions. This would ease the impacts of climate change and reduce, but by no means eliminate, pressures for adaptation to climate change in coastal cities.
2 Global warming and sea-level rise are already occurring at the upper limit of the current IPCC range of scenarios. If this high-end trend were to continue into the future, adaptations would have to be undertaken sooner than otherwise, and with a more urgent effort to integrate much higher levels of infrastructure protection within a larger framework of preparing for radical climate change.

From the present perspective, it appears probable that over the next half century New York City can adapt to sea-level rise and storm surge reasonably well, although at possibly substantial cost. Beyond that period it would be hazardous to predict, given the large uncertainties in the climate and economic models, even for a well-managed and wealthy city such as New York.

References

Aerts, J., Major, D. C., Bowman, M., Dircke, P. and Marfai, M. A. (2009) *Connecting Delta Cities: Coastal Cities, Flood Risk Management, and Adaptation to Climate Change*, Free University of Amsterdam Press, Amsterdam, The Netherlands

ASCE (American Society of Civil Engineers Metropolitan Section) (2009) *Against the Deluge: Storm Surge Barriers to Protect New York City*, Infrastructure Group Seminar, Polytechnic Institute of New York University, New York, 30–31 March

Barron, J. (2007) 'A sudden storm brings a metropolis to its knees', *New York Times, The Metro Section*, 9 August

Bowman, M. J., Colle, B., Flood, R., Hill, D., Wilson, R. E., Buonaiuto, F., Cheng, P. and Zheng, Y. (2005) *Hydrologic Feasibility of Storm Surge Barriers to Protect The Metropolitan New York–New Jersey Region*, Final report to New York Sea Grant, HydroQual, Inc., http://stormy. msrc.sunysb.edu/phase1/Phase%20I%20final%20Report%20Main%20Text.pdf

Cazenave, A. and Llovel, W. (2010) 'Contemporary sea level rise', *Annual Reviews in Marine Science*, vol 2, pp145–173

Dircke, P. T. M., Jongeling, T. H. G. and Jansen, P. L. M. (2009) 'A global overview of navigable storm surge barriers: Suitable gate types for New York from a Dutch perspective', in *ASCE Proceedings of the 2009 Seminar Against the Deluge: Storm Surge Barriers to Protect New York City*, 30–31 March 2009, New York, NY, pp93–102

Ferdaña, Z. A., Newkirk, S., Whelchel, A., Gilmer, B. and Beck, M. W. (2010) 'Adapting to climate change: Building interactive decision support to meet management objectives for coastal conservation and hazard mitigation on Long Island, New York, USA', in A. Pérez, B. Herrera Fernandez and R. Cazzolla Gatti (eds) *Building Resilience to Climate Change: Ecosystem-Based Adaptation and Lessons from the Field*, IUCN, Gland, Switzerland, http:// data.iucn.org/dbtw-wpd/edocs/2010-050.pdf

Hill, D. (ed) (1996) 'The baked apple? Metropolitan New York in the greenhouse', *Annals of the New York Academy of Sciences*, vol 790, p221

Horton, R. and Rosenzweig, C. (2010) 'Climate risk information', in New York City Panel on Climate Change (ed) *Climate Change Adaptation in New York City: Building a Risk Management Response*, New York Academy of Sciences, New York, NY, vol 1196, Appendix A

Horton, R., Herweijer, C., Rosenzweig, C., Liu, J., Gornitz, V. and Ruane, A. (2008) 'Sea level rise projections for current generation CGCMs based on the semi-empirical method', *Geophys. Res. Lett.*, vol 35, pL02715

Horton, R., Bader, D., DeGaetano, A. and Rosenzweig, C. (2011) 'Climate risks in New York State', in NYSERDA ClimAID Team, *Integrated Assessment for Effective Climate Change Adaptation Strategies in New York State*, C. Rosenzweig, W. Solecki, A. DeGaetano, M. O'Grady, S. Hassol and P. Grabhorn (eds) New York State Energy Research and Development Authority (NYSERDA), Albany, NY, Chapter 1

IPCC (Intergovernmental Panel on Climate Change) (2000) *Special Report on Emissions Scenarios*, IPCC, Geneva

IPCC (2007) *Climate Change 2007: The Physical Science Basis. Contribution of Working Group I to the Fourth Assessment Report of the Intergovernmental Panel on Climate Change*, Cambridge University Press, Cambridge, UK

Jacob, K., Deodatis, G., Atlas, J., Whitcomb, M., Lopeman, M., Markogiannaki, O., Kennett, Z., Morla, A., Horton, R., Bader, D., Leichenko, R., Ventura, P. and Klein, Y. (2011) 'Transportation infrastructure', in NYSERDA ClimAID Team, *Integrated Assessment for Effective Climate Change Adaptation Strategies in New York State*, C. Rosenzweig, W. Solecki, A. DeGaetano, M. O'Grady, S. Hassol and P. Grabhorn (eds) New York State Energy Research and Development Authority (NYSERDA), Albany, NY, Chapter 9

Kaufman, L. (2010) 'Front-line city in Virginia tackles rise in sea', *The New York Times*, 25 November

Leichenko, R., Major, D. C., Johnson, K., Patrick, L. and O'Grady, M. (2011) 'An economic analysis of climate change impacts and adaptations in New York State', in NYSERDA ClimAID Team, *Integrated Assessment for Effective Climate Change Adaptation Strategies in New York State*, C. Rosenzweig, W. Solecki, A. DeGaetano, M. O'Grady, S. Hassol and P. Grabhorn (eds) New York State Energy Research and Development Authority (NYSERDA), Albany, NY, Appendix

Major, D. C. (2011) 'Adaptation to climate change in coastal cities, 2050', in W. M. Grayman, D. P. Loucks and L. Saito (eds) *Toward a Sustainable Water Future: Visions for 2050*, EWRI-ASCE, Reston, VA

Major, D. C. and O'Grady, M. (2010) 'Adaptation assessment guidebook', in New York City Panel on Climate Change (ed) *Climate Change Adaptation in New York City: Building a Risk Management Response*, New York Academy of Sciences, New York, NY, vol 1187, Appendix B

Major, D. C., Goldberg, R., Gornitz, V. and Horton, R. (2010) *Methods, Models and Data Used for Sea Level Rise and Storm Surge Estimates for Project Phase I, Sea Level Rise and Storm Surge Frequency and Impacts Component for the Project Demonstrating Ecosystem-based Management on Long Island, NY: Decision Support for Coastal Hazard Mitigation and Biodiversity Conservation*, Columbia University Earth Institute Center for Climate Systems Research, Submission draft, 12 August, New York, NY

MTA (Metropolitan Transportation Authority) (2007) *August 8, 2007 Storm Report*, 20 September

Multihazard Mitigation Council (2005a) *Natural Hazard Mitigation Saves: An Independent Study to Assess the Future Savings from Mitigation Activities, vol 1: Findings, Conclusions, and Recommendations*, www.floods.org/PDF/MMC_Volume1_FindingsConclusionsRecommendations.pdf

Multihazard Mitigation Council (2005b) *Natural Hazard Mitigation Saves: An Independent Study to Assess the Future Savings from Mitigation Activities, vol 2: Study Documentation*, National Institute of Building Sciences, Washington, DC

New York City, Office of the Mayor (2007) *PlaNYC: A Greener, Greater New York*, New York, NY

Nordenson, G., Seavitt, C. and Yarinsky, A. with Cassell, S., Hodges, L., Koch, M., Smith, J., Tantala, M. and Veit, R. (2010) *On the Water: Palisade Bay, New York*, Metropolitan Museum of Art, New York, NY

NPCC (New York City Panel on Climate Change) (2010) *Climate Change Adaptation in New York City: Building a Risk Management Response*, vol 1187, C. Rosenzweig and W. Solecki (eds), New York Academy of Sciences, New York, NY

NRC (National Research Council) (2009) *Informing Decisions in a Changing Climate*, National Academy Press, Washington, DC

NRC, Panel on Adapting to the Impacts of Climate Change, Board on Atmospheric Sciences and Climate, Division on Earth and Life Studies (2010) *America's Climate Choices: Adapting to the Impacts of Climate Change*, NRC, Washington, DC

NYCDEP (New York City Department of Environmental Protection) (2008) *Climate Change Program, Assessment and Action Plan*, New York, NY

NYSERDA ClimAID Team (2011) *Integrated Assessment for Effective Climate Change Adaptation Strategies in New York State*, C. Rosenzweig, W. Solecki, A. DeGaetano, M. O'Grady, S. Hassol and P. Grabhorn (eds) New York State Energy Research and Development Authority (NYSERDA), Albany, NY

Peltier, W. R. (2004) 'Global glacial isostasy and the surface of the ice-age earth: The ICE-5G (VM2) model and GRACE', *Annual Reviews of Earth and Planetary Science*, vol 32, pp111–149

Rahmstorf, S. (2007) 'A semi-empirical approach to projecting future sea-level rise', *Science*, vol 315, pp368–370

Rosenzweig, C. and Solecki, W. (eds) (2001) *Climate Change and a Global City: The Potential Consequences of Climate Variability and Change*, Metro East Coast, Report for the US Global Change Research Program, Columbia Earth Institute, New York, NY

Rosenzweig, C., Major, D. C., Demong, K., Stanton, C., Horton, R. and Stults, M. (2007a) 'Managing climate change risks in New York City's water system: Assessment and adaptation planning', *Mitigation and Adaptation Strategies*, vol 12, no 8, pp1391–1409

Rosenzweig, C., Horton, R., Major, D. C., Gornitz, V. and Jacob, K. (2007b) 'Climate component', in Metropolitan Transportation Authority, *August 8, 2007 Storm Report*, 20 September

Rosenzweig, C., Solecki, W. D., Blake, R., Bowman, M., Faris, C., Gornitz, V., Horton, R., Jacob, K., LeBlanc, A., Leichenko, R., Linkin, M., Major, D., O'Grady, M., Patrick, L., Sussman, E., Yohe, G. and Zimmerman, R. (2011) 'Developing coastal adaptation to climate change in the New York City infrastructure-shed: process, approach, tools, and strategies', *Climatic Change*, vol 106, no 1, pp93–127

Solecki, W., Leichenko, R., Major, D. C., Brady, M., Robinson, D., Barnes, M., Gerbush, M. and Patrick, L. (forthcoming) *Economic Assessment of Climate Change Impacts in New Jersey*, City University of New York, Center for Sustainable Cities, New York, NY

Sussman, E. and Major, D. C. (2010) 'Law and regulation', in New York City Panel on Climate Change (ed) *Climate Change Adaptation in New York City: Building a Risk Management Response*, New York Academy of Sciences, New York, NY, vol 1187, Chapter 5

US Bureau of the Census (undated) *Current Estimates Program*, www.census.gov/popest/estimates.html

Vermeer, M. and Rahmstorf, S. (2009) 'Global sea level linked to global temperature', *Proceedings of the National Academy of Sciences*, vol 106, no 51, pp21527–21532

Walsh, B. (2009) 'The big green apple', *Time*, 2 April

Yohe, G. and Leichenko, R. (2010) 'Adopting a risk-based approach', in New York City Panel on Climate Change (ed) *Climate Change Adaptation in New York City: Building a Risk Management Response*, New York Academy of Sciences, New York, NY, vol 1187, Chapter 2

14

Climate Adaptation in the City of Jakarta

*Philip J. Ward, Muh Aris Marfai, Poerbandono
and Edvin Aldrian*

14.1 Introduction

With a coastline of approximately 81,000km, and more than 17,500 islands, Indonesia is extremely vulnerable to coastal inundation (Marfai and King, 2008a, 2008b). The capital and largest city of Indonesia, Jakarta, is particularly vulnerable to coastal and riverine flooding (Aerts et al, 2009; Yulianto et al, 2009), as well as localized flooding due to intense precipitation events. This vulnerability stems from a complex interaction of several physical and socio-economic parameters.

Flooding and flood management are not new to Jakarta. The region has suffered from flooding ever since the founding of Batavia (the colonial name for the current city of Jakarta) in 1619. Soon after this, a canal system was constructed similar to those of Amsterdam and other Dutch cities at the time, and already in 1725 a dam was built to divert the waters of the Ciliwung westwards through the Westerse Vaart (western canal). Over the course of the last four centuries, many more technical flood management strategies have been designed and/or implemented; but reports of flooding are prevalent throughout this long period (Abeyasekere, 1987). Nevertheless, the impacts of those floods have become more severe during recent decades. In the 21st century alone there have already been several floods with devastating consequences. For example, the floods in 2002 and 2007 caused estimated damages of 450 million Euros and 350 million Euros, respectively (NEDECO, 2002, cited in Caljouw et al, 2005; Steinberg, 2007).

In this chapter we first describe the location and climate characteristics of Jakarta (section 14.2). In section 14.3 we explore the main causes of flooding in Jakarta, both physical and socio-economic, and how these parameters are expected to change in the coming century. In section 14.4 we describe the main socio-economic impacts of flooding in Jakarta. Section 14.5 examines measures that have traditionally been taken to adapt to the flood problem (which mainly focus on reducing flood probabilities by building infrastructural protection) and the city's desire to move towards a more integrated flood risk management approach. In section 14.5 we also present an example of how recent developments in flood damage modelling and mapping in Jakarta could assist in this transition.

14.2 Location and Climate Characteristics

Jakarta is located on the northern coast of West Java (see Figure 14.1). The Special Capital Region of Jakarta (DKI Jakarta: Daerah Khusus Ibukota) covers an area of approximately 662 square kilometres and has a population of about 9 million (BPS Jakarta, 2007). However, during the daytime, commuters from the suburbs enlarge the population by about another third. In terms of physical geography, Jakarta is a lowland area with a relatively flat topography, with slopes ranging from 0° to 2° in the northern and central parts and 0° to 5° in the southern part (Abidin et al, 2001). Regionally speaking, the lowland area around Jakarta has five main landforms:

1   volcanic alluvial fans in the southern part;
2   landforms of marine origin in the northern part adjacent to the coastline;
3   beach ridge landforms in the north-west and north-east parts;
4   (mangrove) swamps in the coastal fringe; and
5   former channels running perpendicular to the coastline (Sampurno, 2001).

According to Yong et al (1995), the Jakarta Basin consists of a 200 to 300m thick sequence of Quaternary deposits that overlies Tertiary sediments. The top sequence is

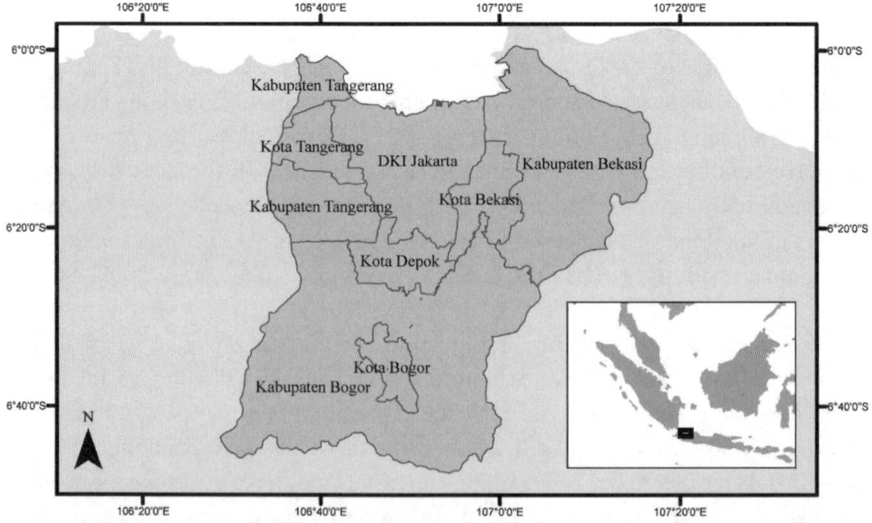

Figure 14.1 *Map showing the districts of Jakarta (western Java, Indonesia)*

thought to be the base of the groundwater basin. Three aquifers are recognized inside a 250m-thick sequence of Quaternary sediment of the Jakarta Basin – namely, the Upper Aquifer, an unconfined aquifer occurring at a depth of less than 40m; the Middle Aquifer, a confined aquifer occurring at a depth between 40m and 140m; and the Lower Aquifer, a confined aquifer occurring at a depth between 140m and 250m.

Jakarta is characterized by a tropical monsoonal climate with a mean annual temperature of approximately 27°C, and a mean temperature variation between 24°C and 29.5°C. The average annual rainfall in Jakarta (at the Tanjung Priok station) over the period of 1978 to 2007 was about 1640mm, with most of the rain (approximately 80 per cent) falling in the wet season between November and May (on average, 1383mm) (see Figure 14.2) (Marfai et al, 2009). There are large differences in the quantity of precipitation in Jakarta and in the mountainous areas in the river catchments upstream from Jakarta, where annual rainfall can be as high as 4500mm. There are 13 rivers that flow into the Jakarta metropolitan area. The Jakarta Bay area is highly dependent upon water flow from upstream. Therefore, there is a need to preserve natural vegetation in the upstream area in order to increase water-holding capacity.

Climate data for Jakarta show the frequent occurrence of high-intensity rainfall events. The most intense events tend to occur around January and February, and can cause major riverine flooding; the major floods of 2002, 2007 and 2008 all occurred at this time of year (Aldrian, 2009). During this period the ocean is as warm as the land and the probability of convection extends into the night, and not only during the afternoon, as is usually the case (Kusumayanti, 2008). Based on hourly rain gauge station observations for the Pondok Betung station to the south-west of Jakarta, we found that the flood events of 2007 and 2008 were associated with morning rainfalls, when the

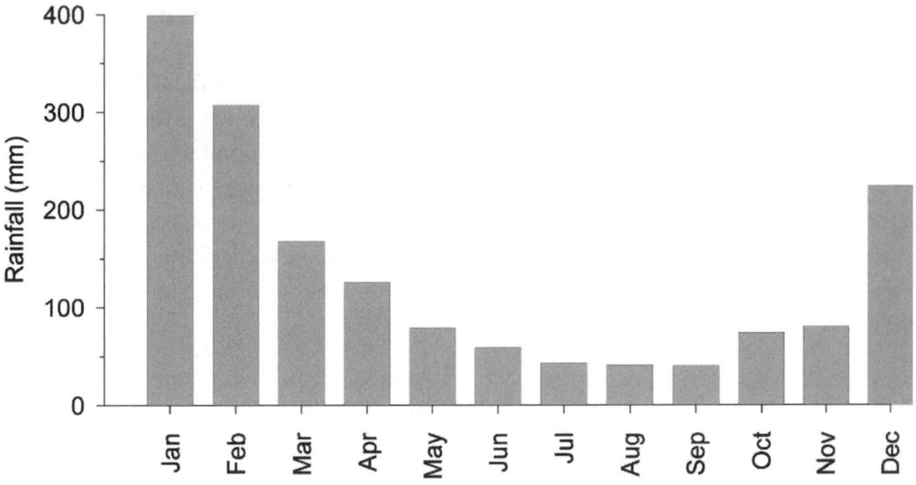

Figure 14.2 *Mean monthly rainfall in Jakarta (Tanjung Priok measuring station) between 1978 and 2007*

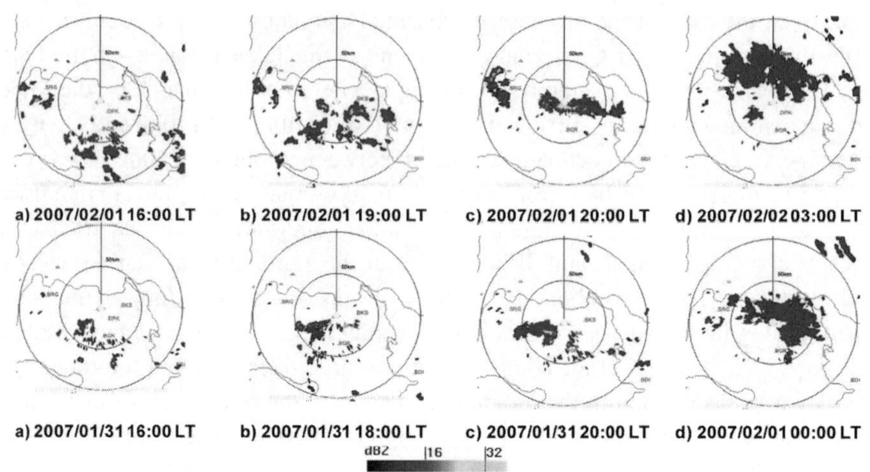

Figure 14.3 *Radar image during the flood event of 2007*

Source: Aldrian (2008)

total precipitation in 24 hours reached as high as 342 and 209mm, respectively, which is in excess of the mean monthly rainfall for this station. During the recent floods, similar rainfall patterns can be observed from weather radar imagery, with a diurnal pattern of intense coastal rainfall in the morning and intense rainfall in the mountainous hinterlands in the afternoon; an example for the 2007 event can be found in Figure 14.3.

14.3 Causes of the Flood Problem in Jakarta

Jakarta is located in the delta of several rivers, the main one being the Ciliwung River, and large parts of the city are close to the current mean sea level. Moreover, the 13 rivers that flow through Jakarta bring another intricacy to the problem. Hence, Jakarta has always been susceptible to flooding. However, several physical and socio-economic factors mean that that vulnerability has increased during recent years, and future trends in those factors are expected to lead to a further increase in vulnerability in the future. In this section, we discuss several of these key factors, but also indicate how these are interlinked and interdependent.

14.3.1 Physical causes of the flood problem in Jakarta

14.3.1.1 Land subsidence

Land subsidence is a serious problem in Jakarta (Abidin et al, 2010) and has four possible causes – namely, groundwater extraction, construction loads, natural consolidation of alluvium soil and geotectonic adjustments (Rismianto and Mak, 1993; Murdohardono and Sudarsono, 1998; Harsolumakso, 2001; Hutasoit, 2001). The first three are believed to be the most dominant (Abidin et al, 2010). The results of three levelling surveys conducted between 1982 and 1997, 11 global positioning system (GPS) surveys between 1997 to 2008, and InSAR technique (a synthetic aperture radar technique for monitoring the Earth's surface deformation) between 2006 and 2007 show that land subsidence in Jakarta has large spatial and temporal variations. Observed subsidence rates are generally approximately 1 to 15cm per year; but can be up to 20 to 25cm per year for certain locations and/or time periods (Abidin et al, 2001, 2004, 2008a, 2008b). The spatial and temporal variations indicate that the sources of land subsidence may also differ spatially. There is a strong indication that land subsidence in the Jakarta area is related to the high volume of groundwater extraction from the middle and lower aquifers, with secondary contributions by building/construction loading and natural consolidation of sedimentary layers. In general, land subsidence in northern Jakarta is more rapid than in the south of the city. Recent estimates of Abidin et al (2010) suggest an average subsidence rate of 4cm per year in northern Jakarta, with more rapid subsidence shown in a number of *cones* of subsidence.

14.3.1.2 Climate change

The mean annual temperature in Indonesia as a whole increased by approximately 0.3°C over the course of the 20th century (Hulme and Sheard, 1999). In the case of Jakarta, there was an increase of about 1.07°C over 100 years in January and of 1.40°C in July (BMKG, or Badan Meteorologi, Klimatologi, dan Geofisika – the Meteorological, Climatological and Geophysical Agency). Similar work indicates that there has been no clear trend (increase or decrease) in the rainfall of Jakarta over the period of 1950 to 1997. Over the same period, average annual rainfall decreased by approximately 2 to 3 per cent across the country, mainly in the wet season from December to February (Cruz et al, 2007). Several climate models project a temperature increase of about 0.1°C to 0.3°C per decade over the 21st century (Hulme and Sheard, 1999). The same projections suggest that annual precipitation will increase across most of Indonesia in the future, although there may be a decrease in Java, where Jakarta is located. However, most studies of South-East Asia suggest that extreme rainfall events will increase in severity and frequency in the 21st century (Cruz et al, 2007). As already stated, the most severe recent flood events in Jakarta occurred as a result of intense rainfall during the January to February period, and increases in this intensity would be expected to increase the severity and/or frequency of riverine floods.

Climate change will also affect Jakarta through sea-level rise. In most parts of coastal Asia, sea-level rise is currently taking place at a rate of approximately 1 to 3mm per year (IPCC, 2007a). Observations based on altimetry satellite detection for South-East Asia also show a mean sea-level rise in the Bay of Jakarta of about 2 to 4mm per annum over the period of 1992 to 2005 (Prijatna and Darmawan, 2005) (see Figure 14.4). Detailed projections of the impacts of climate change on sea-level rise in the Jakarta Bay during the 21st century are not available. However, the observed

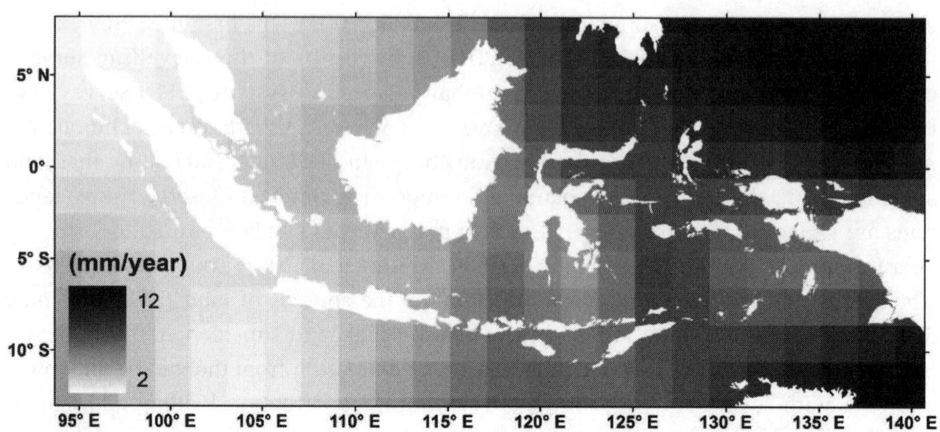

Figure 14.4 *Regional mean annual sea-level rise over the period of 1992 to 2005 based on altimetry satellite detection*

Source: adapted from Prijatna and Darmawan (2005)

changes over recent decades fall within the range of global sea-level rise reported by the Intergovernmental Panel on Climate Change (IPCC). Globally, the IPCC (2007b) projects the *likely* minimum and maximum global mean sea-level rise until 2100 to be between 18 and 59cm. Recent research suggests that South-East Asia will face increased intensities of tropical cyclones in the future, and storm activity will increase in the Indonesian region (APN, 2008), which could lead to increased storm surge heights (Cruz et al, 2007).

14.3.1.3 Low drainage capacity of channels

A major causal mechanism of flooding in Jakarta is the lack of drainage and/or storage capacity in the city's waterways; flooding occurs once the discharge capacity of rivers and drainage canals/sluices has been reached (Deltares, 2009). This lack of drainage has two main features:

1   The capacity of the designed water infrastructure does not have the ability to deal with the amount of water flowing.
2   And/or the actual discharge capacity of the city's waterways is much lower than their design capacity, as they are clogged up by sediment eroded from upstream and by solid waste.

Estimates suggest that of the approximately 23,400 cubic metres of garbage produced per day in Jakarta, only 14,700 cubic metres are disposed of by the City Sanitation Office (Steinberg, 2007). Large amounts of this garbage end up in the city's waterways, clogging up drainage channels. These problems are also compounded by land-use

Figure 14.5 *Accumulation of solid waste in Manggarai Gate, Jakarta*

Source: Ministry of Public Work (2008)

change upstream, which causes increased erosion and delivery of sediments to the river, and by land subsidence which lowers the gradient between the drainage channels/rivers and the outlet. Figure 14.5 shows an example of the garbage and solid waste clogging up drainage channels.

14.3.2 Socio-economic drivers of flooding in Jakarta

The physical changes discussed above are driven by changes in socio-economic factors from the household to the global level. At the global level, socio-economic developments determine the amounts of greenhouse gas emissions, which are a key component of global climate change and sea-level rise (IPCC, 2007b).

Over the last half century, Jakarta's population has risen rapidly from 2.7 million in 1960 to 9 million in 2007 (BPS Jakarta, 2007), and is projected to increase to 25 million by 2050 (Aerts et al, 2009). At the same time, the gross domestic product (GDP) of Indonesia is projected to grow by approximately 4.5 per cent per year between 2005 and 2030.

Past population pressures and economic developments have caused extensive land-use change in the whole of Java (Verburg et al, 1999), particularly in Jakarta (Firman, 2009). During the past three decades, fringe areas have experienced extensive land conversion from prime agricultural land to new urban and industrial areas (Verburg et al, 1999; Firman, 2000). Meanwhile, in the urban centre, many former residential areas have been converted into offices and business spaces, while open green space in Jakarta has greatly decreased from 28.8 per cent of total area in 1984 to an estimated 6.2 per cent in 2007 (Firman, 2009). These land-use changes have two main impacts upon flooding in Jakarta: they change the hydrological conditions of the basin and, therefore, runoff and flood frequency; and land use in flood-prone areas plays a key role in deciding the level of damage that occurs. Model results of Poerbandono et al (2009) and Julian et al (2010) suggest that over the course of the 20th century and early decades of the 21st century, land-use change in Jakarta and its adjacent basins led to an increase in annual runoff of approximately 3 to 9 per cent. Likely causes of this are that upstream deforestation can lead to a decrease in evapotranspiration and reduced soil water retention capacity (Ward et al, 2008), and that urbanization can lead to a decrease in the amount and rate of infiltration into the soil (Haase, 2009).

The increased population and economic development have also played a key role in lowering the drainage capacity of the city's drainage channels due to deforestation that leads to increased soil erosion in the upper catchment and therefore the clogging up of waterways with sediments, and the increase in solid waste mentioned above. Moreover, these socio-economic developments continue to increase pressure on the city's water supply system, with demand continually rising. At present, only 40 per cent of the city's supply is estimated to be piped, with 40 per cent coming from bore-holes and 20 per cent from traditional water vendors. As already discussed, the overuse of groundwater is thought to be one of the major causes of the acute land subsidence problem in Jakarta. Meanwhile, following rapid urban development in Jakarta, slum settlements are also increasing swiftly, especially along the river channels; this condition leads to an increased vulnerability to river flooding.

14.4 Impacts of Flooding upon Jakarta

Recent floods in Jakarta have demonstrated some of the massive impacts in terms of loss of lives and economic damage. The number of fatalities following the floods of 2002 are estimated at about 80 (Caljouw et al, 2005), and in 2007 between 58 (according to BAKORNAS, the National Coordinating Board for the Management of Disaster) and 74 (according to Office of Coordination of Humanitarian Affairs) people were killed. The direct economic damage associated with these floods was also large, estimated at approximately 450 million Euros (NEDECO, 2002, cited in Caljouw et al, 2005) and 350 million Euros (Steinberg, 2007), respectively.

Future sea-level rise is also expected to have several adverse impacts upon farming and coastal livelihoods. For example, in the lower Citarum Basin, the World Bank/DFID (2007) state that sea-level rise could result in the inundation of about 26,000ha of ponds and 10,000ha of cropland, which could result in the loss of 15,000 tonnes of fish, prawn and shrimp output, and about 940,000 tonnes of rice production.

Transportation systems are also hard hit in the event of flooding. At the best of times, Jakarta suffers from chronic congestion, described by many as the city's most significant 'urban nightmare' (Steinberg, 2007). During floods, this problem is exacerbated – for example, during the floods of 2002 and 2007, massive traffic jams hampered evacuation, with major thoroughfares being brought to a complete standstill for days. Train and air services were also badly hit, with about 80 per cent of flights at the Soekarno-Hatta Airport being delayed during the floods in 2002.

The impacts on Jakarta's population are not limited to the mainland; the Seribu Islands, located in the Bay of Jakarta, are also administratively part of Jakarta Province and are extremely vulnerable to inundation as a result of sea-level rise. The Seribu Islands are a collection of approximately 110 vegetated coral cays, with a total population of about 18,500 people (Central Bureau of Statistics) and a dryland area of about 77ha (Central Bureau of Statistics). In the Seribu Islands, most of the dryland area is at a height of just decimetres above mean sea level, and is already vulnerable to inundation during spring tides, even without climate change or extreme storm events. Figure 14.6

Figure 14.6 *The beach wall surrounding the port areas of Pramuka does not prevent the beach from inundation during perigean spring tides (November 2008)*

Source: Poerbandono

shows the height of tidal water on the main administrative island of Seribu (Pramuka) during the perigean spring tide of November 2008, breaching the port's defensive wall. Erosion is also threatening cay beach zones as sea-level rise facilitates erosive waves to expand to higher elevation landward. There is a multiplicative association between sandy beach erosion and sea-level rise, and under plausible global warming, erosion will be exacerbated in the 21st century (Zhang et al, 2004; Church et al, 2008). This loss of living space of cays' inhabitants may affect their livelihood (e.g. Lewis, 1989; Patel, 2006). Growing inhabitation of coral cays contributes to even more stress. Due to the low income levels of the islands' inhabitants and the natural conditions, the Seribu Islands are extremely vulnerable to flooding in the event of high tides.

14.5 Flood Adaptation in Jakarta

The descriptions above demonstrate Jakarta's vulnerabilities to both coastal and riverine flooding. The question that remains is what adaptation measures have traditionally been taken in Jakarta, and what new approaches could be implemented in the future?

14.5.1 Traditional flood management

Traditionally, government policies on flooding emphasized protection against floods based on technical measures (Texier, 2008). Soon after the founding of Batavia (the colonial name for the current city of Jakarta) in 1619, a canal system was constructed similar to those of Amsterdam at the time (De Haan, 1935). In 1725, a dam was built to divert the waters of the Ciliwung westwards through the Westerse Vaart (western canal); but in times of flooding this had to be opened to prevent its destruction (Caljouw et al, 2005). Several other flood-control canals have been built since then, all draining westwards. However, reports of flooding are prevalent throughout the past four centuries (Abeyasekere, 1987).

In 1917, the proposal of Van Breen led to the development of several structural flood defence measures, including the western flood channel (Western Banjir Canal). Other technical measures that were proposed, but not realized at the time, include a large polder along the north coast and the Eastern Banjir Canal (Diposaptono et al, 2004). During the 1940s, a comprehensive irrigation and drainage plan for West Java was drawn up, which included dams, sea-dikes and polders (Caljouw et al, 2005). In 1965, the Government of Indonesia developed a 'master plan for drainage and flood control' (revised in 1973), which was essentially a modification of the Van Breen plan (Diposaptono et al, 2004). In 1984, a master plan was drawn up that extensively mentioned flood control and drainage, again based on structural flood protection measures (Caljouw et al, 2005).

These projects provide technical solutions and assume effective maintenance; some parts of the plans were realized, others were not. In any case, by the time parts of the plans were realized (e.g. the Western Banjir Canal), the population had far outgrown the numbers on which the designs were originally based. To the present day, this system of dealing with flooding remains largely unchanged, with the division of the Ciliwung into two channels – namely, the West Banjir Canal traversing the southern and western parts of the city, and the main stream (Inner City Ciliwung River). In the last year, the construction of the Eastern Banjir Canal has been completed.

Technical measures continue to be taken, including improved maintenance, cleaning and dredging of waterways; improvement and maintenance of technical infrastructure; and improving the discharge capacity of rivers and retention capacity of the soil. During recent years there has also been increasing investment in non-structural measures, such as awareness-raising programmes; law enforcement and early warning and emergency assistance systems; and upper watershed planning and management (Caljouw et al, 2005). Following recent floods, private adaptation measures have also been taken by individuals, such as developing two-storey houses; placing sand bags along the streets; raising the levels of floors in houses; and developers raising real-estate ground levels by several metres (Caljouw et al, 2005; Marfai et al, 2009). While there has clearly been some more attention paid to non-structural measures during recent years, this was not based on an overall strategy or vision.

14.5.2 Moving towards integrated flood risk management

Traditionally, then, flood management in Jakarta focused on providing physical protection against flooding. However, there is a growing recognition amongst practitioners and researchers that as a result of climate change, as well as broader environmental and socio-economic changes, it will be more difficult and expensive in the future to provide such safety standards. Furthermore, it is clear that many of the technical measures have had limited effects due to delays between their design and their implementation. As in other parts of the world, there is therefore a growing interest in developing a more integrated system of flood risk management. In this context, flood risk is defined as the probability of flooding multiplied by the potential consequences of flooding, such as loss of lives and economic damage (e.g. Smith, 1994; Merz et al, 2004; Büchele et al, 2006). In 2008 and 2009, a series of four high-level meetings between experts, decision-makers and stakeholders from Jakarta and The Netherlands were carried out to identify challenges, develop strategies and implement solutions to reduce flood risks in Jakarta. This series of meetings was called the Jakarta Delta Dialogues, and was attended by the governor of DKI Jakarta, the head of the Provincial Planning Agency, the head of the Spatial Planning Department, the director-general of water, and several other representatives of relevant Jakarta institutions. In the following sub-section we elaborate upon how flood damage modelling and risk mapping could assist Jakarta in such a transition from traditional flood management to an integrated flood risk management approach.

14.5.3 Flood damage modelling and flood risk mapping

Several expert reports suggest that a key issue hampering successful flood reduction in Jakarta is the inability of authorities to cope with and plan across policy sectors, such as spatial planning, water resource management, land-use management, urban infrastructure, solid waste management, and housing (NEDECO, 2002, cited in Caljouw et al, 2005). The flood risk management approach is important in this integration, as it considers both the causes and consequences of flooding. An important facet of flood risk management is the preparation of flood risk maps, which show areas with the highest risk of flood damage or loss of lives, and can be used to communicate risk to inhabitants and businesses, who can change their risk perception and potentially lead to risk-reducing behaviour (Botzen et al, 2009). Here we describe a damage assessment model developed by Ward et al (2011) to assist in flood risk mapping and assessment in Jakarta, and show how the tool can be used to produce maps of potential damage exposure for several different scenarios of sea-level rise.

14.5.3.1 Methodology

A geographic information system (GIS)-based inundation model has been set up to produce inundation maps and to calculate maximum potential damage exposure for given coastal flood events under both current environmental conditions and under scenarios of future environmental change (Ward et al, 2011). The GIS-based inundation model produces a map of grid cells that would be inundated for a given coastal flood event (inundation map) – for example, the inundation area that would be caused by a storm surge with a recurrence period of 100 years. The input required by this inundation model are a digital elevation model (DEM), and the difference in sea level between a given flood scenario and current mean sea level (hereinafter referred to as 'floodwater level' and given in metres above current mean sea level, or masl). The output inundation map is then overlaid with a map of land use, and each land use is assigned an economic value, in order to calculate the maximum damage exposure per grid cell. More details on the model can be found in Marfai and King (2008a) and Ward et al (2011).

In this study, we use the model to assess the inundation extent and damage exposure in northern Jakarta under three scenarios:

1   a coastal flood event with a return period of 100 years under current conditions;
2   a coastal flood event with a return period of 100 years assuming a sea-level rise of 59cm (maximum *likely* global sea-level rise of IPCC, 2007b); and
3   a coastal flood event with a return period of 100 years assuming a sea-level rise of 59cm and land subsidence of 360cm (4cm per annum between 2010 and 2100).

14.5.3.2 Input data

The main input required by the model is a DEM. The DEM for the current conditions was generated at a resolution of 5m × 5m from the topographic maps of BAKOSURTANAL (Indonesian Survey and Mapping Coordination Agency) (scale 1:25,000).

A floodwater level (masl) is given as input to assess which land cells would be inundated should such a flood event occur. Estimates of floodwater level associated with a return period of 100 years under current conditions were derived from the Dynamic Interactive Vulnerability Assessment (DIVA) model (Dinas-Coast Consortium, 2006). For the 1/100-year event, the associated floodwater level is estimated at approximately 1.60 masl. Following the method of Nicholls et al (2008), we assume that future sea-level rise will cause this floodwater level to change by the magnitude of the sea-level rise. Hence, the scenario of sea-level rise used here of 59cm would result in an increase of the 1/100-year floodwater level to 2.19 masl.

Each grid cell is also assigned a land use, and each land use is assigned an economic value. This value represents the estimated market value of the buildings and material assets per hectare for each land-use class. By overlaying the inundation maps on the land-use maps, we estimated the total inundated area in each land-use class, and also the total value of exposed assets (damage exposure) per land-use class. It should be noted that this represents the maximum potential damage exposure since the actual damage that would occur is also dependent upon other factors, such as water depth in the inundated area (e.g. Merz et al, 2007), flow velocity (e.g. Kreibich et al, 2009), flood mitigation measures (e.g. Nicholls et al, 2008) and sediment load (e.g. Ward et al, 2009). The land-use map is based on an updated version of detailed topographic maps by the Jakarta local government (Pemerintah Kota Jakarta, Dinas Pertanahan dan Pemetaan) for the year 2004 (based on aerial photography in the year 2003), updated by means of visual image analysis based on an IKONOS image from 2007. For each land-use class, we assigned an average market value of buildings and tangible assets based on interviews, literature review and statistics (DPB, 2002; DGEM, 2004, Marfai and King, 2008b): business areas (2.5 million Euros per hectare); uniform settlements (1.2 million Euros per hectare); un-uniform settlements (1 million Euros per hectare); agriculture (80,000 Euros per hectare); fishponds (95,000 Euros per hectare); and open areas (1700 Euros per hectare).

14.5.3.3 Results

The area that would be inundated in northern Jakarta under the three scenarios is shown in Figure 14.7, and Figure 14.8 shows: (a) the total inundated area per land-use class for each scenario; and (b) the total value of exposed assets per land-use class for each scenario.

Under current conditions, the inundated area for a coastal flood with a return period of 100 years is approximately 3400ha, with a corresponding damage exposure of about 4 billion Euros. Under a scenario of sea-level rise of 59cm, the total damage exposure increases by a factor of 1.7, while the inundated area extends to approximately 6300ha. If we include land subsidence and sea-level rise to 2100, the inundated area is more than double again, with an associated damage exposure of almost 17 billion Euros. Under this scenario, large areas of northern Jakarta would be exposed to devastating flooding. Figure 14.8 shows the importance of land use in flood-prone areas in terms of

a) 1:100 flood event (current)

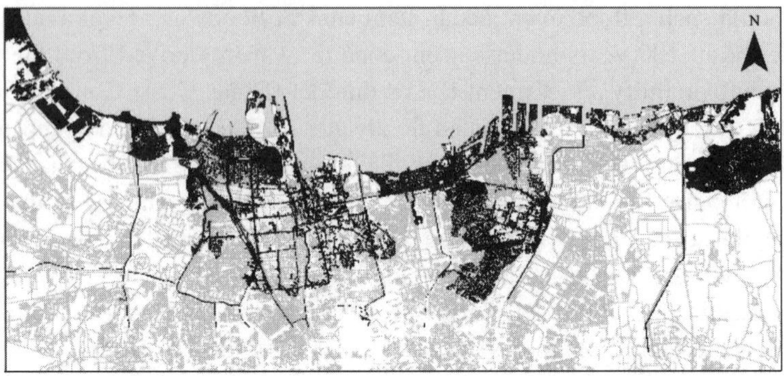

b) 1:100 flood event (sea-level rise)

c) 1:100 flood event (sea-level rise and land subsidence)

Figure 14.7 *Inundation maps for three coastal flooding scenarios: (a) a coastal flood event with a return period of 100 years under current conditions; (b) a coastal flood event with a return period of 100 years assuming a sea-level rise of 59cm; and (c) a coastal flood event with a return period of 100 years assuming a sea-level rise of 59cm and land subsidence of 360cm*

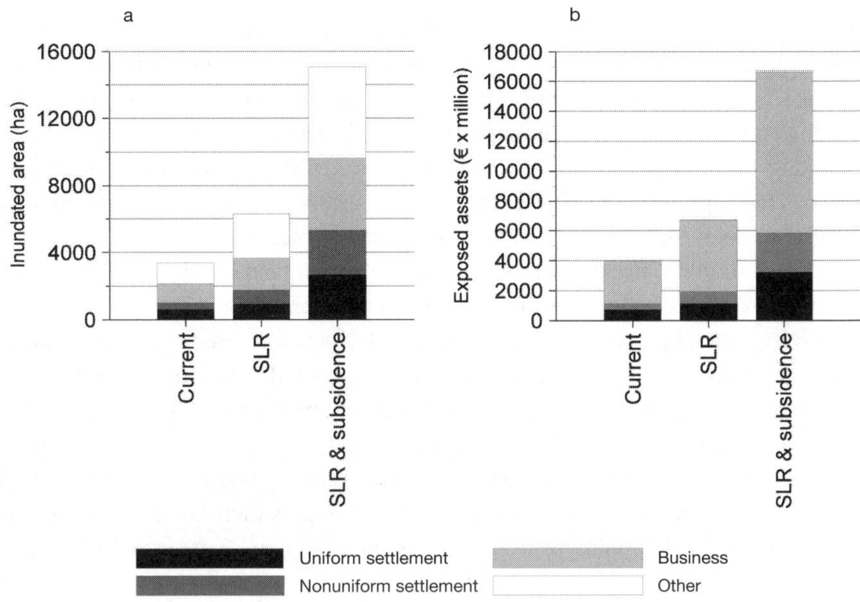

Figure 14.8 *Simulation results for the three inundation scenarios showing: (a) inundated area per land use (hectares); and (b) exposed assets per land use (million Euros)*

the propagation of flood damage exposure. For example, under the scenario of sea-level rise and subsidence, approximately 29 per cent of the total inundated area is of the land-use class *business*. However, in terms of total damage exposure, *business* accounts for about 64 per cent under the same scenario.

Such flood maps are of use in flood risk management since they show the specific areas that would be vulnerable to flooding under future scenarios and give a visualization of the flood problem that can be used to promote stakeholder awareness participation. Moreover, the model can be used to assess the effects of adaptation measures on flood damage, both spatially and at the aggregated level. Hence, the model can be further developed in the coming years to provide stakeholders with a tool for assessing the effectiveness of flood adaptation measures.

14.6 Conclusions

This chapter has shown that flooding is clearly not a new problem in Jakarta; the city has been prone to flooding since it was first established. However, recent and projected changes in several physical and socio-economic factors have meant that both the probability of (major) flooding and the resulting consequences are increasing, and will continue to increase in the future. Traditional flood management in Jakarta has mainly focused on technical measures aimed at reducing the likelihood of flooding. However, due to the city's rapid growth and delays in implementation, many of the city's flood protection measures are now under-designed, even if the possible impacts of climate change are not included. Furthermore, many of the structural defence measures are in need of improved maintenance. Coupled with climate change, there is a realization that the city could benefit greatly from an integrated flood risk management-based approach. This approach considers both structural measures to reduce the probability of flooding (e.g. dikes and sea walls), and non-structural measures to reduce the negative consequences should a flood occur (e.g. spatial planning and building codes). A desire to move towards such an approach has been seen in recent years. High-level workshops have been carried out to identify challenges, develop strategies and implement solutions to reduce flood risks in Jakarta. It is likely that Jakarta will need to use both structural and non-structural measures to reduce risk. We have shown how flood damage models are being developed to assist in flood risk assessments. In the years to come, flood risk assessments can be used to assess the effectiveness and the costs and benefits of new and innovative risk reduction measures, whether they are structural or non-structural in nature. Moreover, the mapping of flood risk can be used to increase stakeholder awareness and involvement.

Acknowledgements

We thank both the KNAW (Royal Netherlands Academy of Arts and Sciences) Mobility Programme (09-MP-10) and Knowledge for Climate (HSINT02; Theme 1) for supporting this chapter.

References

Abeyasekere, S. (1987) *Jakarta: A History*, Oxford University Press, Singapore

Abidin, H. Z., Djaja, R., Darmawan, D., Hadi, S., Akbar, A., Rajiyowiryono, H., Sudibyo, Y., Meilano, I., Kusuma, M.A., Kahar, J. and Subarya, C. (2001) 'Land subsidence of Jakarta (Indonesia) and its geodetic monitoring system', *Natural Hazards*, vol 23, pp365–387, doi:10.1023/A:1011144602064

Abidin, H. Z., Djaja, R., Andreas, H., Gamal, M., Hirose, I. K. and Maruyama, Y. (2004) 'Capabilities and constraints of geodetic techniques for monitoring land subsidence in the urban areas of Indonesia', *Geomatics Research Australia*, vol 81, pp45–58

Abidin, H. Z., Andreas, H., Djaja, R., Darmawa, D. and Gamal, M. (2008a) 'Land subsidence characteristics of Jakarta between 1997 and 2005, as estimated using GPS surveys', *GPS Solutions*, vol 12, pp23–32

Abidin, H. Z., Andreas, H., Gamal, M., Susanti, P., Hutasoit, L., Fukuda, Y., Deguchi, T. and Maruyama, Y. (2008b) 'Land subsidence characteristics of the Jakarta Basin (Indonesia) as estimated from leveling, GPS and InSAR and its relation with groundwater extraction', in *Proceedings of 36th IAH Congress, 26 October–1 November 2008*, Toyama, Japan

Abidin, H. Z., Andreas, H., Gumilar, I., Gamal, M., Susanti, P., Fukuda, Y. and Deguchi, T. (2010) 'Land subsidence in Jakarta Basin (Indonesia): Characteristics, causes and impacts', in *IAHS Book Series: Groundwater System Response to a Changing Climate*, CRC Press/Balkema, The Netherlands

Aerts, J. C. J. H., Major, D. C., Bowman, M. J., Dircke, P., Marfai, M. A., Abidin, H. Z., Ward, P. J., Botzen, W., Bannink, B. A., Nickson, A. and Reeder, T. (2009) *Connecting Delta Cities: Coastal Adaptation, Flood Risk Management and Adaptation to Climate Change*, VU University Press, Amsterdam, The Netherlands

Aldrian, E. (2006) 'Decreasing trends in annual rainfalls over Indonesia: A threat for the national water resource?', *Jurnal Meteorologi dan Geofisika, BMG*, vol 7, no 2, pp40–49

Aldrian, E. (2008) 'Dominant factors of Jakarta's three largest floods', *Jurnal Hidrosfir Indonesia*, vol 3, pp107–114

Aldrian, E. (2009) 'Meteorological conditions during Jakarta's three largest floods', in *Prosiding Seminar Scientific Jurnal Club Tahun*, BMKG, Jakarta, Indonesia

APN (2008) *Integrating Indonesian Capacity for Coastal Zone Management*, Labmath-Indonesia, Bandung, Indonesia

Botzen, W. J. W., Aerts, J. C. J. H. and Van den Bergh, J. C. J. M. (2009) 'Willingness of home-owners to mitigate climate risk through insurance', *Ecological Economics*, vol 68, no 8–9, pp2265–2277

BPS Jakarta (2007) 'Jakarta dalam angka 2007', *Katalog BPS 1403.31*, Badan Pusat Statistik Propinsi DKI Jakarta, Jakarta, Indonesia

Büchele, B., Kreibich, H., Kron, A., Thieken, A., Ihringer, J., Oberle, P., Merz, B. and Nestmann, F. (2006) 'Flood-risk mapping: Contributions towards an enhanced assessment of extreme events and associated risks', *Natural Hazards and Earth System Sciences*, vol 6, pp485–503

Caljouw, M., Nas, P. J. M. and Pratiwo (2005) 'Flooding in Jakarta: Towards a blue city with improved water management', *Bijdragen tot de Taal-, Land- en Volkenkunde (BKI)*, vol 161, pp454–484

Church, J. A., White, N. J., Aarup, T., Wilson, W. S., Woodworth, P. L., Domingues, C. M., Hunter, J. R. and Lambeck, K. (2008) 'Understanding global sea levels: Past, present and future', *Sustainable Science*, vol 3, pp9–22

Cruz, R. V., Harasawa, H., Lal, M., Wu, S., Anokhin, Y., Punsalmaa, B., Honda, Y., Jafari, M., Li, C. and Huu Ninh, N. (2007) 'Asia', in M. L. Parry, O. F. Canziani, J. P. Palutikof, P. J. Van der Linden and C. E. Hanson (eds) *Climate Change 2007: Impacts, Adaptation and Vulnerability. Contribution of Working Group II to the Fourth Assessment Report of the Intergovernmental Panel on Climate Change*, Cambridge University Press, Cambridge, UK

De Haan, F. (1935) *Oud Batavia*, Nix, Bandoeng, Indonesia

Deltares (2009) *Flood Hazard Mapping 2 – Overview*, Report no Q0743.00, Deltares, Delft, The Netherlands

DGEM (2004) *Civil-Society and Inter-Municipal Cooperation for Better Urban Services/Mitigation of Geohazards*, Directorate of Geological and Mining Area Environment (DGME), Department of Energy and Mineral Resources, Jakarta, Indonesia

Dinas-Coast Consortium (2006) *Dynamic Interactive Vulnerability Assessment (DIVA)*, Potsdam Institute for Climate Impact Research, Potsdam, Germany, http://diva.demis.nl, accessed 31 March 2009

Diposaptono, S., Pratikto, W. A. and Mano, A. (2004) 'Flood in Jakarta: Lessons learnt from the 2002 flood', in Y. Goda, W. Kioka and K. Nadaoka (eds) *Asian and Pacific Coasts 2003: Proceedings of the 2nd International Conference*, World Scientific Publishing Co. Ltd., London

DPB (Development Planning Board of Semarang) (2002) *Semarang City Planning 2000–2010*, DPB, Semarang, Indonesia

Firman, T. (2000) 'Rural to urban land conversion in Indonesia during boom and bust periods', *Land Use Policy*, vol 17, pp13–20

Firman, T. (2009) 'The continuity and change in mega-urbanization in Indonesia: A survey of Jakarta-Bandung Region (JBR) development', *Habitat International*, vol 33, pp327–339, doi:10.1016/j.habitatint.2008.08.005

Haase, D. (2009) 'Effects of urbanisation on the water balance: A long-term trajectory', *Environmental Impact Assessment Review*, vol 29, pp211–219, doi:10.1016/j.eiar.2009.01.002

Harsolumakso, A. H. (2001) 'Struktur geologi dan daerah genangan', *Buletin Geologi*, vol 33, pp29–45

Hulme, M. and Sheard, N. (1999) *Climate Change Scenarios for Indonesia*, Climatic Research Unit, Norwich, UK

Hutasoit, L. N. (2001) 'Kemungkinan hubungan antara kompaksi alamiah dengan daerah genangan air di DKI Jakarta', *Buletin Geologi*, vol 33, pp21–28

IPCC (Intergovernmental Panel on Climate Change) (2007a) *Climate Change 2007: Impacts, Adaptation and Vulnerability. Contribution of Working Group II to the Fourth Assessment Report of the Intergovernmental Panel on Climate Change*, Cambridge University Press, Cambridge, UK

IPCC (2007b) *Climate Change 2007: The Physical Science Basis. Contribution of Working Group II to the Fourth Assessment Report of the Intergovernmental Panel on Climate Change*, Cambridge University Press, Cambridge, UK

Julian, M. M., Nishio, F., Poerbandono and Ward, P. J. (2010) 'Simulation of river discharges in major watersheds of north-western Java from 1901 to 2006', *The 4th Indonesia Japan Joint Scientific Symposium 2010*, Bali, Indonesia

Kreibich, H., Piroth, K., Seifert, I., Maiwald, H., Kunert, U., Schwarz, J., Merz, B. and Thieken, A. H. (2009) 'Is flow velocity a significant parameter in flood damage modelling?', *Natural Hazards and Earth System Sciences*, vol 9, pp1679–1692

Kusumayanti, Y. (2008) *Variasi Spasial dan Temporal Hujan Konvektif di Pulau Jawa, Skripsi sarjana*, IPB, Jakarta, Indonesia

Lewis, J. (1989) 'Sea level rise: Some implications for Tuvalu', *The Environmentalist*, vol 9, no 4, pp269–275

Marfai, M. A. and King, L. (2008a) 'Tidal inundation mapping under enhanced land subsidence in Semarang, Central Java Indonesia', *Natural Hazards*, vol 44, pp93–109, doi:10.1007/s11069-007-9144-z

Marfai, M. A. and King, L. (2008b) 'Potential vulnerability implications of coastal inundation due to sea level rise for the coastal zone of Semarang city, Indonesia', *Environmental Geology*, vol 54, pp.1235–1245, doi:10.1007/s00254-007-0906-4

Marfai, M. A., Yulianto, F., Hizabron, D. R., Ward, P. and Aerts, J. (2009) *Preliminary Assessment and Modeling the Effects of Climate Change on Potential Coastal Flood Damage in Jakarta*,

VU University Amsterdam and Gadjah Mada University Yogyakarta, Amsterdam and Yogyakarta, The Netherlands and Indonesia

Merz, B., Kreibich, H., Thieken, A. and Schmidtke, R. (2004) 'Estimation uncertainty of direct monetary flood damage to buildings', *Natural Hazards and Earth System Sciences*, vol 4, pp153–163

Merz, B., Thieken, A. H. and Gocht, M. (2007) 'Flood risk mapping at the local scale: Concepts and challenges', in S. Begum, M. J. F. Stive, and J. W. Hall (eds) *Flood Risk Management in Europe: Innovation in Policy and Practice*, Springer, Dordrecht, The Netherlands

Ministry of Public Work (2008) *Issues and Activities for Flood Mitigation in Directorate General for Human Settlement*, Public Work Department, Jakarta, Indonesia

Murdohardono, D. and Sudarsono, U. (1998) 'Land subsidence monitoring system in Jakarta', in *Proceedings of the Symposium on Japan-Indonesia IDNDR Project: Volcanology, Tectonics, Flood and Sediment Hazards, 21–23 September 1998*, Bandung, Indonesia

Nicholls, R. J., Hanson, S., Herweijer, C., Patmore, N., Hallegatte, S., Corfee-Morlot, J., Château, J. and Muir-Wood, R. (2008) *Ranking Port Cities with High Exposure and Vulnerability to Climate Extremes: Exposure Estimates*, OECD Environment Working Paper no 1 ENV/WKP(2007), OECD, Paris, France, www.oecd-ilibrary.org/environment/ranking-port-cities-with-high-exposure-and-vulnerability-to-climate-extremes_011766488208, accessed 25 December 2009

Patel, S. S. (2006) 'A sinking feeling', *Nature*, vol 440, pp734–736

Poerbandono, Ward, P. J. and Julian, M. M. (2009) 'Set up and calibration of a spatial tool for simulating latest decades' flow discharges of the western Java: Preliminary results and assessments', *ITB Journal of Engineering Science*, vol 41B, no 1, pp50–64

Prijatna, K. and Darmawan, D. (2005) 'Sea level change monitoring in southeast Asian waters', in *Proceedings of the SEAMERGES Final Symposium*, Bangkok, Thailand

Rismianto, D. and Mak, W. (1993) 'Environmental aspects of groundwater extraction in DKI Jakarta: Changing views', in *Proceedings of the 22nd Annual Convention of the Indonesian Association of Geologists*, Bandung, Indonesia

Sampurno (2001) 'Geomorfologi dan daerah genangan DKI Jakarta', *Buletin Geologi*, vol 33, p12

Smith, D. I. (1994) 'Flood damage estimation: A review of urban stage-damage curves and loss functions', *Water SA*, vol 20, pp231–238

Steinberg, F. (2007) 'Jakarta: Environmental problems and sustainability', *Habitat International*, vol 31, pp354–365, doi:10.1016/j.habitatint.2007.06.002

Texier, P. (2008) 'Floods in Jakarta: When the extreme reveals daily structural constraints and mismanagement', *Disaster Management and Prevention*, vol 17, pp358–372

Verburg, P. H., Veldkamp, T. A. and Bouma, J. (1999) 'Land use change under conditions of high population pressure: The case of Java', *Global Environmental Change*, vol 9, pp303–312, doi:10.1016/S0959-3780(99)00175-2

Ward, P. J., Renssen, H., Aerts, J. C. J. H., Van Balen, R. T. and Vandenberghe, J. (2008) 'Strong increases in flood frequency and discharge of the River Meuse over the late Holocene: Impacts of long-term anthropogenic land use change and climate variability', *Hydrology and Earth System Sciences*, vol 12, pp159–175, www.hydrol-earth-syst-sci.net/12/159/2008

Ward, P. J., Van Balen, R. T., Verstraeten, G., Renssen, H. and Vandenberghe, J. (2009) 'The impact of land use and climate change on late Holocene and future suspended sediment yield of the Meuse catchment', *Geomorphology*, vol 103, pp389–400, doi:10.1016/j. geomorph.2008.07.006

Ward, P. J., Marfai, M. A., Yulianto, F., Hizbaron, D. R. and Aerts, J. C. J. H. (2011) 'Coastal inundation and damage exposure estimation: A case study for Jakarta', *Natural Hazards*, vol 56, pp899–916, doi:10.1007/s11069-010-9599-1

World Bank/DFID (2007) 'Executive summary: Indonesia and climate change', *Working Paper on Current Status and Policies*, World Bank/DFID, Washington, DC

Yong, R. N., Turcott, E. and Maathuis, H. (1995) 'Groundwater extraction-induced land subsidence prediction: Bangkok and Jakarta case studies', in *Proceedings of the Fifth International Symposium on Land Subsidence*, IAHS Publication no 234, IAHS, Wallingford, UK

Yulianto, F., Marfai, M. A., Parwati and Suwarsono (2009) 'Simulation model of the overflowing flood of Ciliwung River at kampong Melayu-Bukit Duri, Jakarta, Indonesia', *Journal of Remote Sensing and Digital Image Processing*, vol 6, pp43–53

Zhang, K., Douglas, B. C. and Leatherman, S. P. (2004) 'Global warming and coastal erosion', *Climatic Change*, vol 64, pp41–58

Climate Adaptation and Flood Management in the City of Rotterdam

Piet Dircke, Arnoud Molenaar and Jeroen Aerts

15.1 Introduction

Rotterdam is considered the marine gateway to Western Europe and is situated on the banks of the 'New Meuse' River ('Nieuwe Maas'), one of the channels in the delta formed by the Rivers Rhine and Meuse. Rotterdam has a long history as a port and is currently the second largest port in the world in terms of tonnage. It is, however, also a vulnerable location to floods as most of the city's neighbourhoods are located below sea level. The harbour areas lie several metres above sea level and are all human-made residential areas, which were developed over the last centuries.

An important date in the history of Rotterdam is May 1940, when large parts of the city centre were completely destroyed during a bombardment by the German air force. The centre of Rotterdam, therefore, was completely rebuilt after World War II except for a few buildings such as the town hall and the Laurens Church (see Figure 15.1).

Soon after the end of World War II, and persisting throughout the 1960s, many industrialized cities, such as Rotterdam, experienced a significant peak in birth rates that produced the generation known as the Baby Boomers. In Rotterdam, labour supply growth increased over the 1970s, as the Baby Boom generation entered the labour force, and continued to grow throughout the 1990s. This was also partly the result of an unprecedented number of women entering the workforce, as well as positive net migration rates over the same period. By 2000, The Netherlands' birth rate fell to 1.5 from 3.1 in 1950; over the same period, life expectancy rose by five and eight years for men and women, respectively. Currently, the population of Rotterdam is aging rapidly. For the Randstad conurbation (the urban area from Rotterdam to Amsterdam), population growth is expected to continue to increase, mainly due to the expected net immigration.

The Rotterdam region is situated mostly below sea level (up to –6m); but the city is well protected now from the risks of flooding from the sea and large rivers that flow through The Netherlands by a complex system of dikes, closure dams and storm surge barriers – part of the Dutch Delta Plan. The Delta Works were established after the disastrous 1953 floods that caused 1835 casualties. A main aspect of the Delta Plan was

Figure 15.1 *Laurens Church in Rotterdam in approximately 1684, 1940 and 2009*

to improve the protection of Rotterdam during an extreme storm surge. It was decided to construct the Maeslant Storm Surge Barrier, which protects Rotterdam in the case of an extreme flood event but stays open under normal conditions to allow ships free access to the older port areas, as well as the inland shipping canals behind the barrier.

Today, Rotterdam is a safe city that also wants to protect its citizens against the future impacts of climate change, such as sea-level rise and intensified rainfall. For this, Rotterdam has launched the Rotterdam Climate Initiative to deal with the challenges of climate change in many ways – for instance, with innovative multifunctional flood protection in the city centre, an extensive green roofs programme, the development of urban water plazas, climate-robust buildings, and communities of amphibian or even floating homes.

This chapter provides an overview of recent developments in Rotterdam in the area of climate adaptation and flood risk management. The chapter first provides an overview of Rotterdam in terms of historical and current water management practices (section 15.2). Next, flood risk management in Rotterdam (section 15.3) will be discussed, followed by the climate adaptation plan of Rotterdam (section 15.4). The chapter finally lists specific adaptation measures that have been – or are currently being – implemented in the city of Rotterdam (section 15.5) and concludes with a future outlook (section 15.6).

15.2 Current and Future Flood Risks

Rotterdam has a temperate climate influenced by the North Sea, with moderate temperatures and rainfall all year round. More intensified rainfall and longer dry spells are some of the expected impacts of climate change, in addition to sea-level rise. Heat waves, in which temperatures rise above 30°C already occur now, and will occur more frequently in the future. Summers are moderately hot, with short wet periods. Rainfall is almost equally distributed over the year, with an average annual rainfall of around 790mm. A new precipitation record was set in August 2006, when almost 300mm of precipitation fell in one month, causing extensive flooding and damage in the Rotterdam city area. Winters are relatively wet, with persistent rainfall periods. These periods of excessive rainfall sometimes can cause floods in the river basins of the Rhine and Meuse (Linde et al, 2009)

A large proportion of The Netherlands' economic assets are clustered in the port area of Rotterdam, where the estimated potential damage in the event of flooding is in the order of tens of billions of Euros. Potentially at risk in Rotterdam are industries, energy plants, port facilities, railways, tunnels and container terminals. In addition, a large section of Rotterdam's working population is employed in the port area, and many businesses are highly dependent upon port activities. Moreover, the port area of Rotterdam contains one of the largest petrochemical industries in Europe. Severe economic damage can occur from long-term closures of the port and its industry. Leakages from chemical plants and old underground oil tanks can cause considerable environmental damage and are difficult to clean after the flood event.

15.2.1 Climate change scenarios

The climate scenarios for Rotterdam are largely based on the Royal Netherlands Meteorological Institute (KNMI) 2006 scenarios, also known as the KNMI'06 scenarios. These scenarios are summarized in Table 15.1. The goal of the KNMI'06 scenarios included mapping the uncertainties regarding the future climate. The latter were divided into two major groups:

1 uncertainties regarding future population growth and economic, technological and social developments, along with the accompanying greenhouse gas and particulate emissions;
2 uncertainties due to science having only a partial knowledge of complex climate system processes.

In order to deal with these uncertainties, the KNMI has constructed four scenarios for 2050 (see Figure 15.2), giving a balanced view of a broad range of possible futures.

Table 15.1 *Summary of the KNMI'06 climate scenarios for the year 2050*

	KNMI'06			
	G	*G+*	*W*	*W+*
Summer (June, July, August)				
Average temperature (°C)	+0.9	+1.4	+1.7	+2.8
Warmest summer day (°C)	+1.0	+1.9	+2.1	+3.8
Average precipitation (%)	+3	-10	+6	−19
Number of wet days (%)	−2	−10	−3	−19
Sum of days with precipitation that is exceeded once a decade on average (%)	+13	+5	+27	+10
Potential evaporation (%) (WB21 on annual basis)	+3	+8	+7	+15
Sea-level rise (cm)	15–25	15–25	20–35	20–35
Winter (December, January, February)				
Average temperature (°C)	+0.9	+1.1	+1.8	+2.3
Coldest winter day (°C)	+1.0	+1.5	+2.1	+2.9
Average precipitation (%)	+4	+7	+7	+14
Number of wet days (%)	0	+1	0	+2
Sum of ten days with precipitation that is exceeded once a decade on average (%)	+4	+6	+8	+12
Highest daily wind velocity per year (%)	0	+1	0	+2
Sea-level rise (cm)	15–25	15–25	20–35	20–35

Source: van den Hurk et al (2006)

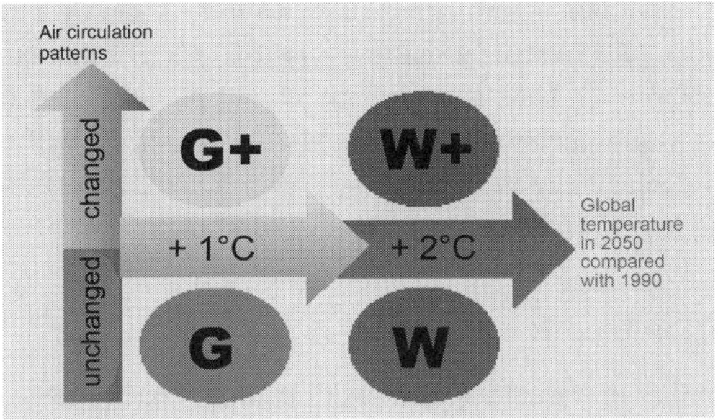

Figure 15.2 *Schematic overview of the four KNMI'06 climate scenarios*

Note: See also Table 15.1.
Source: van den Hurk et al (2006)

Each of these scenarios is plausible, and together they span a large proportion of the range shown in climate models. Emphasis has also been placed upon the variables that are most relevant for policy. For these situations, a picture as complete as possible has been sketched for our future climate.

The KNMI'06 climate scenarios do not describe abrupt climate changes. Climate models still have too many shortcomings in this respect due to a lack of scientific knowledge about these phenomena. Observed indications of rapid changes are also still uncertain. The scenarios therefore do not include any phenomena of uncertain realism. There are differing reasons as to why these extremes vary in nature and why they have not been included in the KNMI'06 scenarios.

15.2.2 Sea-level rise

Sea-level rise is a natural phenomenon, and historical measurements in Rotterdam show an increase in mean sea-level rise of 17 to 22cm over the last 100 years. Prior to the Industrial Revolution, sea-level rise in Rotterdam could mainly be attributed to regional subsidence of the Earth's crust, which is still slowly readjusting to the melting of ice sheets since the end of the last ice age. For Rotterdam, land subsidence accounts for 3 to 4mm per year, mainly due to post-glacial geological processes. The *Fourth Assessment Report* of the Intergovernmental Panel on Climate Change (IPCC, 2007) estimates an increase of global temperature of between 1.1°C and 6.4°C over the next century. As a result, average sea levels could rise by up to 59cm by 2100. There are, however, regional differences in projected sea-level rise, and it is expected that sea levels in the north-east of the Atlantic Ocean will rise 15cm more than the world average by 2100. This can be explained through the weakening of the warm Gulf Stream, gravitational effects and the extra warming of seawater at greater depths. The projected sea-level rise for Rotterdam is estimated at around 50 to 85cm by 2100. The maximum low-probability scenarios indicate a sea-level rise of 1.08 to 1.4m. However, the future behaviour of the large ice sheets in Greenland and Antarctica is largely uncertain. Although it is not well understood how quickly the ice sheets will melt, a theoretical collapse of the Greenland and West and East Antarctica ice sheets through accelerated glacier flow could lead to a rise in sea level of several metres over the next centuries.

15.2.3 Population trends

The expected number of casualties as a result of flooding in the Rotterdam region is regarded as an important indicator of vulnerability. Plate 13 shows the projected effects of growth in urban development in low-lying polders north of Rotterdam by 2040 on the potential number of casualties in the province of South Holland in the case of dike breaches. The expected growth in population in vulnerable areas of South Holland is expected to be much higher than the average population growth in South Holland as a whole (+50 per cent in the areas that could be hit by flooding, compared to +33 per cent for the area as a whole). This is one of the most important factors explaining why the estimated number of fatalities grows at a faster pace than the nation's average population growth. The increase in the potential number of fatalities is primarily caused by the growth of the population in low-lying polders; the limited rise in sea level has a relatively small effect in the low-lying polders. For example, a sea-level rise of 30cm could cause an increase in the fatality rate by as much as 20 per cent, while an 87 per cent growth in the population of Wateringse Veld by 2040 is projected to cause a 156 per cent increase in potential fatalities in the area. The influence of population growth on the fatality rate is therefore considerably greater than the effect of a 30cm sea-level rise.

15.3 Flood Risk Management

15.3.1 The 1953 storm surge

The Dutch flood management system is very much shaped by the storm surge disaster of 1953. During the night of 31 January 1953, a north-westerly windstorm pounded the coast of Zeeland, North Brabant and South Holland, including the city of Rotterdam. The flood protection system failed to prevent the flooding of many parts of the south-western delta of The Netherlands. Many dikes turned out to be too low and too weak, and breached. Large stretches of the east coast of the UK were also inundated. The consequences were disastrous, with 1835 fatalities, 100,000 evacuees, 200,000ha of land being inundated, and large economic losses of about 50 billion Euros (US$70 billion in 2009 values). After the catastrophic event of 1953, a committee was established to design a plan to study the causes of the flood and to develop measures to prevent similar disasters in the future. The Dutch government stated that a major storm surge flood must never happen again; therefore the Delta Plan consisted of the reconstruction of the dikes that failed during the 1953 event, as well as developing the storm surge barriers, resulting in very high coastal protection levels.

Plans were made to seal off the estuaries and connect the islands of South Holland and Zeeland with dams. The combination of these plans and dams was called the Delta Plan. The plan led to the implementation of the Delta Works, which is a series of dams, sluices, dikes and storm surge barriers constructed between 1958 and 1997 in the southwest of The Netherlands. The aim of the Delta Works was to improve flood protection by shortening and strengthening the Dutch coastline, thus reducing the number of dikes that had to be raised. With over 10,550 miles (16,979km) of dikes (1800 miles designated as primary dikes and 8750 miles as secondary dikes) and 300 structures, such as sluices and bridges, the Dutch flood protection system is still one of the most extensive engineering undertakings in the world. The Delta Works reduced the length of the dikes exposed to the sea by approximately 400 to 450 miles (640 to 700km). In most cases, building a barrier or a dam was much easier, faster and cheaper than reinforcing existing dikes.

15.3.2 Flood risk management in Rotterdam

As a result of the Delta Plan and Delta Law, Dutch flood protection standards are currently the highest in the world. Most of the protection system around Rotterdam is designed to withstand a storm that is estimated to occur, on average, once in every 10,000 years. The design surge level is determined at 4m (as a comparison, the 1953 storm generated a surge height of 3m). To protect against such a storm, taking both surge levels and breaking wave heights into account, the average dike along the Rotterdam coast is over 10m in height. This protection level reflects both the number of inhabitants and the economic value of assets within a dike ring; the more people and economic value have to be protected by dikes, the higher the safety standard is. As climate change is expected to increase the frequency and severity of flooding events, these flood probabilities will accordingly increase with sea-level rise. Therefore, reinforcing flood protection is, and will be, an on-going concern in Rotterdam and The Netherlands.

As the port of Rotterdam is of vital economic importance for The Netherlands and Europe, large parts of the port area and the city are protected by the Maeslant Storm Surge Barrier (see also Chapter 10). This barrier, however, was designed to cater for a maximum sea-level rise of about 0.5m, which is below the expected maximum sea-level rise for the year 2100. Both the Port Authority and the city of Rotterdam, together with the national government, are now considering options for coping with the increasing future flood risk due to climate change. The port area is relatively safe, as it is located on man-made land several metres above sea level. However, the area lies outside the dike protection system and is protected only by the barrier. In the future, occasionally, high water levels could be problematic.

As ships must have free access to the port, the city of Rotterdam and the Port Authority have chosen to develop the port area outside the dike protection system and the storm surge barrier at a high elevation, so most of these port areas are well protected against floods. Hence, over the last 100 years, 12,000ha of land have been elevated using fill materials to several metres above sea level. Along with the Palm Islands in Dubai, the most recent enlargement of the port area in Rotterdam (called Maasvlakte 2) is now the largest area of man-made land in the world to be largely surrounded by water (Aerts et al, 2009).

The urban water system has its own challenges, as it is an internal drainage system below sea level. This is a well-designed system of canals, lakes, drainage basins and pumps (Lansen et al, 2010).

15.4 Rotterdam Adaptation Strategy (RAS)

The city of Rotterdam started the Rotterdam Climate Initiative to develop Rotterdam into a climate-neutral city (RCI, 2010). The focus of this plan is on mitigating the emission of greenhouse gases and on strengthening the city's economy through innovative solutions to save energy and store carbon dioxide (CO_2). The major objective is to achieve a 50 per cent reduction in CO_2 emissions by 2025 (compared to the level of emissions in 1990), in conjunction with economic growth. A 50 per cent reduction in CO_2 emissions by 2025 would imply an annual reduction of 30 megatonnes of CO_2 emissions. The founders of the Rotterdam Climate Initiative are the Rotterdam Port Authority, the companies in the industrial port district, the municipality and the environmental protection agency (Rijnmond).

The Rotterdam Climate Proof (RCP) plan focuses on adaptation, and is complementary to the Rotterdam Climate Initiative. The main objective is to make Rotterdam climate proof in 2025. Within the RCP initiative, water management is an important aspect; but water is not only seen as a threat. It is also seen as an asset, a building block for the development of an attractive and economically strong city – hence, it plays a crucial role in the adaptation strategy of Rotterdam. Plate 14 shows a map of Rotterdam and locations where additional investments in the water system are needed to become climate proof in 2030.

An important aspect of adaptation planning in Rotterdam is to understand how the regional and national government plans for adaptation intersect, and, hence, the requirements and boundary conditions that they set for the city of Rotterdam. In 2008, for example, the Veerman Committee, a special national committee set up to study the challenges that The Netherlands face as a result of future climate changes, presented new recommendations on flood management and adaptation to climate change for the whole of The Netherlands (Deltacommissie, 2008; Kabat et al, 2009).

The new Veerman Plan indicates a need to build special flood protection around the port and city of Rotterdam. Increasing sea levels will result in more frequent closures of the Maeslant Barrier, which can lead to an increased risk of flooding from the large rivers flowing through Rotterdam that cannot discharge freely into the North Sea under these conditions. A number of different scenarios are under consideration in an open debate known as Rijnmond Closable but Open: should Rotterdam be protected by an extended and complex system of barriers, dams and gates that keep the sea out, or should Rotterdam embrace the sea, opt for a new balance in the water system and allow the salt water to flow freely in and out of the city?

15.4.1 Objectives of the RAS

The RCP team is developing a Rotterdam Adaptation Strategy (RAS), which has several challenges and measures (see Table 15.2):

- *Flood protection.* Sea-level rise will result in increased flood risk. According to Dutch law, the flood protection system will have to be reinforced. All quays and dikes that are not high enough will be elevated and reinforced during the coming years. However, in the long term, additional reinforcement may be needed. For this reason, space needs to be reserved now for the possible upgrading of the flood protection system in the future.
- *Architecture and spatial planning.* One of the challenges is to find alternative options that both enhance flood protection and add value to the attractiveness of the city. To achieve this, spatial planners, architects and flood protection authorities should join forces in looking for these alternative adaptation options. Traditional solutions are considered inadequate in this respect. In the city centre and the old neighbourhoods, for example, it is not possible to tackle the problems of water storage by constructing extra facilities. The costs are exorbitant and existing buildings cannot simply be demolished. Innovations such as green roofs, 'water plazas' and alternative forms of water storage are therefore essential for the further development of the city. The city also plans to develop new suburban centres outside the dike system. The challenge for spatial planners and architects is to find areas that are able to absorb high flood levels. The Rotterdam port area is situated several metres above sea level. Some of the older parts of the port are currently not in use for port activities and are potentially suitable for urban development, as long as architectural planning accommodates the possibility of flood risks in the future.
- *Rainwater storage/updating the sewage system.* The intensity and frequency of extreme rainfall events will increase in the future; there is a risk that the current sewage system may not be able to treat and drain the surplus water. In practice, rainwater usually drains away via the sewers; increasing amounts of rainfall already lead to problems with the existing sewage system. One possible way to avoid these problems is to collect the rainwater and allow it to drain away in a system other than the sewers, separating the dirty 'black' wastewater from the relatively clean 'brown' rainwater. However, this system should not affect public health, the quality of the

Table 15.2 *Rotterdam Adaptation Strategy: List and classification of potential measures*

	Minimizing chance	*Minimizing consequences*	*Stimulating recovery*
Region	Improve surge barrier*	Set up early warning system and design evacuation plan* Create cooling recreational areas close to the city***	Elevate energy infrastructure*
City	Improve dikes* Avoid vulnerable land-use functions in vulnerable areas* Increase the amount of public green space* Create space for innovative water storage**	Elevate infrastructure* Use water structure to transport goods and people* Inform people of what to do during times of excessive heat***	Arrange back-up systems (water, etc.)* Set up emergency shelters*** Arrange help desk for damage-related questions**
Quarter/ street	Integrate buildings with dikes* Combine numerous small gardens to create one big green oasis***	Create safe havens* Create shade*** Raise pavements/lower streets**	Adapt public space*** Install pumps**
Building	Build on mounds* Construct green roofs and green façades*** Less paving and more vegetation in gardens**	Install wet- or dry-proof ground floor***	Plant wet-proof vegetation***

Notes: * Measures related to water safety; ** measures related to urban water management; *** measures related to the urban climate.

groundwater or groundwater levels. Sewage pipes generally last for about 50 years; the reconstruction would take several decades. One option is to look for locations for the (temporary) storage of rainwater.

- *Learning by doing.* Continuously develop (scientific) knowledge while immediately applying results to practical situations.
- *Communication.* Exchange knowledge and stimulate cooperation with universities, companies and other cities, and inform the general public to show what the city is doing and how we can turn climate change from a threat into an opportunity. On an international level, RAS may prove to be of significant value to other delta cities, and Rotterdam will invest in knowledge exchange with other delta cities through the Connecting Delta Cities (CDC) initiative.

15.4.2 Acceptation and implementation

A crucial element for the RAS is to involve stakeholders and communities in the region and to implement the strategy. At the same time, different topics must be prioritized based on progressive insight and the on-going development of the city. Another important action is to increase awareness among policy-makers and city planners, and to integrate the need to strengthen the city's adaptation capability with mainstream city development planning. Finally, Rotterdam is also working on designing a monitoring system, as well as a climate barometer that will translate all of its efforts and actions into smart results and will help to assess the extent of Rotterdam's success in closing the gap between its ambitions and an ever-changing climate.

15.5 Examples of Adaptation Measures

15.5.1 Rainwater storage: Updating the storm-water sewerage system

For Rotterdam, climate change will result in more prolonged periods of drought and more heavy showers, both during the summer and winter periods. Precipitation is expected to increase by 7 to 28 per cent in winter. There is a risk that the current sewerage system may not be able to treat and drain this surplus water. The Rotterdam area requires an additional 400,000 to 600,000 cubic metres of storm-water storage space. At least 80ha of extra lakes and canals would be needed to provide this storage in open water. In the city centre, open water areas are used for storing extra water by retrofitting ponds in city parks or adjusting canals to store more water. In the case of an extreme precipitation event, water levels may rise without inundating surrounding areas. Green roofs slow the rate of roof runoff and can retain significant amounts of rainwater. After a rainfall event, green roofs gradually release water back into the atmosphere via evaporation. In order to stimulate citizens, public and private organizations, the city of Rotterdam has a large government support programme in place for partial subsidizing of the development of green roofs. Another example is the construction of an underground parking garage, below which the sloping entrance to a storm-water storage basin is connected to the city's storm-water and sewerage system. The capacity of the reservoir is large enough to store 50 per cent of the expected volume of rainwater that falls in one storm on Rotterdam city centre (10,000 cubic metres). Furthermore, a water plaza in Rotterdam can store water during times of peak rain events, but is used as a playground in normal conditions (Figure 15.3).

Figure 15.3 *A water plaza in Rotterdam, which can store water in times of peak rain events but is used as a playground during normal conditions*

Source: De Urbanisten, Rotterdam, The Netherlands

15.5.2 Adaptive housing

'Water-robust' building strategies are considered a promising and sustainable way to deal with the challenges of climate change. Adaptive waterfront development is seen as a solution to address these different issues and requirements. Rotterdam plans to develop 1600ha of waterfront locations in the old harbour area in the centre of the city, called Stadshavens. The properties are built on elevated land outside the dikes that has been raised during the last 100 years. Through adaptive architecture, this new neighbourhood can be made climate proof and serve as a high-quality waterfront-located living area. This requires new ways of developing buildings that, for example, allow water to move through the neighbourhood in the event of a flood without causing casualties or damage to assets.

Another example is the 'floating pavilion' (see Figure 15.4). The pavilion will temporarily house the Rotterdam National Water Centre. As the water level rises, the floating pavilion, a visible icon and landmark for the city, will automatically rise accordingly. This makes the pavilion an example of climate change-resilient building and of Rotterdam's commitment to a sustainable and climate-resilient water management strategy. In addition to sustainable technology on the inside, such as heating and

Figure 15.4 *A floating conference and exhibition centre in Rotterdam, which rises with elevated water levels*

Source: RCP Rotterdam Climate Proof, City of Rotterdam, The Netherlands

cooling systems that use solar energy, climatic zones and a separate filter installation, the outside is also sustainable as the pavilion is flexible and can moor at any location.

15.6 Future Outlook: Rotterdam Smart Delta City of the Future

Rotterdam is the perfect living showcase for climate change adaptation in The Netherlands and an inspiring example for delta cities worldwide. Rotterdam is currently developing innovative technologies to become a smart delta city of the future. One element of a smart delta city is a real-time operational information system for the management of (water) infrastructure and the development of control mechanisms for the Rotterdam Adaptation Strategy. The system pools various data flows on water and climate in the region, creating an accurate dynamic picture, and enables administrators and operators to respond adequately to threats such as floods, drought and changes in water conditions that may cause damage.

15.6.1 Innovation and collaboration

Trendsetting research, innovative knowledge development and decisive implementation of the suggested measures will result in strong economic incentives. For this, Rotterdam has invested in a new scientific campus which specializes in sustainable products, water management and climate change. Rotterdam has imperative reasons to highlight its qualities as a water knowledge city on an international level. Accordingly, the city was present at the world expo 2010 in Shanghai through its Rotterdam Water City pavilion and the city also organized, in cooperation with the Dutch Knowledge for Climate programme, the international Rotterdam Deltas in Times of Climate Change conference during 2010.

15.6.2 Flood control 2015

One of the major projects in this field is called Flood Control 2015, a Dutch public–private consortium of nine water specialists and experts in the field of water management, information and communications technology (ICT) and crisis management (see www.floodcontrol2015.com). Flood Control 2015 is currently researching the feasibility of a smart flood control system for Rotterdam. Smart gaming, a demonstrator flood control room, decision support systems, application of sensor technology in dikes, and many other tools are under consideration and may be integrated within a new smart flood protection system for Rotterdam. At the Rotterdam University of Applied Sciences, smart flood control is one of the topics that can be studied on the new water management course. In collaboration with private firms, students are currently developing a Rotterdam flood management game in an innovation lab.

15.6.3 Connecting delta cities

Rotterdam, as an affiliate member of the C40 initiative, is heading the Connecting Delta Cities initiative. The objective is to establish a network of delta cities, active in the field of climate change adaptation, that will exchange knowledge and share best practices. Under this initiative, the delta cities of Rotterdam, New York, Jakarta, London, New Orleans, Ho Chi Minh City, Hong Kong and Tokyo have decided to join forces, starting with an overview of best practices in these cities (Aerts et al, 2009; Dircke et al, 2010). Other cities, including Melbourne, Shanghai and Copenhagen, are keen to join this initiative. Cooperation with other delta networks is also increasing. This extensive and well-organized CDC network of knowledge-driven and ambitious delta cities is one of the best ways to deal with the challenges of climate change, and to turn them into opportunities for safe and attractive delta cities.

References

Aerts, J. C. J. H., Major, D. C., Bowman, M. J., Dircke, P., Marfai, M. A., Abidin, H. Z., Ward, P. J., Botzen, W. J. W., Bannink, B., Nickson, A. and Reeder, T. (2009) *Connecting Delta Cities: Coastal Adaptation, Flood Risk Management and Adaptation to Climate Change,* VU University Press, Amsterdam, The Netherlands

Deltacommissie (2008) *Samen Werken met Water,* Deltacommissie, The Netherlands

Dircke, P., Aerts, J. and Molenaar, A. (2010) *Connecting Delta Cities: Sharing Knowledge and Working on Adaptation to Climate Change,* City of Rotterdam Press, Rotterdam, The Netherlands

IPCC (Intergovernmental Panel on Climate Change) (2007) *Climate Change 2007: The Physical Science Basis. Contribution of Working Group I to the Fourth Assessment Report of the Intergovernmental Panel on Climate Change,* Cambridge University Press, Cambridge, UK

Kabat, P., Fresco, L. O., Stive, M., Veerman, C., van Alphen, J., Parmet, B., Hazeleger, W. and Katsman, C. (2009) 'Dutch coasts in transition', *Nature Geoscience,* vol 2, pp450–452, doi:10.1038/ngeo572

Lansen, A. J., Jonkman, S. N., van der Meer, R. A. E. and van Barneveld, N. (2010) *Flood Risks in Unembanked Areas,* Knowledge for Climate KvK report KvK022/2010, Rotterdam, The Netherlands

Linde, A., Aerts, J., Bakker, A. and Kwadijk, J. (2009) 'Simulating river discharge under climate change using different scenarios and modeling methods, with a focus on flood-peak probabilities', *Water Resources Research,* vol 44, doi:10.1029/2007WR006168

RCI (Rotterdam Climate Initiative) (2010) *Rotterdam Climate Initiative,* www.rotterdamclimateinitiative.nl

van den Hurk, B., Klein Tank, A., Lenderink, G., van Ulden, A., van Oldenborgh, G. J., Katsman, C., van den Brink, H., Keller, F., Bessembinder, J., Burgers, G., Komen, G., Hazeleger W. and Drijfhout, S. (2006) *KNMI Climate Change Scenarios 2006 for The Netherlands,* KNMI Scientific Report WR 2006-01, KNMI, De Bilt, The Netherlands

Index

Note: page numbers referring to figures are given in *italics*; page numbers referring to tables are given in **bold** type.